AF598031

NEUROMETHODS

Series Editor
Wolfgang Walz
University of Saskatchewan
Saskatoon, SK, Canada

For further volumes:
http://www.springer.com/series/7657

Microdialysis Techniques in Neuroscience

Edited by

Giuseppe Di Giovanni

Faculty of Medicine and Surgery, Department of Physiology and Biochemistry, University of Malta, Msida, Malta

Vincenzo Di Matteo

Consorzio "Mario Negri" Sud, Santa Maria Imbaro, Chieti, Italy

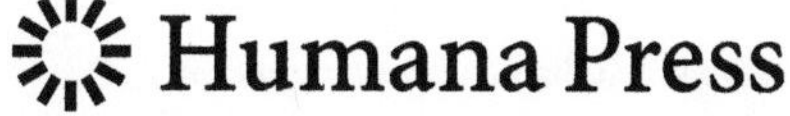

Editor-in-Chief
Giuseppe Di Giovanni
Faculty of Medicine and Surgery
Department of Physiology and Biochemistry
University of Malta
Msida, Malta

Assistant Editor
Vincenzo Di Matteo
Consorzio "Mario Negri" Sud
Santa Maria Imbaro
Chieti, Italy

ISSN 0893-2336 ISSN 1940-6045 (electronic)
ISBN 978-1-62703-172-1 ISBN 978-1-62703-173-8 (eBook)
DOI 10.1007/978-1-62703-173-8
Springer New York Heidelberg Dordrecht London

Library of Congress Control Number: 2012949563

Printed on acid-free paper

Humana Press is a brand of Springer
Springer is part of Springer Science+Business Media (www.springer.com)

Dedication

To Ennio Esposito

Who died unexpectedly on the 23rd of October 2011. We will miss a brilliant and influential scientist with many fresh ideas, who was supportive and understanding, and a good friend.

Giuseppe and Vincenzo

Foreword

It is not surprising that it was the great experimenter and behavioral physiologist Jose Delgado who constructed the first "Dialytrode" for collecting amino acids from the monkey brain (1). However, the dialysis bag at the tip of his probe did not allow for very effective exchange of substances and was used intermittently to collect the dialysate. At the same time I worked as a histologist at the Karolinska institute in Stockholm, mapping the monoamine neurons with the Falck–Hillarp technique. After 10 years in front of the microscope I was dreaming of a way to capture neurotransmitter release in real time—not only looking at the beautiful green fluorescent nerve terminals.

One day I saw the cross section of a capillary in the brightly green fluorescent brain parenchyma. It struck me that the capillary was just like a dialysis tube embedded in the tissue. If I had such a tube, chances were that I would be able to extract neurotransmitters from the extracellular liquid and bring them out of the tissue to be analyzed and quantified over time!

As you know good ideas in science must come at the right time to be realized. The same year the Dow Company had introduced the first "hollow fiber" dialysis tubes with perfect dimensions for being implanted into an animal brain. We got hold of some fibers and implanted them horizontally through the two caudate nuclei of the rat brain calling it the first real "brain wash" technique.

In order to capture the release of dopamine we needed to label the nerve terminals by perfusing the dialysis tubes with labeled tyrosine and then recover the release of labeled dopamine (2). However, time was once again in our favor. In 1973, Kissinger et al. (3) described the first HPLC technique with electrochemical detection that soon became commercially available, and we were finally able to monitor endogenous dopamine release over time even in awake, behaving animals. We set out to develop a more practical needle probe with a dialysis tube at its end, and once it worked one might say that "the rest is history."

My colleagues Trevor Sharp and Tyra Zetterström have summarized these exciting years and the contributions from scientists all over the world in a great review article (4).

Prof. Di Giovanni and Dr. Di Matteo in their book present microdialysis as it stands today. It has advanced from being a "brain wash" technique to a highly sophisticated "universal biosensor" applicable to essentially all tissues in animals as well as humans. It has developed hand in hand with new analytical techniques offering a remarkable increase in sensitivity, and I dare to say that today there is hardly any small molecular substance from the tissues that cannot be captured and analyzed.

The chapters in this book speak for themselves. They range from new techniques for both detection and quantification of the release of several different neurotransmitters in vitro and in vivo, even in freely moving animals, to sophisticated use of reverse dialysis and the application of microdialysis in pharmacokinetic studies. It covers the use in human neurointensive care and is on its way to becoming a routine diagnostic technique, as well as being an experimental technique applicable to the study of human brain physiology, e.g., in Parkinson patients. I must admit that thinking back on the day exactly 50 years ago when I first stepped into my laboratory at the Karolinska institute I feel a sting of envy. How great it was to spend ones time exploring animal and human physiology and pharmacology with such a versatile and universal technique as microdialysis!

Stockholm, Sweden ***Urban Ungerstedt***

References

1. Delgado JM, et al. (1972) Dialytrode for long term intracerebral perfusion in awake monkeys. Arch Int Pharmacodyn Ther 198:9–21
2. Ungerstedt U, Pycock C (1974) Functional correlates of dopamine neurotransmission. Bull Schweiz Akad Med Wiss 30:44–55
3. Kissinger PT, et al. (1973) An electrochemical detector for liquid chromatography with picogram sensitivity. Anal Lett 6:465–477
4. Sharp T, Zetterstöm T (2007) What did we learn from microdialysis? In: Westerink BHC, Cremers TIFH (Eds) Handbook of microdialysis, vol. 16. Elsevier, Oxford, UK, pp. 5–16

Preface to the Series

Under the guidance of its founders Alan Boulton and Glen Baker, the Neuromethods series by Humana Press has been very successful since the first volume appeared in 1985. In about 17 years, 37 volumes have been published. In 2006, Springer Science + Business Media made a renewed commitment to this series. The new program will focus on methods that are either unique to the nervous system and excitable cells or which need special consideration to be applied to the neurosciences. The program will strike a balance between recent and exciting developments like those concerning new animal models of disease, imaging, in vivo methods, and more established techniques. These include immunocytochemistry and electrophysiological technologies. New trainees in neurosciences still need a sound footing in these older methods in order to apply a critical approach to their results. The careful application of methods is probably the most important step in the process of scientific inquiry. In the past, new methodologies led the way in developing new disciplines in the biological and medical sciences. For example, Physiology emerged out of Anatomy in the 19th century by harnessing new methods based on the newly discovered phenomenon of electricity. Nowadays, the relationships between disciplines and methods are more complex. Methods are now widely shared between disciplines and research areas. New developments in electronic publishing also make it possible for scientists to download chapters or protocols selectively within a very short time of encountering them. This new approach has been taken into account in the design of individual volumes and chapters in this series.

Wolfgang Walz

Preface

The development of microdialysis, a minimally invasive technique, by Ungersted about 35 years ago has changed radically the way we study neurochemistry in vivo. Thanks to the substantial modification of a dialysis membrane attached to the end of Delgado's "dialytrode," Ungersted and Pycock in the 1970s succeeded in giving the first demonstration of the feasibility of using brain dialysis for measuring neurotransmitter release in vivo, measuring amphetamine-induced release of dopamine in the rat striatum. Interesting details about the development of this innovative technique can be found in the Foreword that Prof. Ungersted has provided to this volume, for which we are very grateful. Since its first application, microdialysis has become incredibly popular to study brain function and has been applied with success in different fields from psychopharmacology, neurobiology, and physiology in animals and also humans. The success of such a technique in neuroscience research is principally due to its capability to provide information on basal and stimulated levels of extracellular neurotransmitters offering the opportunity to single out the role of various receptor subtypes in regulation of synaptic and extrasynaptic neurotransmitter release and metabolism in discrete nuclei. Moreover, the development of different analytical methods, first of all high performance liquid chromatography (HPLC), has made this technique versatile, cheap, and easy to use routinely. Other different analytical methods have been coupled to microdialysis. They can be divided in non-separation-based methods, which allow the detection of one analyte at a time, in contrast to separation-based methods that can be used for the detection of multiple analytes in each sample. Due to limited spatial and temporal resolution, conventional microdialysis is best suited for sampling extrasynaptic pools of neurotransmitters and neuromodulators. Nevertheless, advances in probe design, fluid collection and handling, as well as analytical techniques are making way for breakthrough advances in microdialysis moving the membrane closer to the synapse.

This volume, with contributions from leading experts in their fields, follows a tradition set by the *Neuromethods* series by focusing on the practical aspects of microdialysis in animals and humans, highlighting current technical limitations, and providing a vision of what is yet to come for the determination of the most disparate compounds in the brain. The book has been organized to display the research topics to which modern microdialysis methods have been applied. The 16 chapters each contain an introduction that gives a broad overview of a focused topic, followed by an extensive protocol on how the experiments are performed along with invaluable practical advice. We hope that this text will be a valuable reference for students, neuroscientists, and physicians for the use of microdialysis in the study of brain functions and its clinical applications.

I would like to thank my friend and colleague, Vincenzo Di Matteo, for his help in producing this book, and Springer and its publishing editor for this series, Wolfgang Walz, for helping us in driving the book's development and eventual publication.

Finally, I would like to express our sincere appreciation to all the authors who have responded very willingly and contributed their time and expertise in preparing their individual contribution to a consistently high standard.

Msida, Malta *Giuseppe Di Giovanni*

Contents

Contributors

PETER J. BAAB • *Pediatric Research, University of Maryland, Baltimore, MD, USA*
CHIARA BALLINI • *Dipartimento di Farmacologia Preclinica e Clinica, Università degli Studi di Firenze, Florence, Italy*
FACUNDO MARTÍN BERTERA • *Departamento de Farmacología, Facultad de Farmacia y Bioquímica, Instituto de Fisiopatología y Bioquímica Clínica, Universidad de Buenos Aires, Junín, Buenos Aires, Argentina*
LORIA BIANCHI • *Dipartimento di Farmacologia Preclinica e Clinica, Università degli Studi di Firenze, Florence, Italy*
ANOUK BORG • *Neurocritical Care Unit, The National Hospital for Neurology and Neurosurgery, University College London Hospitals, London, UK*
MARÍA JOSÉ CANO-CEBRIÁN • *Department of Pharmacy and Pharmaceutical Technology, University of Valencia, Valencia, Spain*
CHENG-FU CHEN • *Department of Mechanical Engineering, University of Alaska Fairbanks, Fairbanks, AK, USA*
MARIA ALESSANDRA COLIVICCHI • *Dipartimento di Farmacologia Preclinica e Clinica, Università degli Studi di Firenze, Florence, Italy*
LAURA DELLA CORTE • *Dipartimento di Farmacologia Preclinica e Clinica, Università degli Studi di Firenze, Florence, Italy*
GIUSEPPE DI GIOVANNI • *Faculty of Medicine and Surgery, Department of Physiology and Biochemistry, University of Malta, Msida MSD, Malta*
VINCENZO DI MATTEO • *Consorzio "Mario Negri" Sud, Santa Maria Imbaro, Chieti, Italy*
KELLY L. DREW • *Department of Chemistry and Biochemistry, University of Alaska Fairbanks, Fairbanks, AK, USA*
WOLFHARDT FREINBICHLER • *Dipartimento di Farmacologia Preclinica e Clinica, Università degli Studi di Firenze, Florence, Italy*
SALVATORE GALATI • *Neurology Department, Neurocenter (EOC) of Southern Switzerland, Lugano, Switzerland*
LUIS GRANERO • *Department of Pharmacy and Pharmaceutical Technology, University of Valencia, Valencia, Spain*
TOM K. GREEN • *Department of Chemistry and Biochemistry, University of Alaska Fairbanks, Fairbanks, AK, USA*
NEIL D. HERSHEY • *Departments of Chemistry, University of Michigan, Ann Arbor, MI, USA*
CHRISTIAN HÖCHT • *Departamento de Farmacología, Facultad de Farmacia y Bioquímica, Instituto de Fisiopatología y Bioquímica Clínica, Universidad de Buenos Aires, Junín, Buenos Aires, Argentina*
ROBERTO W. INVERNIZZI • *Laboratory of Neurochemistry and Behavior, Department of Neuroscience, Istituto di Ricerche Farmacologiche "Mario Negri", Milan, Italy*

JAN KEHR • *Department of Physiology and Pharmacology, Karolinska Institutet, Stockholm, Sweden; Pronexus Analytical AB, Stockholm, Sweden*
ROBERT T. KENNEDY • *Departments of Chemistry and Pharmacology, University of Michigan, Ann Arbor, MI, USA*
YUQING LIN • *Beijing National Laboratory for Molecular Sciences, Key Laboratory of Analytical Chemistry for Living Biosystems, Institute of Chemistry, The Chinese Academy of Sciences (CAS), Beijing, China*
OMAR S. MABROUK • *Departments of Chemistry and Pharmacology, University of Michigan, Ann Arbor, MI, USA*
LANQUN MAO • *Beijing National Laboratory for Molecular Sciences, Key Laboratory of Analytical Chemistry for Living Biosystems, Institute of Chemistry, The Chinese Academy of Sciences (CAS), Beijing, China*
MAURO PESSIA • *Section of Human Physiology, University of Perugia School of Medicine, Perugia, Italy*
MASSIMO PIERUCCI • *Department of Physiology and Biochemistry, University of Malta, Msida MSD, Malta*
JAN POELAERT • *Department of Anesthesiology & Perioperative Medicine, Acute and Chronic Pain Therapy, Universitair Ziekenhuis Brussel, Brussel, Belgium*
ANA POLACHE • *Department of Pharmacy and Pharmaceutical Technology, University of Valencia, Valencia, Spain*
GARY M. POLLACK • *Department of Pharmaceutical Sciences, College of Pharmacy, Washington State University, Spokane, WA, USA*
JEANELLE PORTELLI • *Department of Pharmaceutical Chemistry, Drug Analysis and Drug Information, Center for Neurosciences, Vrije Universiteit Brussel, Brussels, Belgium*
BRIAN T. RASLEY • *Department of Chemistry and Biochemistry, University of Alaska Fairbanks, Fairbanks, AK, USA*
LIN SHI • *Department of Anesthesiology & Perioperative Medicine, Acute and Chronic Pain Therapy, Universitair Ziekenhuis Brussel, Brussel, Belgium*
MARTIN SMITH • *Neurocritical Care Unit, The National Hospital for Neurology and Neurosurgery, University College London Hospitals, London, UK*
ILSE SMOLDERS • *Department of Pharmaceutical Chemistry, Drug Analysis and Drug Information, Center for Neurosciences, Vrije Universiteit Brussel, Brussels, Belgium*
CHIARA STEFANINI • *Dipartimento di Farmacologia Preclinica e Clinica, Università degli Studi di Firenze, Florence, Italy*
KRISTIAN E. SWEARINGEN • *Institute for Systems Biology, Seattle, WA, USA*
CARLOS ALBERTO TAIRA • *Departamento de Farmacología, Facultad de Farmacia y Bioquímica, Instituto de Fisiopatología y Bioquímica Clínica, Universidad de Buenos Aires, Junín, Buenos Aires, Argentina*
KEITH F. TIPTON • *Department of Biochemistry, Trinity College Dublin, Dublin, Ireland*
VINCENT UMBRAIN • *Department of Anesthesiology & Perioperative Medicine, Acute and Chronic Pain Therapy, Universitair Ziekenhuis Brussel, Brussel, Belgium*
BENJAMIN P.E. WARLICK • *School of Molecular and Cellular Biology, University of Illinois at Urbana-Champaign, Urbana, IL, USA*

TAKASHI YOSHITAKE • *Department of Physiology and Pharmacology, Karolinska Institutet, Stockholm, Sweden*
ZIPIN ZHANG • *Key Laboratory of Analytical Chemistry for Living Biosystems, Beijing National Laboratory for Molecular Sciences, Institute of Chemistry, The Chinese Academy of Sciences (CAS), Beijing, China*
CAROL L. ZIELKE • *Pediatric Research, University of Maryland, Baltimore, MD, USA*
H. RONALD ZIELKE • *Pediatric Research, University of Maryland, Baltimore, MD, USA*
TEODORO ZORNOZA • *Department of Pharmacy and Pharmaceutical Technology, University of Valencia, Valencia, Spain*

Chapter 1

Cerebral Microdialysis: Research Technique or Clinical Tool?

Anouk Borg and Martin Smith

Abstract

Cerebral microdialysis is a well-established laboratory tool that is now widely used as a bedside monitor of brain tissue biochemistry during neurointensive care. With its ability to create a facsimile of brain tissue extracellular fluid (ECF) and characterize metabolic and biochemical changes, cerebral microdialysis is able to elucidate pathophysiological processes after brain injury and provide objective endpoints for clinical interventions and research. Microdialysis allows early recognition of cerebral hypoxia/ischemia and bioenergetic failure by monitoring changes in brain ECF glucose, lactate, pyruvate, glycerol, and glutamate concentrations. However, the sensitivity and specificity of microdialysis markers of ischemia and bioenergetic failure are not well characterized and there are no data to confirm whether microdialysis-guided therapy can influence outcome. The development of a system providing rapid analysis in "real time" is crucial to maximize the clinical applicability of the microdialysis technique.

Key words: Brain injury, Brain monitoring, Brain metabolism, Cerebral microdialysis, Neurointensive care

1. Introduction

The first animal experiments using microdialysis were carried out during the 1960s. Bito et al. implanted dialysis sacks into the cerebral hemispheres of dogs and 10 days later removed them and analyzed the fluid that they contained (6% dextran) for the concentration of amino acids (1). In subsequent years the microdialysis technique underwent several improvements and began to be applied in humans where it rapidly became a popular bioanalytical sampling tool in clinical brain research (2–4). In 1995 commercially available microdialysis instrumentation was introduced for clinical use (CMA Microdialysis AB, now M Dialysis AB, Sweden) and this facilitated the translation of a laboratory-based research technique into the clinic.

Giuseppe Di Giovanni and Vincenzo Di Matteo (eds.), *Microdialysis Techniques in Neuroscience*, Neuromethods, vol. 75,
DOI 10.1007/978-1-62703-173-8_1,

It is in neurointensive care that microdialysis has found a particular niche because of its ability to monitor endogenous substances in brain tissue extracellular fluid (ECF). Trends in various biochemical substances can aid in the understanding of the pathophysiology of acute brain injury and identifying secondary (ischemic) insults early, and thereby guide clinical management. Cerebral microdialysis (CM) is now a key component of multimodal monitoring techniques during neurointensive care.

2. Principles of Cerebral Microdialysis

The microdialysis technique makes use of the dialysis principle whereby diffusion of a substance across a semipermeable membrane occurs along its concentration gradient. The final concentration of substances in the dialysate depends on the balance between substrate delivery to and uptake from the brain ECF and also on several other factors related to the technique itself (see Sect. 2.2). The principles of microdialysis have been described in detail elsewhere (5) and only a brief summary of those issues that are particularly relevant to its clinical applications will be discussed.

2.1. The Microdialysis Catheter

The scope of CM is to sample and measure brain interstitial biochemical concentrations in vivo. Diffusion drives the passage of molecules along their concentration gradient from the brain ECF across the dialysis membrane. In order to maintain the concentration gradient, the membrane is continuously flushed on its inner side with a solute (the perfusate) that lacks the substance of interest in the ECF. This arrangement is found in a microdialysis catheter which consists of a thin (0.6 mm diameter) double-lumen probe, with parallel inlet and outlet tubes, lined at its tip with a dialysis membrane. The perfusion fluid, which is isotonic to the tissue interstitium, enters the microdialysis catheter via the inlet tube and its flow rate is controlled by a miniature pump. Perfusate passes along the catheter to the semipermeable polyamide dialysis membrane covered tip across which exchange of molecules between interstitial and perfusion fluid takes place. At the distal end of the catheter the fluid, now the dialysate, passes via the outlet tube to a small collecting chamber. In the clinical setting the dialysate is collected into small microvials that are changed regularly, usually every hour, and placed in a bedside microdialysis monitor where the dialysate is analyzed (Fig. 1). Samples can also be stored for off-line analysis.

2.2. Recovery of Measured Substances

Unless there is total equilibration between the brain ECF and perfusate, the concentration of a given substance in the dialysate will be lower than its actual concentration in the ECF. The relationship between dialysate and ECF concentrations is termed the relative

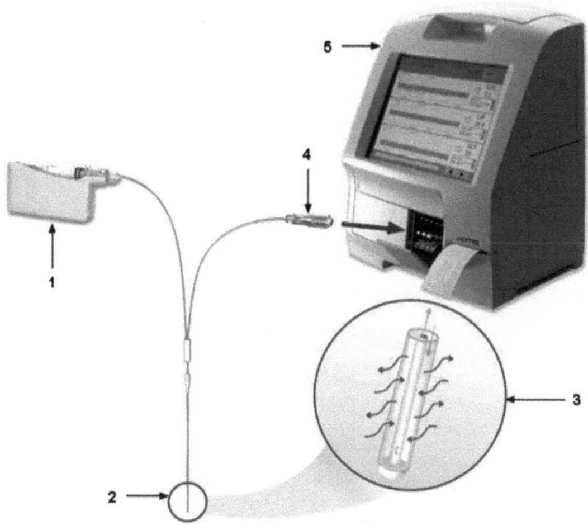

Fig. 1. Components of a clinical microdialysis system. (*1*) Microdialysis pump. (*2*) Microdialysis catheter. (*3*) Microdialysis catheter tip showing exchange of molecules across the dialysis membrane. (*4*) Microvial for collection of the microdialysate. (*5*) Bedside analyzer. Reproduced with permission from M Dialysis AB, Sweden.

recovery and is defined as the dialysate/interstitial concentration ratio expressed as a percentage:

$$\text{Relative recovery} = \frac{C_{\text{md}}}{C_{\text{ecf}}} = 1 - \exp - K_0 \cdot \frac{A}{F},$$

where C_{md} and C_{ecf} are the concentration of the substance in the microdialysate and ECF, respectively, K_0 the average mass transfer coefficient, A the membrane surface area, and F the dialysate flow rate (2).

Since full equilibration does not occur across the microdialysis membrane, the concentration of the measured substance in the dialysate is never equal to the true interstitial concentration. The interstitial concentration can be calculated using one of several models (see Sect. 3.4) but, as it is trends in tissue biochemistry (rather than absolute values) that are used to guide patient management, there is usually no requirement to determine actual interstitial concentration during clinical applications on the neuro-intensive care unit.

The relative recovery of a particular substance depends on several factors, including three equipment-dependent factors—perfusion flow rate, size of the dialysis membrane, and membrane pore size.

1. A low perfusion flow rate allows more time for equilibration across the dialysis membrane and for the concentration in the dialysate to approach that of the interstitial fluid. The standard perfusion flow rate applied in clinical CM is 0.3 μL/min and this allows hourly sampling with good recovery rates for commonly measured substances. Increased temporal resolution can

be achieved using higher flow rates but at the expense of lower relative recovery. For example, flow rates up to 1.0 μL/min allow sampling every 10–15 min but with a recovery of only 30% (6).

2. A higher recovery is also achieved if the area of the dialysis membrane is increased. Commercially available catheters are a standard 0.6 mm diameter, so membrane area is in practice determined by catheter length. Those designed for use in the brain have a membrane length of 10 mm and, with a perfusion flow of 0.3 μL/min, a relative recovery of around 70% for commonly measured variables (6). Increasing the catheter length to 30 mm delivers almost 100% recovery but decreases anatomical specificity.
3. Another important characteristic of the microdialysis membrane that affects recovery is its pore size, which is in turn determined by the material used for its manufacture. 20 and 100 kDa molecular weight cutoff catheters are commercially available for clinical use and have similar recovery of commonly measured variables (7). However, 100 kDa cutoff catheters also permit sampling of macromolecules, such as cytokines and other proteins, and are widely used for research purposes (8).

As previously noted, the concentration of a substance in the dialysate also depends on its supply to the tissue being monitored and its local uptake or release from cells. These in vivo issues lead to unknown variations in recovery but are usually considered irrelevant during clinical use.

2.3. Perfusate Composition

It is important that the composition of the perfusate closely mimics that of the brain ECF since this will otherwise influence recovery. The perfusate solution should be isotonic with the tissue interstitium and contain adequate levels of cations, in particular Na^+, K^+, Ca^{2+}, and Mg^{2+}, to prevent depletion from surrounding tissue. It should also have a similar pH and temperature and, for research purposes, a low protein content to facilitate high-performance liquid chromatography without the need for prior deproteinization of the dialysate. 0.9% saline and Ringer's solution have both been used as perfusates and have similar recovery of glutamate, glucose, lactate, and pyruvate (6). However, 0.9% saline contains no calcium (Ringer's solution contains 2 mM calcium), so the use of 0.9% saline as perfusate leads to depletion of calcium in the ECF. Since the release of neurotransmitters such as dopamine is calcium dependent, calcium levels in the perfusate may influence the ECF concentration of some neurotransmitters and therefore their recovery (9). For this reason, Ringer's solution is a preferable perfusion fluid to 0.9% saline. Commercially available artificial cerebrospinal fluid (CSF) preparations (Perfusion Fluid CNS, M Dialysis AB) are also available and widely used in the clinical setting.

3. Technical Considerations

With the advent of commercially available bedside analyzers, point-of-care testing of brain tissue biochemistry has become a reality during neurointensive care. More recently, online analysis that provides almost continuous data (every 30 s), albeit a limited number of substrates, has been used in the research setting.

3.1. Bedside Analysis

3.1.1. Commercially Available Equipment

Collected samples of dialysate can be measured at the bedside using a commercial microdialysis analyzer with a temporal resolution (every 20–60 min) that is sufficient for routine clinical use during neurointensive care. The CMA 600 microdialysis analyzer (M Dialysis AB) can measure four analytes from a single sample and is able to process the small dialysate volumes obtained during routine clinical use. Reagents for glucose, lactate, pyruvate, glycerol, glutamate, and urea are available and are measured using completely automated enzymatic phosphorylation reaction techniques and spectrophotometric detection (10). When the reagent reacts with an individual analyte a colored substance of a particular bandwidth is generated and this is detected photometrically. Absorbance measurements are made with a single-beam filter photometer and 375 and 520 nm filters. The photometer uses a capillary flow-through cuvette with a volume of 2 μL. It takes approximately 90 s to analyze each substance and the result is displayed on a screen and saved onto a hard disk. A new generation of bedside analyzers, the ISCUS^{flex} (M Dialysis AB), was introduced in 2008 and uses similar technology to its predecessors but is substantially smaller, has batch processing capability, and permits monitoring and data display on up to eight patients simultaneously. Following bedside analysis, dialysate can be stored prior to remote off-site analysis for an infinite range of substrates using enzyme spectrophotometry and high-performance liquid chromatography.

Urea levels have been used to monitor the in vivo performance of the microdialysis catheter. Urea is an endogenous compound that is evenly distributed amongst different body compartments, so any variation in the brain ECF concentration of a biomarker that is not accompanied by a change in urea can be considered to reflect a true biological change in the brain (11). Urea levels are also independent of brain injury.

3.1.2. Continuous Cerebral Microdialysis

Continuous or online CM is now in use in several centers for research purposes. This technique continuously and automatically takes online measurements of the analytes of interest without the need for collection of dialysate into vials at specific time intervals. Instead the dialysate is analyzed electrochemically in real time using microchip electrophoresis, and analyte concentrations can be estimated every 30–60 s. Higher flow rates are required (2 μL/min) to allow such frequent measurements since the minimum volume required for analysis is 1 μL (12). The advantage of using rapid sampling CM is that short-lived pathological events can be detected. However, there are currently only a limited number of substances that can be analyzed using this method.

3.2. Catheter Placement

Because the volume of tissue that can be monitored by a microdialysis catheter is limited to a few cubic millimeters around its tip, only local tissue biochemistry is monitored. It is therefore crucial that the catheter tip is accurately placed in the tissue of interest (13). The ideal location of the microdialysis catheter was debated at a meeting of microdialysis experts in Stockholm in 2004 when it was suggested that, in order to detect the early biochemical changes that are associated with brain injury, the catheter should be placed in the tissue most at risk of secondary insults (14). In traumatic brain injury (TBI) this is the area of tissue adjacent to a focal lesion (pericontusional area) and in aneurysmal subarachnoid hemorrhage (SAH) the parent vascular territory, i.e., the area most at risk from vasospasm.

A microdialysis catheter can be inserted via a cranial access device or during a surgical procedure when it is brought out through the craniotomy and tunneled subcutaneously. The latter allows placement under direct vision and the area of interest can therefore be more accurately identified. Whichever method is used, the dura should be incised prior to insertion in order to prevent damage to the microdialysis membrane during catheter placement. Because CM data can only be reliably interpreted if the exact location of the catheter, and its proximity to abnormal tissue, is known, this should be verified by imaging. Commercially available catheters have an incorporated gold tip to allow visualization on computed tomography.

3.3. Insertion Artifacts

The insertion of the microdialysis catheter causes disturbance to the surrounding brain parenchyma because of temporary disruption to the blood–brain barrier, decreased regional cerebral blood flow, electrolyte shifts between intra- and extracellular compartments, and a local inflammatory response (2). A "run-in" period of around 1 h, during which microdialysis values are unreliable, should therefore be allowed after catheter placement (14).

3.4. Quantitative Analysis

Although the trends in the level of an analyte are sufficient in most clinical circumstances, some applications (e.g., drug studies) require an assessment of actual ECF concentration. As well as taking into account factors that influence relative recovery as discussed earlier (see Sect. 2.2), analytical models must be used if assessment of actual ECF concentration is required.

The extrapolation-to zero-flow model is most commonly used to determine true interstitial concentration. It makes use of the different dialysate concentrations obtained by varying the flow rate

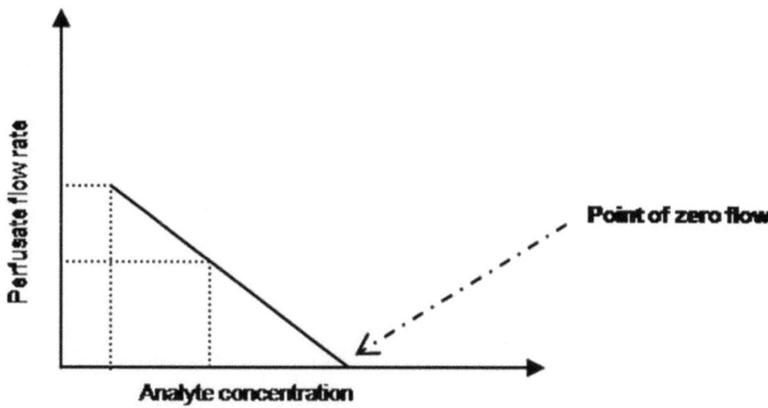

Fig. 2. A natural log plot showing the exponential relationship between perfusate flow rate and analyte concentration. The measured concentration at different flow rates can be extrapolated to that at zero flow, when full equilibration across the dialysis membrane can be assumed. This represents the true tissue concentration.

of the perfusate and then using logistic regression analysis to extrapolate the concentration that would be obtained at zero flow (Fig. 2), i.e., when full equilibration across the dialysis membrane can be assured (15). The method assumes that, at zero flow, dialysis sampling does not influence ECF analyte concentrations.

The no net flux method is an alternative steady-state model that involves adding the analyte in question to the perfusion fluid at varying concentrations and then calculating the difference in the concentration between perfusate and dialysate (16). When this difference is zero, the concentration in the perfusate is considered to be the true ECF concentration.

4. Bedside Biomarkers of Ischemia and Cell Damage

Fluctuations in the delivered levels of oxygen and glucose affect the metabolic state of brain cells and the resultant metabolic products that are released into the ECF. By analyzing changes in the interstitial compartment, CM can monitor how brain tissue reacts to pathophysiological events and responds to treatment. Each sampled substance can act as a marker of a particular cellular process associated with hypoxia, ischemia, and cellular energy failure (Fig. 3). Consensus guidelines suggest that lactate, pyruvate, glucose, glycerol, and glutamate are the most useful biochemical variables in the clinical setting (5, 14, 17, 18).

4.1. Lactate, Pyruvate, and the Lactate: Pyruvate Ratio

Glucose is metabolized to pyruvate in a process that yields two molecules of adenosine 5′-triphosphate (ATP) for each molecule of glucose. Under aerobic conditions, the majority of pyruvate then enters the highly efficient, energy-producing citric acid cycle where

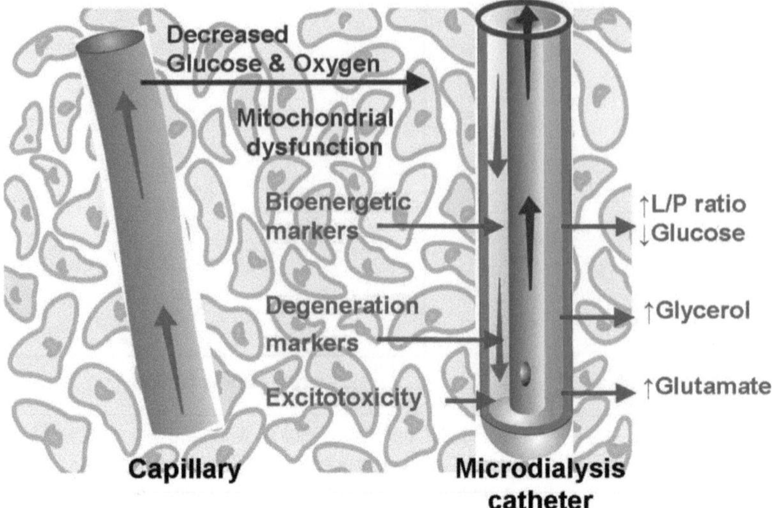

Fig. 3. Schematic representation of a blood capillary and microdialysis catheter in brain tissue. The concentration of substrate in the collected fluid (the microdialysate) is related to the balance between substrate delivery to, and uptake/excretion from, the brain extracellular fluid as well as to pathophysiological perturbations. Reproduced with permission from Br J Anaesth 2006; 97:18–25.

subsequent metabolism, through electron complex-mediated reduction of oxygen, yields another 36 molecules of ATP. However, in the absence of oxygen, pyruvate undergoes anaerobic metabolism to lactic acid (lactate in solution) outside the citric acid cycle, with a low yield of ATP. Mitochondrial failure, which is common after brain injury, also results in higher lactate concentrations because of the inability of cells to utilize delivered oxygen and glucose. An increase in the lactate concentration also occurs secondary to increased cellular metabolism but, in this case, pyruvate levels will also rise. The ECF lactate-to-pyruvate ratio (LPR) therefore provides more useful clinical information than the absolute levels of lactate and pyruvate in isolation. Two types of elevated LPR, related to the mechanism of the physiological perturbations described above, are therefore described (19). Type 1 changes are associated with reduced ECF pyruvate and elevated lactate concentrations secondary to classic ischemia, whereas the type 2 pattern occurs when reduced pyruvate is the predominant metabolic perturbation, a situation that reflects impairment of the glycolytic pathway in the presence of adequate (or reduced) glucose supply (20, 21). LPR is a sensitive biomarker of tissue ischemia and elevations correlate well with clinical events such as intracranial pressure (ICP) plateau waves (22), as well as with the severity of brain injury and outcome (23, 24). Because lactate and pyruvate have very similar molecular weights, the LPR is independent of catheter recovery in vivo (25).

The LPR in normal subjects is around 20, and levels >25 are often considered to be an early sign of metabolic abnormalities (18). However, a tissue hypoxic threshold for an abnormal LPR has not been clearly established (26) and different thresholds for abnormality, ranging from 20 to 40, are applied in clinical studies.

4.2. Glucose

Although glucose is the main energy source of the brain, there are no substantial cerebral stores of glucose and a continued supply is essential for maintenance of normal cerebral metabolism. The balance between glucose delivery and demand is altered after brain injury and brain ECF glucose is an important biomarker of altered cerebral metabolism. Low brain ECF glucose concentration can be a reflection of decreased cerebral glucose delivery because of hypoperfusion or systemic hypoglycemia, or hyperglycolysis secondary to hypoxia/ischemia-induced anaerobic metabolism. CM glucose should therefore be interpreted in light of the LPR, systemic glucose concentration, and, where possible, brain tissue PO_2 (18). An increased LPR in combination with low brain glucose is a sign of severe hypoxia/ischemia (27) and associated with poor outcome after TBI (28) and SAH (29). Whether preventing or reversing these metabolic abnormalities has an effect on outcome remains to be established.

The lactate-to-glucose ratio (LGR) is also a sensitive marker of hypoxia/ischemia and, in general, reflects sustained lactate production driven by hypoxia/ischemia-induced hyperglycolysis (30).

4.3. Glycerol

Glycerol is a breakdown product of phospholipids and therefore a useful biomarker of cell membrane degradation (31). Impaired cellular metabolism results in activation of phospholipases, cell membrane breakdown, and release of glycerol and fatty acids into the interstitial fluid. Glycerol is water soluble, easier to measure than fatty acids, and extensively studied after brain injury. Glycerol levels are very low in normal brain tissue (32) and a steep rise occurs following TBI (33, 34). However, since glycerol can leak from the plasma through the blood–brain barrier, high brain ECF glycerol levels may also be due to systemic factors such as triglyceride breakdown (5). High brain ECF glycerol levels are associated with cerebral ischemia and secondary adverse events after brain injury, with levels usually peaking within the first 24 h (33).

4.4. Glutamate

Brain ECF glutamate levels increase in the presence of ischemia (27, 35). The presence of a high concentration of interstitial glutamate was thought to cause excitotoxicity via calcium influx through glutamate-mediated ion channels, and studies have repeatedly demonstrated beneficial effects of glutamate receptor blockade on neurons in culture (36). However the concept of glutamate-induced excitotoxicity as a mechanism leading to secondary brain injury has been challenged because the levels of extracellular glutamate observed in vivo are not high enough to produce the depolarizations that are observed in experimental models (37). In any case, glutamate measured with CM does not necessarily reflect the glutamate levels at the synaptic cleft, where the effect on glutamate receptors is thought to occur.

Although the exact mechanism of glutamate-induced cell damage is still unclear, several studies have shown that high CM glutamate levels are associated with poor outcome after TBI (27, 38) and SAH (25, 39). Furthermore a low glutamate level (<10 μmol/L) is associated with a good outcome after brain injury (40). Further studies are required to determine whether CM-monitored glutamate is sensitive enough to be used as a reliable primary endpoint in studying the effects of interventions after ABI.

4.5. Interpretation of Biochemical Variables

Analysis of the temporal profile of biochemical markers after brain injury reveals fluctuating values, reflecting the changing metabolic state. This makes it difficult to interpret "one off" measurements in isolation and the "LTC-method" has been proposed as a way of interpreting microdialysis variables:

1. The *level* of measured microdialysis variable is compared to the physiological range (Table 1).
2. The *trend* of the measured variable over recent hours is reviewed.
3. A *comparison* is made between microdialysis and other physiological variables, such as ICP, cerebral perfusion pressure (CPP), and brain tissue PO_2.

Table 1
Baseline concentrations of the biochemical markers that are commonly measured during neurointensive care

Analyte	Normal value ± SD Reinstrup et al. (32)	Normal value ± SD Schulz et al. (30)
Glucose (mmol l^{-1})	1.7 ± 0.9	2.1 ± 0.2
Lactate (mmol l^{-1})	2.9 ± 0.9	3.1 ± 0.3
Pyruvate (μmol l^{-1})	166 ± 47	151 ± 12
Lactate:pyruvate ratio	23 ± 4	19 ± 2
Glycerol (μmol l^{-1})	82 ± 44	82 ± 12
Glutamate (μmol l^{-1})	16 ± 16	14.0 ± 3.3

These values are derived from microdialysis samples obtained from uninjured regions of human brain during wakefulness (30, 32)

5. Clinical Applications

Acute brain injury is frequently exacerbated by secondary events that lead to secondary brain injury and adverse clinical outcome. The initial injury activates an auto-destructive cascade of metabolic, immunological, and biochemical changes that ultimately results in irreversible cell damage or death (41). The pathophysiological processes of secondary brain injury are incompletely understood, but failing cellular metabolism, calcium overload, increased production of free radicals, and release of neurotoxic levels of excitatory amino acids are implicated. The neurointensive care management of acute brain injury is aimed at preventing or minimizing the burden of secondary injury. Monitoring brain tissue biochemistry has the potential to identify impending or early-onset secondary injury, and allows timely and individualized treatment strategies that might potentially improve outcome (18).

CM has been extensively studied during the neurointensive care management of TBI, SAH, ischemic stroke, and epilepsy. In addition the technique has relevance for intra-operative monitoring, and during pharmacokinetic studies involving drug research and development.

5.1. Traumatic Brain Injury

TBI is the leading cause of death and disability in children and young adults and secondary brain injury, which continues to evolve after the initial insult, plays a major role in its poor outcome (42). Several linked mechanisms, including ionic, metabolic, inflammatory, and immunological changes, are activated by the primary insult and lead to a progressive sequence of events that result in secondary (predominantly ischemic) injury and subsequent axonal death (41). Brain monitoring can identify some of these pathological processes and is now a standard of care in many neurointensive care units. Multimodal monitoring in brain injury goes beyond the measurement of systemic physiology, such as cardiorespiratory variables, but seeks also to determine the state of the injured brain with techniques such as ICP, cerebral oxygenation (jugular venous oxygen saturation and brain tissue PO_2), cerebral blood flow monitoring, and, more recently, analysis of brain tissue biochemistry using CM (43). Since energy failure after TBI can be related to mitochondrial dysfunction in the absence of vascular or tissue hypoxia, clinical monitors that identify only ischemia (such as brain tissue PO_2 and jugular venous oximetry) are unable to monitor the entire pathophysiological process. The addition of CM to the multimodal monitoring array therefore brings additional and important information that may be used to guide clinical management. The importance of catheter placement has been discussed previously but its particular relevance during monitoring in TBI is illustrated in Fig. 4.

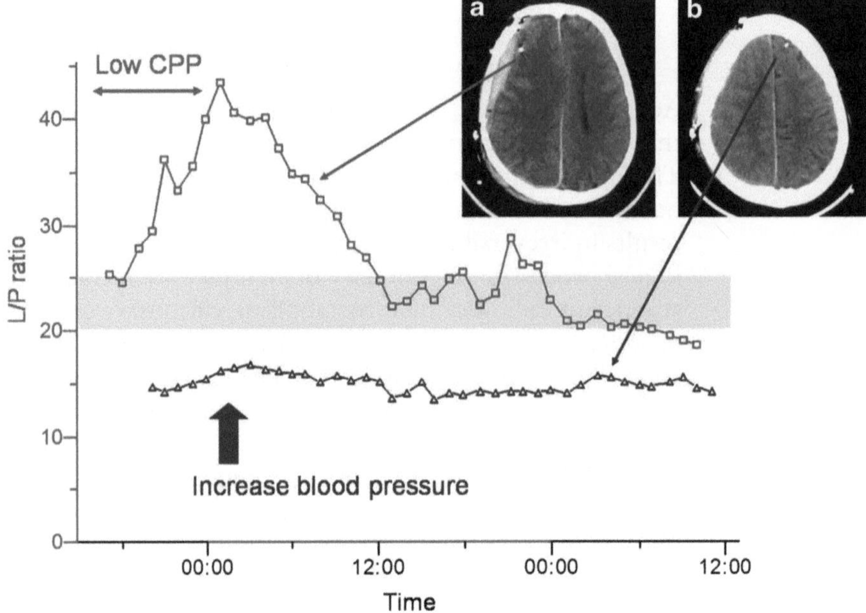

Fig. 4. Changes in lactate:pyruvate (L/P) ratio in "at-risk" (**a**) and normal (**b**) brain tissue during a period of low and normal cerebral perfusion pressure. The normal range for the L/P ratio is shown by the *shaded* area. Note the rise in L/P ratio in the "at-risk" tissue during a period of cerebral hypoperfusion which returns to normal following restoration of adequate cerebral perfusion. This is in contrast to the normal values that are measured throughout by the catheter in normal brain. Modified with permission from Br J Anaesth 2006; 97:18–25.

5.1.1. Typical Changes in Biomarkers

Various studies have demonstrated correlation between adverse clinical events, such as high ICP, low CPP and systemic hypoxemia, and abnormalities in brain ECF lactate, glycerol, glucose, and LPR after TBI. Brain ECF glucose concentration is typically reduced after injury and often associated with severe tissue hypoxia/ischemia and poor outcome (44). However, regional hyperglycolysis rather than critical supply of glucose and oxygen has been implicated as the cause of cerebral hypoglycemia in some circumstances (21). A brain ECF glucose concentration consistently less than 0.66 mmol/L in the first 50 h post TBI is associated with poor outcome and may occur independently of tissue ischemia (28). However, there are no data confirming whether interventions targeted to increase brain glucose influence outcome.

An increase in LPR correlates with the severity of injury and is also associated with a poor outcome after TBI (45, 46). Elevations in LPR may precede a rise in ICP, thereby providing early warning of brain energy crisis, potentially before irreversible tissue damage has occurred (47). Whether this can be exploited to bring forward the window for therapeutic intervention remains to be seen. An increase in LPR has also been shown to be detrimental in normal-appearing post-traumatic frontal lobe. In one study, persistently elevated LPR was found to be predictive of tissue atrophy after 6 months (48). The predilection of atrophy in the frontal lobes, as opposed to global brain atrophy, is unclear but might be due to regional variance in ICP and the tendency for contusional injury in the frontal lobes. Given that the LPR often remains elevated for a considerable period of time after TBI, the window of opportunity for neuroprotection might be longer than previously thought.

Brain ECF glycerol concentration is typically elevated in the first 24 h after TBI, with four- or eightfold increases reported in severe focal or global ischemia, respectively (30). Glycerol concentration declines during the ensuing 72 h, with subsequent rises being associated with adverse secondary ischemic events including intracranial hypertension, low CPP, low brain tissue PO_2, and seizures (33). However, not all secondary insults are associated with secondary increases in glycerol concentration and most studies show no correlation between glycerol levels and outcome after TBI. The clinical role of monitoring brain ECF glycerol therefore remains uncertain. Brain ECF glutamate levels may increase to levels 6–20 times normal in the first few hours after TBI in some patients (23). These high levels usually persist for 12–18 h but in some patients may be sustained for many days. Again, the clinical relevance of these findings is uncertain.

5.1.2. Guiding Treatment

There is an accumulating body of evidence to support the application of CM to guide treatment after TBI.

Cerebral Perfusion Pressure

Defining and maintaining an optimal CPP is a key component of the neurointensive care management of severe TBI, and the impact of different levels of CPP on markers of glucose metabolism has been widely investigated. In a study in which microdialysis catheters were placed in “at-risk” tissue (adjacent to peri-contusional areas) and in non-affected frontal (“normal”) tissue, lactate concentration was higher in “at-risk” compared to “normal” brain tissue (49). Lactate and LPR increased when CPP fell below 50 mmHg in “at-risk” tissue, suggesting that CM has potential to identify the safe lower limit of CPP. However, other studies have suggested that elevation in LPR is unrelated to CPP values that are customarily considered to be adequate, reinforcing the importance of using physiological data from multiple sources to guide individualized patient management (50).

Systemic Glucose Management

Both systemic hyperglycemia and hypoglycemia are detrimental to the injured brain but the optimal target for systemic glucose control after ABI is unclear. Although blood glucose concentration influences brain metabolism after ABI, brain ECF and blood glucose concentrations are not always related; both cerebral low and high glucose episodes can occur independently of blood glucose concentration. Systemic hyperglycemia (glucose >15 mmol/L) is widely associated with increased LPR after TBI (51, 52), but control of elevated blood glucose with insulin infusion can be associated with an increased incidence of cerebral metabolic distress (53).

Tight systemic glucose control (4.4–6.7 mmol/L) has been associated with a greater prevalence of cerebral hypoglycemia (65 vs. 36%) and brain energy crisis (25 vs. 17%) than intermediate level control (6.8–10.0 mmol/L) (54). In this study, cerebral glucose concentration was significantly lower in non-survivors than in survivors and brain energy crisis (low brain ECF glucose in association with elevated LPR) was associated with significantly increased hospital mortality. Therefore, although hyperglycemia is associated with unfavorable outcome after TBI, its treatment might itself adversely affect the injured brain (55).

Hypothermia

The application of therapeutic hypothermia has been widely studied after TBI and recent evidence suggests that it might have preferential effects in "at-risk" brain tissue. In one study, LPR and LGR, and glycerol concentration, were significantly decreased in peri-lesional tissue during hypothermia, raising the possibility that biochemical variables might provide potential therapeutic targets for induced hypothermia after brain injury (56).

Seizures

Non-convulsive seizures are common after TBI and result in elevated LPR, possibly reflecting a situation where seizure-induced increases in tissue energy demands are not being adequately met (57). Cortical spreading depolarization (CSD) is increasingly recognized as a cause of secondary brain injury, has been identified in up to 50% of patients after TBI, and is an independent prognostic factor (58). Since CSD lasts only 2–5 min on average, online CM is the only method that has the temporal resolution to monitor the neurometabolic effects of spreading depressions. Depletion of brain ECF glucose and a rise in lactate concentration have been observed using online CM, suggesting a local energy imbalance caused by an insufficient glucose supply as a result of the CSD (59). The severity of the glucose depletion is proportional to the number of depolarizations, and it is likely that a vicious cycle is established whereby CSD leads to glucose depletion and bioenergetic distress, which in turn lead to further depolarization waves.

5.1.3. Outcome

Recently, a large observational study of 223 patients suggested that brain ECF metabolic markers are independently associated with outcome after severe TBI (60). Outcome was assessed at 6 months post injury and, averaged over the total monitoring period, levels of glutamate and LPR were significantly higher in patients who died. During the initial 72 h of monitoring, median glycerol levels were higher in those who died and LPR and lactate levels lower in patients with a favorable outcome. Brain ECF glucose and LPR were significant independent positive predictors of mortality, whereas pyruvate was an independent negative predictor. As the authors of this study observed, it remains to be established whether treatment-related improvement in biochemistry translates into improved outcome.

5.2. Subarachnoid Hemorrhage

The early case fatality of SAH has fallen in the last decade because of early intervention to secure the ruptured aneurysm and improved recognition and management of complications. The latter has partly been driven by advanced monitoring techniques, including CM, that are able to detect the tissue sequelae of cerebral vasospasm, one of the major causes of death and disability after SAH. Significantly elevated levels of lactate and pyruvate reflect the hypermetabolic state of the injured brain after SAH and are associated with global edema on the initial CT scan (61). Low brain ECF glucose and high LPR are associated with a poor outcome after SAH (29, 62).

CM allows early detection of the metabolic changes that are associated with vasospasm-related ischemia. There is usually a biphasic disturbance in brain tissue biochemistry after SAH, with the degree of the disturbance on day 1 reflecting the severity of the initial hemorrhage (40). The biochemical abnormalities then return towards normal but secondary elevations in the LPR, lactate, and glutamate concentrations, and reductions in glucose, are seen in some patients between days 5 and 10 and are related to vasospasm-induced ischemia (63). In one study, patients with an uneventful clinical course after SAH had stable biomarker levels whereas those with clinical signs of cerebral ischemia had elevations in lactate and glutamate levels (64). Patients who develop a delayed ischemic neurological deficit (DIND) have significantly higher lactate, glutamate, and LPR compared with those who remain asymptomatic (30). Ischemia-related biochemical abnormalities can be reversed following the institution of standard treatments for vasospasm, suggesting that they might be a target for treatment.

The biochemical changes associated with cerebral ischemia may precede the onset of the clinical symptoms of vasospasm. In one study of 42 patients with SAH, an ischemic biochemical pattern, defined in this study as a greater than 20% increase in the LPR and LGR from baseline followed by a 20% increase in glycerol concentration, was identified in 17 of 18 patients who developed a DIND and in only 3 of 24 who did not (65). The ischemic pattern preceded the onset of DIND in all 17 cases and the mean delay from the peak in the LPR or LGR to the occurrence of the DIND was 23 h (range 4–50 h). Identification of early changes in brain tissue biochemistry potentially brings forward the treatment window for cerebral vasospasm but the clinical significance of these findings remains to be determined.

5.3. Epilepsy

When microdialysis is carried out in conjunction with electroencephalography (EEG) monitoring, the effect of seizure activity on the ECF composition can be analyzed. Elevations of glutamate (up to 84-fold) were associated with spontaneous seizure activity in a study of four patients undergoing hippocampal EEG and CM monitoring during temporal lobe epilepsy surgery (66). A rise in the level of glutamate has been shown to precede seizure onset (67) and a transient rise in lactate has been observed during secondary generalized seizures (68).

Exogenous compounds can also be analyzed with microdialysis and the application of CM to monitor antiepileptic drug efficacy has been explored (see Sect. 6.4). Central nervous system (CNS) drug pharmacokinetics can be better studied with CM since plasma concentrations do not necessarily reflect local concentrations in the brain. Monitoring antiepileptic drug concentration with microdialysis can provide decisive information about the ideal dose and/or drug delivery method (69).

6. Future Directions of Clinical Microdialysis

The majority of early clinical CM research involved bedside measurement of analytes but these represent only a tiny proportion of the potential markers of brain pathology and cellular metabolic function. Since CM provides a facsimile of brain ECF that contains all molecules small enough to pass through the microdialysis membrane, any pathological process that results in changes in the biochemical composition of the ECF can theoretically be monitored. The advent of commercially available high-molecular-weight-cutoff microdialysis membranes (100 KD) has allowed macromolecules to be sampled in addition to standard bedside variables.

6.1. Cytokines

It is well known that there is an intense immunological and inflammatory response after human brain injury and that this is detrimental and leads to further, and possibly, irreversible cell damage. There is therefore an increasing interest in the role of inflammatory mediators such as cytokines in brain injury pathophysiology. Recently the temporal profile of 42 cytokines has been analyzed following TBI (70). The levels of 16, including tumor necrosis factor, interleukin (IL)7, IL8, and IL1β, peaked during day 1, whereas the levels of four cytokines, including IL1(ra) and IL6, peaked on day 2, and 19 cytokines peaked between days 4 and 5. There was a large variation in cytokine response between different patients with similar severity and patterns of brain injury and also marked heterogeneity between brain and systemic cytokine concentrations, suggesting that plasma markers cannot reliably be used to quantify cerebral changes. A 6-h period of sample collection is currently required because relatively large dialysate volumes are needed for cytokine analysis. Therefore, although cytokine levels measured using CM present a potential therapeutic target, the methodology needs to be refined before this becomes a reality.

6.2. Other Novel Biomarkers

6.2.1. S100β

The calcium-binding astrocyte protein S100β is released during brain injury and elevated plasma and CSF levels have been associated with secondary injury. It is possible to measure brain ECF S100β using CM and this might provide a more sensitive and specific monitoring of secondary injurious processes than measurement in plasma or CSF (71).

6.2.2. Nitric Oxide

Nitric Oxide (NO) plays a key role in normal CNS function as well as in pathological processes. NO is a reactive molecule and difficult to detect directly, but endogenous production can be estimated by measurement of the concentration of its downstream metabolites nitrate and nitrite—NO(x). Increasing NO(x) is associated with decreasing LPR and lactate, and increasing glucose, suggesting that higher concentrations of NO are associated with more favorable metabolism in injured brains (72). Dialysate NO(x) levels have also been shown to be significantly lower in patients with reduced cerebral blood flow after head injury (73).

6.2.3. Amino Acids

As well as the bedside measurement of glutamate, the temporal profiles of multiple interstitial amino acids have been studied using CM. Increases in non-transmitter amino acids are seen early after SAH and large increases (up to 1,350-fold) in gamma-amino-butyric acid, glutamate, and aspartate concentration have also been identified during cerebral ischemia (74). *N*-Acetylaspartate (NAA) is a neuronal marker present in high concentrations in the CNS and synthesized almost exclusively in the mitochondrion of neurons. CM studies have confirmed that brain ECF NAA concentration is more than 30% lower in non-survivors than survivors of TBI (75). A non-recoverable fall in NAA occurs in non-survivors from day 4 onwards and is associated with a rise in LPR and glycerol concentration. NAA is therefore a potential candidate marker for monitoring therapeutic strategies aimed at preserving mitochondrial function after ABI.

6.3. Proteomic Research

Application of proteomics to CM is an exciting field of research with potential to provide new insights into the pathophysiology of brain injury. Using 100 kDa catheters and a combination of electrophoresis and mass spectrometry, ten proteins that are not present in CSF following acute ischemic stroke have been identified (76). Metabolic distress after TBI has also been shown to be associated with a differential proteome indicating cellular destruction (77). A study investigating a proteome-wide screening identified increases in several isoforms of glyceraldehyde-3-phosphate dehydrogenase in patients with symptomatic vasospasm after SAH, on average 3.8 days before the onset of clinical symptoms (78).

6.4. Clinical Drug Trials

The estimation of drug concentration with CM allows evaluation of the effects of new neuro-protective drugs at the clinically relevant target site. Furthermore, in vivo microdialysis measures the free level of a drug and it is this unbound fraction that is pharmacologically active. CM therefore provides an excellent platform for drug delivery studies (79).

Unless a drug is able to reach sufficient levels within the tissue of interest, a good clinical response in unlikely. Direct assessment of intratumoral drug concentration is important in the development of new brain cancer drugs as well as in assessing efficacy in clinical trials. CM has been used in animal studies to investigate the intratumoral dose–response relationship for cisplatin and demonstrated variable penetration of the drug into the tumor that was in turn associated with a variable tumor response to treatment (80). By altering the constituents of the perfusate, the microdialysis catheter can be used for stereotactic drug delivery to specific structural targets in the brain. This is called retrodialysis and currently remains a research technique (81).

Acknowledgement

M.S. is partly funded by the Department of Health's National Institute for Health Research funding scheme via the UCLH/UCL Biomedical Research Centre.

References

1. Bito L, Davson H, Levin E et al (1966) The concentrations of free amino acids and other electrolytes in cerebrospinal fluid, in vivo dialysate of brain, and blood plasma of the dog. J Neurochem 13:1057–1067
2. Benveniste H, Huttemeier PC (1990) Microdialysis–theory and application. Prog Neurobiol 35:195–215
3. Lindefors N, Amberg G, Ungerstedt U (1989) Intracerebral microdialysis: I. Experimental studies of diffusion kinetics. J Pharmacol Methods 22:141–156
4. Ungerstedt U (1991) Microdialysis–principles and applications for studies in animals and man. J Intern Med 230:365–373
5. Hillered L, Vespa PM, Hovda DA (2005) Translational neurochemical research in acute human brain injury: the current status and potential future for cerebral microdialysis. J Neurotrauma 22:3–41
6. Hutchinson PJ, O'Connell MT, Al-Rawi PG et al (2000) Clinical cerebral microdialysis: a methodological study. J Neurosurg 93:37–43
7. Hutchinson PJ, O'Connell MT, Nortje J et al (2005) Cerebral microdialysis methodology–evaluation of 20 kDa and 100 kDa catheters. Physiol Meas 26:423–428
8. Hillman J, Aneman O, Anderson C et al (2005) A microdialysis technique for routine measurement of macromolecules in the injured human brain. Neurosurgery 56:1264–1268
9. Westerink BH, De Vries JB (1988) Characterization of in vivo dopamine release as determined by brain microdialysis after acute and subchronic implantations: methodological aspects. J Neurochem 51:683–687
10. Nordstrom CH (2010) Cerebral energy metabolism and microdialysis in neurocritical care. Childs Nerv Syst 26:465–472
11. Ronne-Engstrom E, Cesarini KG, Enblad P et al (2001) Intracerebral microdialysis in neurointensive care: the use of urea as an endogenous reference compound. J Neurosurg 94:397–402
12. Parkin M, Hopwood S, Jones DA et al (2005) Dynamic changes in brain glucose and lactate in pericontusional areas of the human cerebral cortex, monitored with rapid sampling on-line microdialysis: relationship with depolarisation-like events. J Cereb Blood Flow Metab 25: 402–413

13. Engstrom M, Polito A, Reinstrup P et al (2005) Intracerebral microdialysis in severe brain trauma: the importance of catheter location. J Neurosurg 102:460–469
14. Bellander BM, Cantais E, Enblad P et al (2004) Consensus meeting on microdialysis in neurointensive care. Intensive Care Med 30: 2166–2169
15. Krebs-Kraft D, Frantz K, Parent M (2007) In vivo microdialysis: a method for sampling extracellular fluid in discrete brain regions. In: Lajtha A (ed) Handbook of neurochemistry and molecular neurobiology, practical neurochemistry methods. Springer, New York
16. Lonnroth P, Jansson PA, Smith U (1987) A microdialysis method allowing characterization of intercellular water space in humans. Am J Physiol 253:E228–E231
17. Goodman JC, Robertson CS (2009) Microdialysis: is it ready for prime time? Curr Opin Crit Care 15:110–117
18. Tisdall MM, Smith M (2006) Cerebral microdialysis: research technique or clinical tool. Br J Anaesth 97:18–25
19. Hillered L, Persson L, Nilsson P et al (2006) Continuous monitoring of cerebral metabolism in traumatic brain injury: a focus on cerebral microdialysis. Curr Opin Crit Care 12:112–118
20. Hlatky R, Valadka AB, Goodman JC et al (2004) Patterns of energy substrates during ischemia measured in the brain by microdialysis. J Neurotrauma 21:894–906
21. Vespa P, Bergsneider M, Hattori N et al (2005) Metabolic crisis without brain ischemia is common after traumatic brain injury: a combined microdialysis and positron emission tomography study. J Cereb Blood Flow Metab 25: 763–774
22. Persson L, Hillered L (1992) Chemical monitoring of neurosurgical intensive care patients using intracerebral microdialysis. J Neurosurg 76:72–80
23. Hutchinson PJ, Gupta AK, Fryer TF et al (2002) Correlation between cerebral blood flow, substrate delivery, and metabolism in head injury: a combined microdialysis and triple oxygen positron emission tomography study. J Cereb Blood Flow Metab 22:735–745
24. Stahl N, Mellergard P, Hallstrom A et al (2001) Intracerebral microdialysis and bedside biochemical analysis in patients with fatal traumatic brain lesions. Acta Anaesthesiol Scand 45: 977–985
25. Persson L, Valtysson J, Enblad P et al (1996) Neurochemical monitoring using intracerebral microdialysis in patients with subarachnoid hemorrhage. J Neurosurg 84:606–616
26. Johnston AJ, Steiner LA, Coles JP et al (2005) Effect of cerebral perfusion pressure augmentation on regional oxygenation and metabolism after head injury. Crit Care Med 33:189–195
27. Hutchinson PJ, Al-Rawi PG, O'Connell MT et al (2000) Biochemical changes related to hypoxia during cerebral aneurysm surgery: combined microdialysis and tissue oxygen monitoring: case report. Neurosurgery 46:201–205
28. Vespa PM, McArthur D, O'Phelan K et al (2003) Persistently low extracellular glucose correlates with poor outcome 6 months after human traumatic brain injury despite a lack of increased lactate: a microdialysis study. J Cereb Blood Flow Metab 23:865–877
29. Kett-White R, Hutchinson PJ, Al-Rawi PG et al (2002) Adverse cerebral events detected after subarachnoid hemorrhage using brain oxygen and microdialysis probes. Neurosurgery 50:1213–1221
30. Schulz MK, Wang LP, Tange M, Bjerre P (2000) Cerebral microdialysis monitoring: determination of normal and ischemic cerebral metabolisms in patients with aneurysmal subarachnoid hemorrhage. J Neurosurg 93:808–814
31. Hillered L, Valtysson J, Enblad P, Persson L (1998) Interstitial glycerol as a marker for membrane phospholipid degradation in the acutely injured human brain. J Neurol Neurosurg Psychiatry 64:486–491
32. Reinstrup P, Stahl N, Mellergard P et al (2000) Intracerebral microdialysis in clinical practice: baseline values for chemical markers during wakefulness, anesthesia, and neurosurgery. Neurosurgery 47:701–709
33. Clausen T, Alves OL, Reinert M et al (2005) Association between elevated brain tissue glycerol levels and poor outcome following severe traumatic brain injury. J Neurosurg 103:233–238
34. Marklund N, Salci K, Lewen A, Hillered L (1997) Glycerol as a marker for post-traumatic membrane phospholipid degradation in rat brain. Neuroreport 8:1457–1461
35. Bullock R, Zauner A, Myseros JS et al (1995) Evidence for prolonged release of excitatory amino acids in severe human head trauma. Relationship to clinical events. Ann N Y Acad Sci 765:290–297
36. Rothman S (1984) Synaptic release of excitatory amino acid neurotransmitter mediates anoxic neuronal death. J Neurosci 4:1884–1891
37. Obrenovitch TP (1999) High extracellular glutamate and neuronal death in neurological disorders. Cause, contribution or consequence? Ann N Y Acad Sci 890:273–286
38. Koura SS, Doppenberg EM, Marmarou A et al (1998) Relationship between excitatory amino

acid release and outcome after severe human head injury. Acta Neurochir Suppl 71:244–246

39. Runnerstam M, von EC, Nystrom B et al (1997) Extracellular glial fibrillary acidic protein and amino acids in brain regions of patients with subarachnoid hemorrhage–correlation with level of consciousness and site of bleeding. Neurol Res 19:361–368
40. Staub F, Graf R, Gabel P et al (2000) Multiple interstitial substances measured by microdialysis in patients with subarachnoid hemorrhage. Neurosurgery 47:1106–1115
41. Teasdale GM, Graham DI (1998) Craniocerebral trauma: protection and retrieval of the neuronal population after injury. Neurosurgery 43: 723–737
42. Langlois JA, Rutland-Brown W, Wald MM (2006) The epidemiology and impact of traumatic brain injury: a brief overview. J Head Trauma Rehabil 21:375–378
43. Tisdall MM, Smith M (2007) Multimodal monitoring in traumatic brain injury: current status and future directions. Br J Anaesth 99: 61–67
44. Goodman JC, Valadka AB, Gopinath SP et al (1999) Extracellular lactate and glucose alterations in the brain after head injury measured by microdialysis. Crit Care Med 27:1965–1973
45. Hutchinson PJ, Al-Rawi PG, O'Connell MT et al (2000) On-line monitoring of substrate delivery and brain metabolism in head injury. Acta Neurochir Suppl 76:431–435
46. Zauner A, Doppenberg EM, Woodward JJ et al (1997) Continuous monitoring of cerebral substrate delivery and clearance: initial experience in 24 patients with severe acute brain injuries. Neurosurgery 41:1082–1091
47. Belli A, Sen J, Petzold A et al (2008) Metabolic failure precedes intracranial pressure rises in traumatic brain injury: a microdialysis study. Acta Neurochir (Wien) 150:461–469
48. Marcoux J, McArthur DA, Miller C et al (2008) Persistent metabolic crisis as measured by elevated cerebral microdialysis lactate-pyruvate ratio predicts chronic frontal lobe brain atrophy after traumatic brain injury. Crit Care Med 36:2871–2877
49. Nordstrom CH, Reinstrup P, Xu W et al (2003) Assessment of the lower limit for cerebral perfusion pressure in severe head injuries by bedside monitoring of regional energy metabolism. Anesthesiology 98:809–814
50. Vespa PM, O'Phelan K, McArthur D et al (2007) Pericontusional brain tissue exhibits persistent elevation of lactate/pyruvate ratio independent of cerebral perfusion pressure. Crit Care Med 35:1153–1160
51. Diaz-Parejo P, Stahl N, Xu W et al (2003) Cerebral energy metabolism during transient hyperglycemia in patients with severe brain trauma. Intensive Care Med 29:544–550
52. Kerner A, Schlenk F, Sakowitz O et al (2007) Impact of hyperglycemia on neurological deficits and extracellular glucose levels in aneurysmal subarachnoid hemorrhage patients. Neurol Res 29:647–653
53. Vespa P, Boonyaputthikul R, McArthur DL et al (2006) Intensive insulin therapy reduces microdialysis glucose values without altering glucose utilization or improving the lactate/pyruvate ratio after traumatic brain injury. Crit Care Med 34:850–856
54. Oddo M, Schmidt JM, Carrera E et al (2008) Impact of tight glycemic control on cerebral glucose metabolism after severe brain injury: a microdialysis study. Crit Care Med 36:3233–3238
55. Oddo M, Schmidt JM, Mayer SA, Chiolero RL (2008) Glucose control after severe brain injury. Curr Opin Clin Nutr Metab Care 11:134–139
56. Wang Q, Li AL, Zhi DS, Huang HL (2007) Effect of mild hypothermia on glucose metabolism and glycerol of brain tissue in patients with severe traumatic brain injury. Chin J Traumatol 10:246–249
57. Vespa PM, Miller C, McArthur D et al (2007) Nonconvulsive electrographic seizures after traumatic brain injury result in a delayed, prolonged increase in intracranial pressure and metabolic crisis. Crit Care Med 35:2830–2836
58. Hartings JA, Strong AJ, Fabricius M et al (2009) Spreading depolarizations and late secondary insults after traumatic brain injury. J Neurotrauma 26:1857–1866
59. Feuerstein D, Manning A, Hashemi P et al (2010) Dynamic metabolic response to multiple spreading depolarizations in patients with acute brain injury: an online microdialysis study. J Cereb Blood Flow Metab 30:1343–1355
60. Timofeev I, Carpenter KL, Nortje J et al (2011) Cerebral extracellular chemistry and outcome following traumatic brain injury: a microdialysis study of 223 patients. Brain 134:484–494
61. Zetterling M, Hallberg L, Hillered L et al (2010) Brain energy metabolism in patients with spontaneous subarachnoid hemorrhage and global cerebral edema. Neurosurgery 66:1102–1110
62. Cesarini KG, Enblad P, Ronne-Engstrom E et al (2002) Early cerebral hyperglycolysis after subarachnoid haemorrhage correlates with favourable outcome. Acta Neurochir (Wien) 144:1121–1131

63. Sarrafzadeh AS, Haux D, Ludemann L et al (2004) Cerebral ischemia in aneurysmal subarachnoid hemorrhage: a correlative microdialysis-PET study. Stroke 35:638–643
64. Nilsson OG, Brandt L, Ungerstedt U, Saveland H (1999) Bedside detection of brain ischemia using intracerebral microdialysis: subarachnoid hemorrhage and delayed ischemic deterioration. Neurosurgery 45:1176–1184
65. Skjoth-Rasmussen J, Schulz M, Kristensen SR, Bjerre P (2004) Delayed neurological deficits detected by an ischemic pattern in the extracellular cerebral metabolites in patients with aneurysmal subarachnoid hemorrhage. J Neurosurg 100:8–15
66. Thomas PM, Phillips JP, Delanty N, O'Connor WT (2003) Elevated extracellular levels of glutamate, aspartate and gamma-aminobutyric acid within the intraoperative, spontaneously epileptiform human hippocampus. Epilepsy Res 54:73–79
67. During MJ, Spencer DD (1993) Extracellular hippocampal glutamate and spontaneous seizure in the conscious human brain. Lancet 341:1607–1610
68. During MJ, Fried I, Leone P et al (1994) Direct measurement of extracellular lactate in the human hippocampus during spontaneous seizures. J Neurochem 62:2356–2361
69. Clinckers R, Smolders I, Vermoesen K et al (2009) Prediction of antiepileptic drug efficacy: the use of intracerebral microdialysis to monitor biophase concentrations. Expert Opin Drug Metab Toxicol 5:1267–1277
70. Helmy A, Carpenter KL, Menon DK et al (2011) The cytokine response to human traumatic brain injury: temporal profiles and evidence for cerebral parenchymal production. J Cereb Blood Flow Metab 31:658–670
71. Sen J, Belli A, Petzold A et al (2005) Extracellular fluid S100B in the injured brain: a future surrogate marker of acute brain injury? Acta Neurochir (Wien) 147:897–900
72. Carpenter KL, Timofeev I, Al-Rawi PG et al (2008) Nitric oxide in acute brain injury: a pilot study of NO(x) concentrations in human brain microdialysates and their relationship with energy metabolism. Acta Neurochir Suppl 102:207–213
73. Hlatky R, Goodman JC, Valadka AB, Robertson CS (2003) Role of nitric oxide in cerebral blood flow abnormalities after traumatic brain injury. J Cereb Blood Flow Metab 23:582–588
74. Hutchinson PJ, O'Connell MT, Al-Rawi PG et al (2002) Increases in GABA concentrations during cerebral ischaemia: a microdialysis study of extracellular amino acids. J Neurol Neurosurg Psychiatry 72:99–105
75. Belli A, Sen J, Petzold A et al (2006) Extracellular N-acetylaspartate depletion in traumatic brain injury. J Neurochem 96:861–869
76. Cuadrado E, Rosell A, Colome N et al (2010) The proteome of human brain after ischemic stroke. J Neuropathol Exp Neurol 69: 1105–1115
77. Lakshmanan R, Loo JA, Drake T et al (2010) Metabolic crisis after traumatic brain injury is associated with a novel microdialysis proteome. Neurocrit Care 12:324–336
78. Maurer MH, Haux D, Sakowitz OW et al (2007) Identification of early markers for symptomatic vasospasm in human cerebral microdialysate after subarachnoid hemorrhage: preliminary results of a proteome-wide screening. J Cereb Blood Flow Metab 27:1675–1683
79. Helmy A, Carpenter KL, Hutchinson PJ (2007) Microdialysis in the human brain and its potential role in the development and clinical assessment of drugs. Curr Med Chem 14:1525–1537
80. Zamboni WC, Gervais AC, Egorin MJ et al (2002) Inter- and intratumoral disposition of platinum in solid tumors after administration of cisplatin. Clin Cancer Res 8:2992–2999
81. Blakeley J, Portnow J (2010) Microdialysis for assessing intratumoral drug disposition in brain cancers: a tool for rational drug development. Expert Opin Drug Metab Toxicol 6: 1477–1491

Chapter 2

In Vivo Microdialysis to Study Striatal Dopaminergic Neurodegeneration

Giuseppe Di Giovanni, Massimo Pierucci, Mauro Pessia, and Vincenzo Di Matteo

Abstract

Microdialysis cerebral technique has been widely employed in order to study neurotransmitter release. This technique presents numerous advantages such as it allows work with sample in vivo from freely moving animals. Different drugs in different points implanted probes in several brain areas can be infused simultaneously by means of microdialysis. Parkinson's disease (PD) is a progressive neurodegenerative disorder that is primarily characterized by the degeneration of dopamine (DA) neurons in the nigrostriatal system, which in turn produces profound neurochemical changes within the basal ganglia, representing the neural substrate for Parkinsonian motor symptoms. Over the years, a broad variety of experimental models of the disease have been developed and applied in diverse animal species. The two most common toxin models used employ 6-hydroxydopamine (6-OHDA) and the 1-methyl-4-phenyl-1,2,3,6-tetrahydropyridine/1-methyl-4-phenilpyridinium ion (MPTP/MPP$^+$), either given systemically or locally applied into the nigrostriatal pathway, to resemble PD features in animals. Both neurotoxins selectively and rapidly destroy catecholaminergic neurons, although with different mechanisms. Since in vivo microdialysis coupled to high-performance liquid chromatography (HPLC) is an established technique for studying physiological, pharmacological, and pathological changes of a wide range of low molecular weight substances in the brain extracellular fluid, here we describe a rapid and simple microdialysis technique that allows the direct quantitative study of the damage produced by 6-OHDA and MPP$^+$ toxins on dopaminergic (DAergic) striatal terminals of rat brain.

Key words: Parkinson's disease, In vivo microdialysis, Corpus striatum, MPP$^+$, 6-OHDA, ROS

1. Introduction

Parkinson's disease (PD) is a progressive neurodegenerative disorder that is primarily characterized by the degeneration of dopamine (DA) neurons in the nigrostriatal system, which in turn produces profound neurochemical changes within the basal ganglia (Fig. 1), representing the neural substrate for Parkinsonian motor

Giuseppe Di Giovanni and Vincenzo Di Matteo (eds.), *Microdialysis Techniques in Neuroscience*, Neuromethods, vol. 75,
DOI 10.1007/978-1-62703-173-8_2,

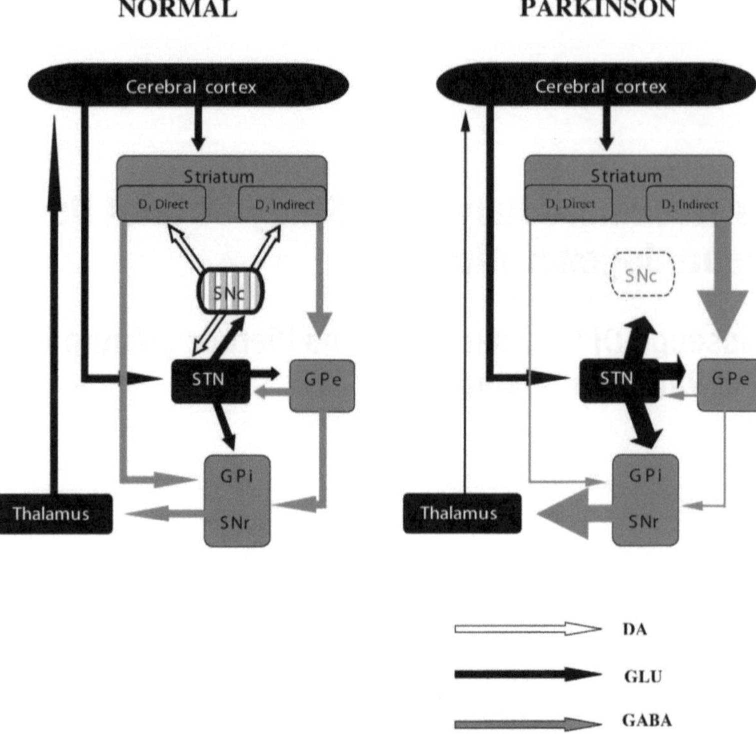

Fig. 1. Simplified diagram illustrating the changes occurring in the basal ganglia functional organization in Parkinson's disease, with respect to normal condition. Relative thickness of *arrows* indicate the degrees of activation of the transmitter pathways. The basal ganglia participate in larger circuits that also include cortex and thalamus. The striatum is the principal input structure of the basal ganglia, and the internal segment of the globus pallidus (GPi) and the substantia nigra pars reticulata (SNr) are the major output structures, projecting toward the thalamus and brainstem. According to conventional anatomical models, basal ganglia input and output structures are linked via a monosynaptic "direct" pathway and a polysynaptic "indirect" pathway that involves the external pallidal segment (GPe) and the subthalamic nucleus (STN). Dopamine released from terminals of the nigrostriatal (SNc) projection is thought to modulate basal ganglia activity by inhibiting activity along the "indirect" pathway through stimulation of dopamine D2 receptors and enhancing activity along the "direct" pathway by the stimulation of the dopamine D1 receptor. The same model has been applied to explain aspects of the pathophysiology of parkinsonism. Loss of striatal dopamine is believed to result in increased striatal inhibition of GPe, leading to disinhibition of STN neurons and to increased basal ganglia output from GPi and SNr. Increased and altered basal ganglia output to the thalamus is thought to disturb cortical processing, which is ultimately responsible for the development of many of the Parkinsonian motor signs. Dopamine (DA) *unfilled arrows*, glutamate (GLU) *black arrows*, γ-aminobutyric acid (GABA) *gray arrows*.

symptoms (1). The pathogenesis of the disease is still not completely understood, but environmental and genetic factors are thought to play relevant roles. Important factors include formation of free radicals, impaired mitochondrial activity, increased sensitivity to apoptosis, excitoxicity, and inflammation (1, 2). It appears clear that understanding the etiopathogenesis of PD, the modalities whereby the neurodegenerative process begins and progresses, is fundamental for the development of drugs to slow or prevent the progression of PD. Most of the progress in this field has been gained thanks to the toxin models of PD. Over the years, a broad variety of experimental models of the disease have been developed and applied in diverse animal species. The two most common toxin models used employ the 6-hydroxydopamine (6-OHDA) and the 1-methyl-4-phenyl-1,2,3,6-tetrahydropyridine/1-methyl-4-phenilpyridinium ion (MPTP/MPP^+), given either systemically or locally applied into the nigrostriatal pathway, to resemble PD features in animals. Both neurotoxins selectively and rapidly destroy catecolaminergic neurons, though with different mechanisms (3, 4). In this regard, in vivo microdialysis, a well-established method for monitoring the extracellular levels of neurotransmitters in the CNS (5) allows online estimates of neurotransmitters in living animals and is a suitable method for monitoring the extracellular levels of neurotransmitters during local administration of pharmacological agents. Different doses of a drug or a combination of agonists and antagonists can be administered in the same experiment without adding any fluid to extracellular spaces (6). Therefore, researchers have extensively used this method as a tool for the study of pathophysiologic changes in chemical processes of Parkinsonian brain, and particularly, the infusion of drugs (toxins) through the microdialysis cannula (Fig. 2) (*retrodialysis*), has permitted the development of some interesting animal models of PD as well as contemporary study of the effects of unilateral lesions of nigrostriatal dopamine neurons (7). Here we will describe a rapid and simple microdialysis technique that allows the direct quantitative study of the damage produced by 6-OHDA and/or MPP^+ toxins on DAergic striatal terminals of rat brain.

This rapid sampling procedure is completed in 2 days: in the first day of the experiment a fixed toxin (e.g., MPP^+ or 6-OHDA) dissolved in Ringer solution is perfused through the microdialysis probe, in a target area of the striatum for a short time (10–15 min) with a fixed concentration (generally 1–10 mM, or more, depending from the used substance), to induce neurodegeneration of the nigrostriatal system. Forty-eight hours after (day 2), the amount of DA released by the perfusion (challenge) of a second dose of MPP^+ could be indicative of the damage produced by a previous perfusion of a toxic compound, since it could be proportional to the number of remaining DAergic terminals (Fig. 3) and the decreases in DA overflow reflecting DA nerve terminal degeneration (8). Indeed, the massive DA extracellular output after the first MPP^+ or 6-OHDA perfusion is an index of DAergic cell disruption (9, 10). Thus, this method is suitable to study different drug-induced DAergic toxicity in the nigrostriatal system and also to estimate the quantitative damage induced by these toxins.

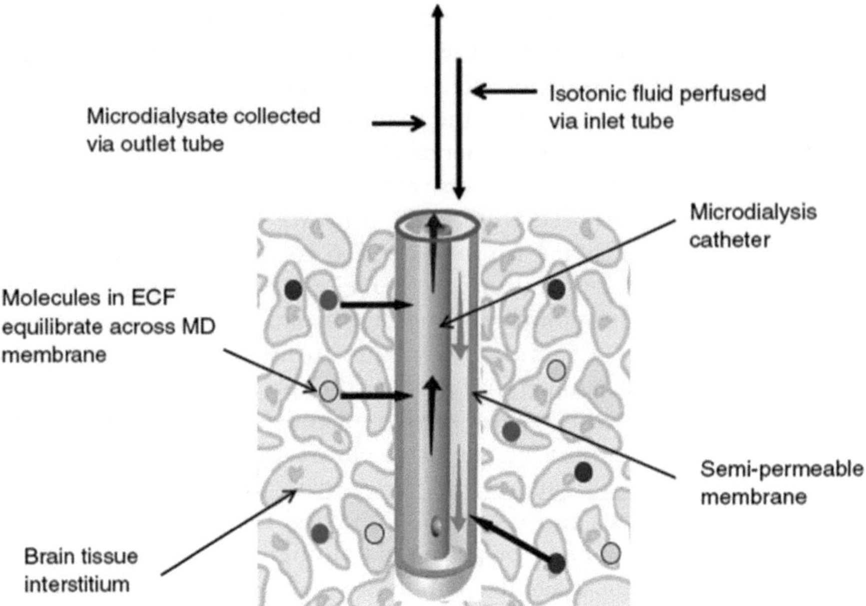

Fig. 2. The Microdialysis probe: A microdialysis (MD) probe is usually constructed as a concentric tube where the perfusion fluid enters through an inner tube, flows to its distal end, exits the tube, and enters the space between the inner tube and the outer dialysis membrane. The direction of flow is now reversed and the fluid moves toward the proximal end of the probe. This is where the "dialysis" takes place, i.e., the diffusion of molecules between the extracellular fluid (ECF) and the perfusion fluid. Vice versa, by reverse dialysis is possible the introduction of a substance into the extracellular space via the microdialysis probe. The inclusion of a higher amount of a drug in the perfusate allows the drug to diffuse through the microdialysis membrane to the tissue. This technique not only allows the local administration of a substance but also permits the simultaneous sampling of the extracellular levels of endogenous compounds.

1.1. Brief Description of Microdialysis Process

As already mentioned, microdialysis is a technique designed to monitor the chemistry of extracellular spaces in living tissue and allows monitoring of neurotransmitters released from practically any region of the brain. It consists of the filtration of water-soluble substances in extracellular fluid through a dialysis membrane (Fig. 2) into a perfusion fluid that is collected and then analyzed for the substances of interest. With this technique, extracellular neurotransmitter levels and other molecules equilibrate with the solution flowing through a dialysis probe implanted in discrete brain areas. Usually microdialysis is coupled with high performance liquid chromatography (HPLC), making it possible to detect extracellular levels of many compounds, from small neurotransmitters to larger peptides (5, 6). The core of microdialysis is the dialysis probe (Fig. 2) designed to mimic a blood capillary. When a physiological salt solution (artificial cerebral fluid solution, aCSF) is slowly pumped through the microdialysis probe, the solution equilibrates with the surrounding extracellular tissue fluid. After a while, it will then contain a representative proportion of the tissue fluid's molecules, and the microdialysate is extracted and later analyzed in the laboratory, usually by HPLC. The body of any "freely moving" awake microdialysis system consists of a dual-channel microdialysis

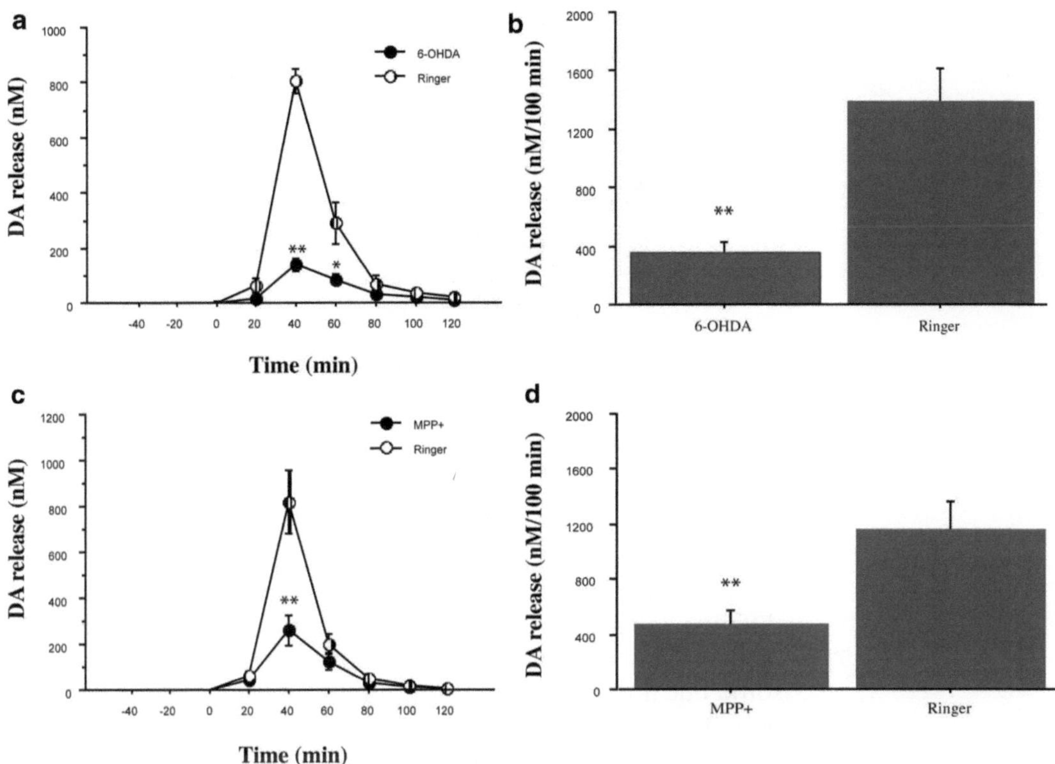

Fig. 3. The 2-day test-challenge microdialysis method. Time course of the effect of 15 min 1 mM MPP+ perfusion 24 h later after perfusion of 6-OHDA (1 mM for 15 min) (**a**), and 1 mM-15 min MPP+ (**c**) on extracellular dopamine output in the corpus striatum. Each data point represents mean ± SEM of absolute levels of DA, without considering probe recovery. Statistical analysis shows that MPP+ and 6-OHDA perfused in the first day of the experiment induced neurodegeneration of nigrostriatal pathway as shown by the decrease in DA release after MPP+ challenge in the second day (one-way ANOVA, followed by Fisher's PLSD *post hoc* test: $^{*}p<0.05$; $^{**}p<0.01$ 6-OHDA vs. Ringer (Control group), and $^{**}p<0.01$ MPP+ vs. Ringer). In (**b**)–(**d**) histograms represent the total DA output measured as the sum of five (100 min) consecutive samples after the same conditions of (**a**) and (**c**) ($^{**}p<0.01$ 6-OHDA (**b**) and MPP+ (**d**) vs. Ringer). Modified from (13).

swivel that has a quartz-lined center and side channels to minimize dead volume and prevent neurotransmitter oxidation (Fig. 4). Dialysate is typically infused through one channel of the swivel, removed through the other, and then collected with a fraction collector or into a microvial. The head block tether and lever arm are necessary to minimize stress on microdialysis probes. The counterbalanced lever arms generally move vertically and horizontally with the animal to prevent slack in the tether. Most of the lever arms use a mass as the counterbalance, the animal can generally stay in a round polycarbonate container that prevents it from damaging the probes (Fig. 4).

A microdialysis probe is usually constructed as a concentric tube (Fig. 2). The perfusion fluid, delivered by a syringe microdialysis pump (Fig. 4) at low flow rate (perfusion rates of 0.3–3 μl/min are typically used depending upon the volume, sample collection time, and analytical sensitivity needed, with typical sample collection times ranging from 1 to 20 min). It enters through an inner tube,

Fig. 4. Photograph of "in vivo freely moving" microdialysis setup. It is formed by a dual-channel microdialysis swivel, head block tether, and lever arm. The aCSF is delivered into the microdialysis probe by a microinjection syringe pump.

flows to its distal end, exits the tube, and enters the space between the inner tube and the outer dialysis membrane, which may be of different lengths, depending on the brain region analyzed. The dialysis membrane is semipermeable and permits free transport of some but not all solutes. Permeability is typically limited to compounds with molecular masses <20,000 Da. The direction of flow is now reversed and the fluid moves toward the near end of the probe (see Fig. 2). Dialysis (the diffusion of molecules between the extracellular and perfusion fluids) takes place at the end of the probe, in the membrane space. It is important to realize that dialysis is an exchange of molecules in both directions. The difference in concentration through the membrane governs the direction of the gradient. It is possible to collect an endogenous compound at the same time that an exogenous compound such as a drug is introduced into the tissue. The gradient of a particular compound into the membrane depends on the difference in concentration between the perfusate and the extracellular fluid and also on the flow velocity inside the microdialysis probe. The absolute recovery (mol/time unit) of a substance from the tissue or of substances entering the tissue from the probe depends on (1) the flow rate of the perfusion fluid, (2) the length of the membrane, (3) the "cut-off" of the dialysis membrane, and (4) the diffusion coefficient of the compound through the extracellular fluid, all important factors to consider for a microdialysis experiment. The advantages and limitations of microdialysis techniques have been reviewed in detail (5, 6, 11). Moreover, this is an ideal technique because multiple sampling of the brain and several regions in a living free-moving animal is possible. No clean-up procedures such as extraction and homogenization of tissue are required before analysis. Behavioral and pharmacology studies can be contemporaneously performed. Neurochemistry around the brain probes can be continuously monitored. Various drugs can be locally administered in the specific brain areas, without having to inject additional volumes.

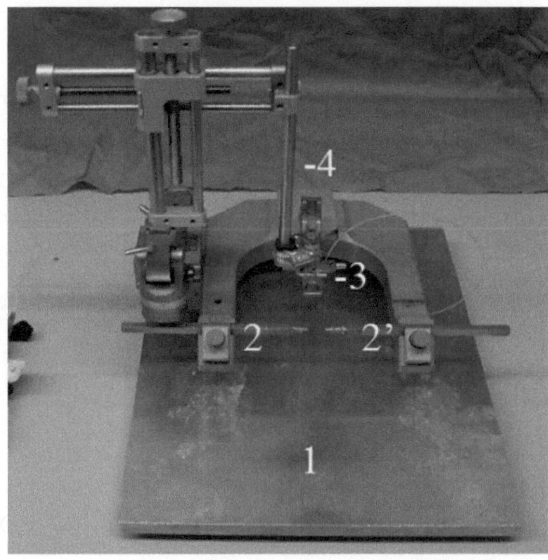

Fig. 5. A typical stereotaxic apparatus: The animal is positioned on the heat-controlled blanket (1), the left (2), and right (2′) ear bars and the incisor adaptor with the nose clamp (3) are used to fix the animal's head, and the probe holder (4) is controlled in the *x*, *y*, and *z* planes by the stereotaxic arm.

2. Materials

2.1. For Microdialysis

Male Sprague–Dawley rats (weighing 340–380 g) are obtained from Charles River, Calco, Italy.

Anesthetic: Chloral hydrate.

Stereotaxic atlas for rat (12).

Stereotaxic frame (David Kopf Instruments, Tujunga, USA) (Fig. 5).

Microdialysis guide cannula for CMA/12 probe (Carnegie Medicin, Stockholm, Sweden) (Fig. 4).

Microdialysis probe (CMA/12, 3 mm length, 500 μm outer diameter, Carnegie Medicin, Stockholm, Sweden) (Fig. 4).

Scalpel.

Spatula.

Dental or hand drill.

Jeweler's forceps and screwdriver.

3/16 in. bone stainless steel screws.

Methylacrylic cement.

Fluorinated ethylene propylene (FEP) tubing 0.12 mm inner diameter (Carnegie Medicin, Stockholm, Sweden).

Tubing Adaptors (for connecting FEP Tubing to the probe, swivel, liquid switch, and syringes; Carnegie Medicin, Stockholm, Sweden).

CMA 110 Liquid Switch (Carnegie Medicin, Stockholm, Sweden) (Fig. 4).

Two-channel liquid swivel (Carnegie Medicin, Stockholm, Sweden) (Fig. 4).

CMA/120 system for freely moving animals (Carnegie Medicin, Stockholm, Sweden) (Fig. 4).

Glass Luer-Lock gas-tight microsyringes (Hamilton).

Microperfusion pump (Harvard Apparatus syringe infusion pump 22, USA) (Fig. 4).

Collection plastic vials (e.g., 0.2-ml microcentrifuge tubes) (Fig. 4).

2.2. Artificial Cerebrospinal Fluid

Na_2HPO_4.

NaH_2PO_4.

$MgCl_2$.

$CaCl_2$.

KCl.

NaCl.

H_2O.

GS (0.22 μm) Millipore glass filters.

2.3. Drugs

1-Methyl-4-phenylpiridinium iodide (MPP^+) from Sigma-Aldrich, St. Louis, MO, USA.

6-Hydroxydopamine hydrobromide (6-OHDA) from Aldrich Chem. Co. Milwakee. WI, USA.

2.4. For HPLC Analysis

HPLC with electrochemical detection (HPLC/EC) system consisting of

HPLC dual-piston pump (LC-10 ADvp pump, Shimadzu Italia, Milano) equipped with pulse dampener.

Manual injector (20-μl sample loop/gas-tight HPLC glass microsyringe).

Coulochem detector (Coulochem II, ESA, Bedford, MA, USA) equipped with a dual electrode analytic cell (model 5014; ESA, Bedford, MA, USA).

Computer software program.

Hypersil column (C18, 4.6×150 mm, 5 μm, Sigma-Aldrich Chemicals, USA).

10 μM Monoamine standard working solutions (in 0.1 N percloric acid and 0.05% $Na_2S_2O_5$) and dilutions (in aCSF) bracketing those estimated for the samples:

4-Hydroxybenzoic acid (4-HBA).

3,4 Dihydroxybenzoic acid (3,4-DHBA).

Dopamine hydrochloride (DA).

Dihydroxyphenilacetic acid (DOPAC).

HPLC-grade water.

Filtration apparatus consisting of

Borosilicate glass vacuum flask.

Fritted filter support and spring clamp.

0.22-μm Filter discs (e.g., Millipore).

Vacuum source.

Ultrapure helium with regulator and Teflon tubing.

The mobile phase is composed of 70 mM NaH_2PO_4, 0.1 mM Na_2EDTA, 0.7 mM triethylamine, 0.1 mM octylsulfonic acid, and 10% methanol, adjusted to pH 4.8 with orthophosphoric acid.

Ascorbate.

2.5. Tyrosine Hydroxylase Immuno histochemistry

4% Paraformaldehyde.

1 M Phosphate buffer, pH 7.4 (PBS).

Sucrose in PBS,

Isopentane –45°C.

Cryostat microtome,

Gelatine-coated slices.

0.6% Hydrogen peroxide.

Methanol.

Goat serum.

Humid chamber.

Policlonal rabbit anti-βTH (Abcam Limited Cambridge, UK) 0.15 μg/μl diluted 1:200 in PBS Triton-X-100.

Biotinylated goat anti-rabbit IgG (Vector, 1:200).

Avidin Biotinylated horseradish peroxidase macromolecular Complex solution (Vectastain® ABC system, Vector Labs, Inc., Burlingame, CA).

Standard (Vector) diaminobenzidine/hydrogen peroxidase chromogen (peroxidase substrate kit DAB; Vector Labs, Inc., Burlingame, CA).

3. Methods

3.1. Surgery for Microdialysis

For short- or long-term measurements of neurochemicals in the conscious animal, a guide cannula is typically implanted above the region of interest. This provides support and protection of the more delicate dialysis probe and is a means to introduce the probe without the trauma of a just-completed surgery. Due to trauma resulting from the introduction of the probe or the probe/cannula assembly into the tissue, the probe is left in situ for some hours prior to the start of experiments. During this period, the probe is perfused with physiological aCSF.

Male Sprague–Dawley rats (Charles River, Calco, Italy) weighing 340–380 g are used. The animals are kept at constant room temperature (21 ± 1°C) and relative humidity (60 ± 5%) under a regular light/dark schedule (light 8.00–20.00 h). Food and water are freely available. Procedures involving animals and their care are conducted in conformity with the institutional guidelines that are in compliance with national (D.L. n. 116, G.U., suppl. 40, 18 Febbraio 1992) and international laws and policies (*EEC Council Directive* 86/609, OJ L 358,1, Dec. 12, 1987; *NIH Guide for the Care and Use of Laboratory Animals*, NIH Publication N. 85-23, 1985 and Guidelines for the Use of Animals in Biomedical Research, Thromb. Haemost. 58, 1078-1084, 1987).

1. Prior to surgery, all surgical instruments are autoclaved or sterilized chemically before the probe implantation, a check is made for leaks and air bubbles. A final check is made to ensure that the active length of the probe extends the appropriate distance from the guide cannula tip.
2. The rat is gently restrained and injected intraperitoneally (i.p.) with chloral hydrate (400 mg/kg, i.p.). Other anesthetics can be used.
3. The animal is mounted in the ear bars of the stereotaxic apparatus (Fig. 5). If the animal is correctly positioned, only movement in the dorsoventral plane should be possible (see Note 1).
4. A midline incision is made using the scalpel blade to pierce the skin.
5. Fascia and connective tissue are scraped away. The entire area is dried with a cotton sponge or Q-tips (see Note 2). The bregma is identified (intersection of the coronal suture with midline) and used to determine the vertical zero coordinate.
6. The guide cannula is attached into the stereotaxic holder and straighten (Fig. 5).
7. The cannula is lowered so that it is centered on the bregma. The anteroposterior (AP) and lateral (L) coordinates are read from the scale and added or subtracted from the stereotaxic coordinates of striatum.
8. By using a 3-Dimensional map of the brain (stereotaxic atlas) along with vernier scales on the apparatus, the guide cannula is placed over the left corpus striatum (AP = 0.8, L = 3.0, V = −3.0 from the dura surface and respect to the bregma), according to the atlas of Paxinos and Watson 1986 (12) (see Note 3).
9. Dental or hand drill with a burr sized is appropriately placed for the diameter of the guide cannula or probe assembly on an

electrode carrier or manipulator that has been modified to accommodate the width of drill. AP and L coordinates may be adjusted. Drill until the dura is punctured, using an up-and-down motion (see Note 4).

10. Three additional holes are drilled around the cannula/probe hole; drill screw holes may be done far enough so as not to interfere with cannula insertion. Using jeweler's forceps and a screwdriver, 3/16-in. bone screws are inserted into the holes (see Note 5).
11. The cannula is slowly lowered into the target striatal region until the cannula's base just rests on the dura surface. The ventral (V) coordinate is readed and the cannula lowered to –3.0 from the dura surface. Then, the skull is permanently fixed to with stainless steel screws and methylacrylic cement.
12. The cement is allowed to dry and the holder is carefully removed. The cannula is secured in place by applying additional dental acrylic cement.
13. aCSF is prepared: 147 mM Na^+, 2.7 mM K^+, 1 mM Mg^{2+}, 1.2 mM Ca^{2+}, 154.1 mM Cl^-, adjusted to pH 7.4 with 2 mM sodium phosphate buffer. The Ringer solution must be filtered through type GS (0.22 μm) Millipore glass filters before use (see Note 6).
14. The microsyringe pump is prepared, filled with aCSF, and attached to the pump.
15. Probe is prepared (see Note 7), and connected to the swivel and syringe by FEP tubing and tubing adaptors (Fig. 4).
16. The dummy cannula is removed from rat guide cannula and the probe is slowly inserted into the striatum (see Note 8).
17. After the probe is inserted, the tether is attached to the animal and placed in the testing cage CMA/120 system for freely moving animals (see Note 9). Balance arm and liquid swivel are connected via a wire guide and spring clamp to a collar that is inserted over the neck of the rat, prior to insertion of the probe. Probe inflow and outflow tubing are secured to the wire guide, which pulls the liquid swivel/balance arm (Fig. 4).
18. Following probe implantation, the rat is allowed to recovery from the surgery and the tissue from the physical damage rendered by probe implantation. The probe is flushed at 0.3 μl/min overnight with aCSF.

3.2. First Day of Microdialysis: Lesion of Nigrostriatal DA System

The experiments are performed 24 h (day 1) after probe implantation, in awake, freely moving animals.

1. The corpus striatum is perfused at a flow rate of 1 μl/min, and every 20 min samples of perfusate are collected in plastic vials and immediately assayed by HPLC with electrochemical detection (see below Sect. 3.5) (Fig. 7).

2. After establishing a steady baseline level in three consecutive samples (control value), MPP^+ iodide or 6-OHDA (1 mM) (see Note 10), dissolved in aCSF, is perfused for 10 min through the microdialysis probe in the rat striatum to induce neurodegeneration of the nigrostriatal DAergic system, and then with the normal aCSF for further 2 h at a flow rate of 1 μl/min (see Note 11).
3. Afterwards the probe is flushed at 0.3 μl/min until the next day of the experiment.

It is important to note that any pharmacological treatment to prevent neurodegeneration can be done 1 h before neurotoxins (prevention studies). Moreover, drugs can be systemically given after the first day of the experiment, to study its possible post-lesion therapeutic effects. In this case the evaluation of neuronal integrity experiment (day 2 step) must be done at least after a week (see Note 12).

On the first day of the experiment it is important to verify that the drug treatment does not cause any significant effect on basal DA release in the rat striata, and does not modify the extracellular DA release induced by the neurotoxins, which indicate that it does not affect the DA-uptake mechanism or DA metabolism (13, 14).

3.3. Second Day of Microdialysis

To check the integrity of the DA system, after establishing a steady baseline level in three consecutive samples (control value), 1 mM MPP^+ (1 ml/min) is perfused for 10 min 1 day after (day 2) the first infusion of MPP^+ or 6-OHDA. Then the perfusion fluid is switched to normal aCSF for 2 h, until the end of experiment. The challenge with MPP^+ produces a maximal DA extracellular output 40 min after its infusion (Fig. 3). In a first experimental group, MPP^+ perfusion produced a maximal DA increase of 819 ± 136 nmol in the non-lesioned rats (aCSF), 260 ± 66 in the MPP^+ lesioned rats (Fig. 3c, d). In a second group, MPP^+ perfusion produced a maximal DA increase of 805 ± 42 nmol in the non-lesioned (aCSF), 139 ± 22 in the 6-OHDA lesioned rats (Fig. 3a, b). The limited rise of DA following the MPP^+ reperfusion is due to toxins-induced neuronal loss in the lesioned groups. The total amounts of DA, released in 100 min, were reduced in MPP^+- or 6-OHDA-lesioned rats, challenged with MPP^+ on day 2, as compared to controls ($p < 0.01$ vs. controls at Fisher's PLSD test, in both groups) (Fig. 3) (13).

3.4. Microdialysis to Study Oxidative Stress

Oxidative stress is defined as an imbalance between the production of reactive oxygen species (ROS) and a biological system's ability to readily detoxify the reactive intermediates and/or easily repair the resulting damage. Increasing evidence suggests that reactive oxygen species are involved in the pathophysiology of PD. The microdialysis technique is suitable for measuring radical formation in the extracellular space in vivo. After probe implantation and obtaining a stable baseline, the probe can be used to simultaneously deliver the stimulating agent (e.g., 6-OHDA or MPP^+)

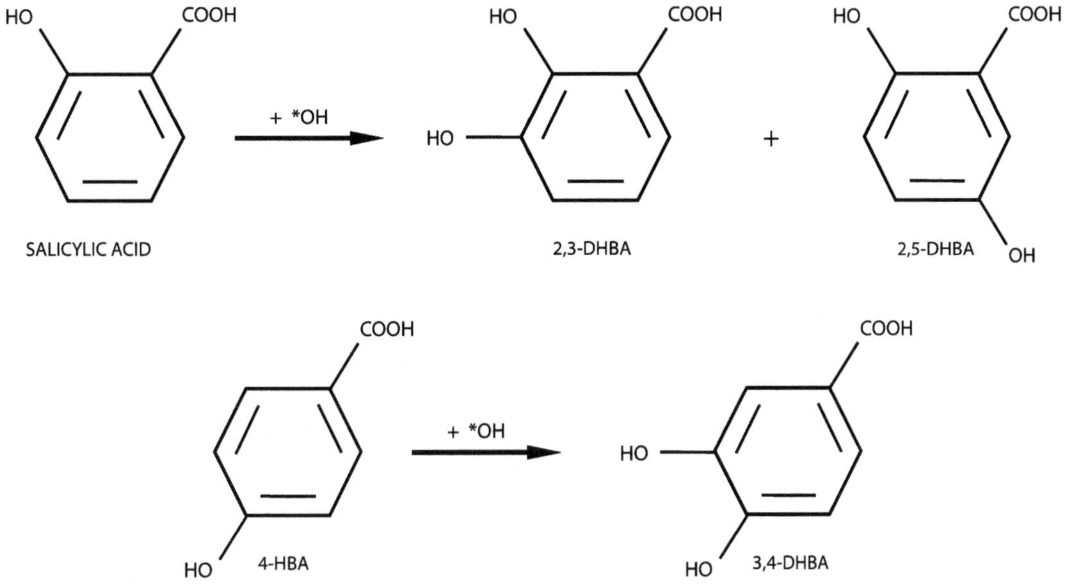

Fig. 6. Reaction scheme of salicylate (2-HBA) with the hydroxyl radicals (•OH) to form 2,3- and 2,5-dihydroxybenzoate (2,3- and 2,5-DHBA) and 4-hydroxybenzoic acid (4-HBA) hydroxylation that forms the 3,4-dihydroxybenzoic acid (3,4-DHBA).

and the trapping agent (see below) for the radical determination, without any purification steps before HPLC analysis. In addition, the implanted microdialysis probe allows reliable recording for many hours, and the animal serves as its own control. Due to the high reactivity and short half-life of ROS in biological tissue, indirect methods for in vivo detection have been widely used. The method for in vivo measurement of the hydroxyl radicals (·OH) is based on the ability of the ·OH to attack the benzene rings of aromatic molecules (Fig. 6). The salicylate (2-HBA) trapping technique is the most widely used technique for ROS detection in vivo and has previously been employed in experimental animal models of PD with microdialysis and HPLC coupled with an electrochemical detector (ECD) (15–17). However, when ROS react with salicylate two adducts: 2,3- and 2,5-DHBA are formed (Fig. 6). Concerns have been raised with regard to the specificity of the salicylate trapping technique since enzymatic formation of 2,5-DHBA occurs at extracerebral sites by the cytochrome p450 system in the liver after systemic administration. In addition, 2,3-DHBA is metabolized in the rat intestine, making it an unspecific marker (18–20). Salicylate may also chelate divalent metal ions (21), thus affecting hydroxyl radical formation. Finally, salicylate has been shown to affect the inducible nitric oxide synthase (iNOS) (22). Therefore, when a trapping method for ROS detection is employed a very careful evaluation of possible sources of spontaneous ROS formation may be made and these sources may be eliminated as far as possible. Plausible factors affecting hydroxylation

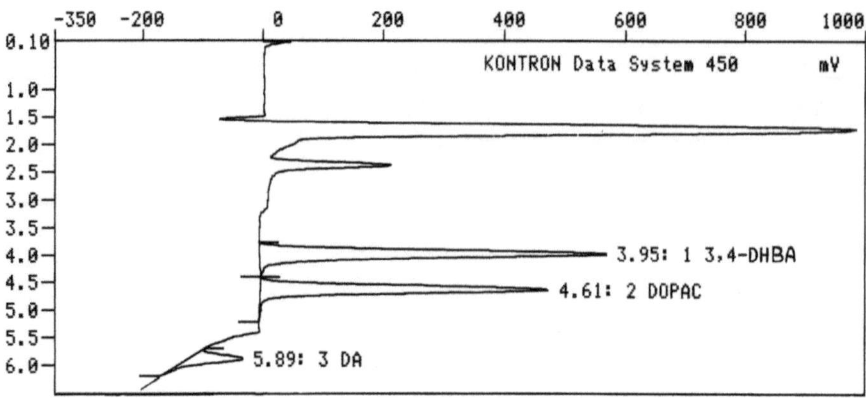

Fig. 7. Chromatogram of 4-hydroxybenzoic acid (4-HBA), dihydroxyphenyl acetic acid (DOPAC), and dopamine (DA).

of DHBAs using microdialysis technique include time, use of metal ions, and reuse of microdialysis probes. The liquid switch perfusion with an artificial physiological solution may possibly corrode some metal surfaces. Reusing the microdialysis probe resulted in increased DHBA production (23). Therefore, to reduce artifactual exogenous hydroxyl free radical production, all stainless steel may be eliminated from the microdialysis setup and HPLC system, the perfusing solutions may kept cold, shielded from the light and freshly prepared just before use. A sensitive and specific hydroxybenzoate (4-hydroxybenzoic acid; 4-HBA) hydroxylation assay is also used in conjunction with HPLC–ECD to monitor hydroxyl radicals (·OH) production (21, 23). 4-HBA is known to trap hydroxyl radicals with equal efficacy as salicylate (21). It offers a number of advantages compared to the use of salycilate: firstly, only one isomer is formed by hydroxylation of 4-HBA, the 3,4-DHBA (Fig. 7), therefore the signal due to ·OH attack is not split between two compounds (2,3- and 2,5-DHBA), decreasing the sensitivity of the assay, as in the case of salycilate (Fig. 7). Secondly, the systemic production of 3,4-DHBA following administration of a large bolus of 4-HBA in vivo is negligible compared to the in vivo enzymatic hydroxylation of salycilate. Lastly, salicylate is an effective chelator of divalent metal ions, including Fe^{2+}. As Fe^{2+} catalyzes the decomposition of H_2O_2 to OH, this may result in an over estimation of apparent ·OH production due to its increased local formation (21, 23). On the other hand, the formation of 3,4-DHBA may reflect either the production of hydroxyl radicals or other highly reactive species such as peroxynitrite produced by the reaction between superoxide anions and nitric oxide (24). Thus, the formation of 3,4-DHBA is likely to reflect ROS production but is not specific for any single radical species.

We here describe the hydroxybenzoate (4-hydroxybenzoic acid) hydroxylation assay, for the above reasons.

1. Starting from the day after the surgery, 0.5 mmol/l 4-hydroxybenzoic acid is added to aCSF solution and perfused continuously.
2. Brain dialysates are collected every 20 min, and immediately assayed for determination of 3,4-dihydroxybenzoic acid (3,4-DHBA) (see below Sect. 3.5) (Fig. 7).
3. After a stable baseline, to induce exogenous hydroxyl radicals stimulation, the perfusion fluid is switched for 10 min to a fluid containing 1 mM MPP^+ or 6-OHDA (or other toxins able to induce oxidative stress).
4. Then the perfusion fluid is switched back to the previous perfusion fluid (normal aCSF + 4-HBA), for further 2 h.
5. In control experiments, only aCSF with 4-HBA is used.

3.5. HPLC Analysis

Dialysate samples are analyzed by reversed-phase HPLC coupled with electrochemical detection (Fig. 7) (13, 14).

1. 2 L of mobile phase composed of 70 mM NaH_2PO_4, 0.1 mM Na_2EDTA, 0.7 mM triethylamine, 0.1 mM octylsulfonic acid, and 10% methanol, adjusted to pH 4.8 with orthophosphoric acid (see Note 13) are prepared.
2. The mobile phase solution is filtered through 0.22-μm filter under vacuum, then degassed by passing a gentle stream of ultrapure helium through the solution for 10 min at a pressure of 5–10 psi, using Teflon tubing (see Note 14).
3. Any air bubbles are purged from the mobile phase inlet lines. The HPLC/EC system is assembled connecting the mobile phase to the inlet lines. The flow rate is setted at 1 ml/min and it is allowed several hours (e.g., 14–16 h) for equilibration of the mobile phase and column.
4. Cell potential of the first electrode is adjusted at −175 mV and the second at +175 mV, respectively.
5. The electrode sensitivity is adjusted to 500 μA for DOPAC and to 5 nA for DA (see Note 15).
6. 20 μl of 10–1 nM or less of DA and DOPAC standard working solution are injected to determine retention times and standard curves, used for the linear regression to determine analyte concentration (see Note 16).
7. A linear regression analysis of the standards is performed and the resulting linear regression equation is used to determine actual concentrations in samples.

3.6. Immunohistochemistry of Tyrosine Hydroxylase

At the end of the experiments, the rats are perfused transcardially under deep anesthesia (chloral hydrate) with physiological saline (180–200 ml), followed by 4% paraformaldehyde in phosphate buffer, pH 7.4 (180–200 ml).

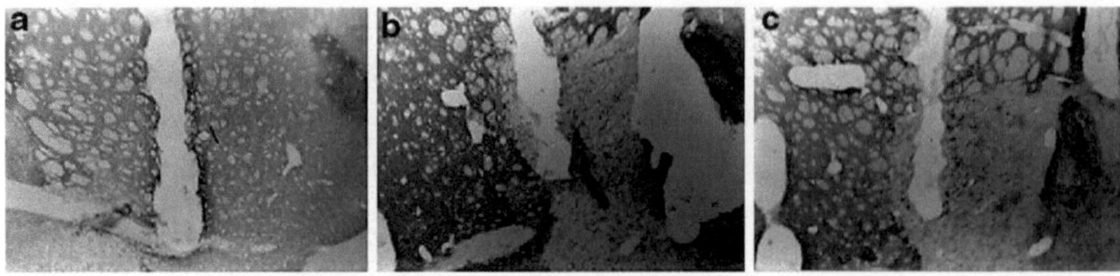

Fig. 8. Photomicrograph of coronal sections (cryostat-cut sections) through striatum (around the microdialysis probe's membrane) after the immunostaining. (**a**) Rat perfused with Ringer's solution. (**b**) Rat perfused with MPP^+. (**c**) Rat perfused with 6-OHDA (magnification 32×). Adapted from (13).

Brains are removed and cryoprotected by serial passages in sucrose in PBS, pH 7.4; first in 10% sucrose for 24 h and then in 30% sucrose for 2–5 days. The brains are then frozen in isopentane −45°C and then stored at −80°C. Coronal sections (30 μm thickness) are cut using a cryostat microtome, mounted in gelatine-coated slices and stained for tyrosine hydroxylase (TH) immunohistochemistry. All experiments are done at room temperature. Sections are washed and then treated with 0.6% hydrogen peroxide in methanol for 30 min, washed again, and incubated in a solution containing PBS and 50% goat serum for 30 min in a humid chamber. Slices are drained and further incubated with policlonal rabbit anti-βTH (Abcam Limited Cambridge, UK) 0.15 μg/μl diluted 1:200 in PBS containing 1.5% goat serum and 0.1% Triton-X-100 over night at 4°C. Sections are then incubated for 2 h with biotinylated goat anti-rabbit IgG (Vector, 1:200) followed by a second 1-h incubation with Avidin Biotinylated horseradish peroxidase macromolecular Complex solution (Vectastain® ABC system, Vector). The peroxidase is then visualized with a standard (Vector) diaminobenzidine/hydrogen peroxidase chromogen reaction for 5 min. The area which shows reduction in TH immunoreactivity around the cannula is identified under a microscope (Fig. 8) (13).

3.7. Data Analysis

Dopamine and DOPAC content in each sample is expressed as the absolute dialysate levels in nmol/l (without considering probe recovery). Data correspond to mean ± SEM values of absolute DA and DOPAC levels obtained in each experimental group. 3,4-DHBA content in each sample is expressed as a percentage of the average baseline level calculated from three fractions collected before neurotoxins infusion. Data are expressed as mean ± SEM of the percentage obtained in each experimental group. Data are analyzed by one-way analysis of variance (ANOVA) with repeated measures, followed by the Fisher's protected least significance difference *post hoc* test (Fisher's PLSD) to allow multiple comparisons between groups. All statistical analyses are performed with StatView™ version 5.0.1 (SAS Institute, Inc., Cary, NC, USA) (13, 14).

4. Notes

1. Stereotaxic surgery requires that the animal be firmly mounted into the stereotaxic instrument. No horizontal movement of the head should occur. The presence of such indicates improper mounting of the animal. Ear bars may require tightening or reinsertion.
2. Dilute hydrogen peroxide can be applied to facilitate drying of the skull and aid visualization of bregma. This step is critical for the proper adherence of dental acrylic to the skull at the end of this surgical procedure.
3. A stereotaxic atlas provides coordinates for the anterior–posterior (AP), lateral (L), and dorsal ventral (DV) planes. Accurate placement depends on adherence to the reference points used for atlas preparation. Depending on the atlas employed, the position of the head on the incisor bar may be either above or below the interaural line. In addition, brain size may vary depending on strain, age, or sex. Pilot studies should be conducted to validate or adjust the coordinates if necessary.
4. The size of the drill bit should be that which will allow lowering of the cannula into the brain without catching the bone. Caution should be exercised in drilling so that the meninges and underlying brain tissue are not damaged.
5. The screws serve to anchor the cannula/probe assembly when dental acrylic is applied. The screws should be rigid but should not extend to the dura.
6. For 500 ml aCSF: 2 mM Na_2HPO_4 (500 ml) and 2 mM NaH_2PO_4 (250 ml) should be prepared. The aCSF salts are dissolved in about 350–400 ml of 2 mM Na_2HPO_4 and adjusted to pH 7.4 with 2 mM NaH_2PO_4, then bringed to 500 ml with H_2O or 2 mM sodium phosphate buffer pH 7.4.
7. Prior to use, any remaining glycerol from dialysis membrane is removed by soaking in 70% ethanol for ~5 min. The probe is perfused for several minutes with distilled water or perfusion fluid (flow rate 2 μl/min) and checked for leaks or air bubbles inside the membrane.
8. Probe insertion requires a steady hand since it is easy to damage the delicate membrane. Lack of outflow, after probe insertion, is often due to occlusion of the outflow tubing. This can be overcome by cutting the tubing. If this does not result in flow, the system (i.e., tubing and pump connectors) should be checked for leaks. If flow still does not occur, the probe is probably at fault. If such is the case, the probe should be removed and inspected for leaks or tears. It is possible to reinsert the probe; however, an ample amount of time must then elapse before sampling can begin.

9. The outflow line must exit the testing cage in such a way that the experimenter can remove the collection vial without disturbing the animal.
10. To ensure maximal reproducibility, MPP^+ or 6-OHDA (10×) dissolved in 0.1% ascorbate/saline are prepared in advance, filtered through 0.22 μm filters, aliquoted, and stored at −20°C. Immediately before use, aliquots gently defrosted, are diluted in normal aCSF to the final concentration of 1 mM each and loaded into a designated microdialysis syringe, protected from light.
11. A liquid switch (CMA 110 Liquid Switch) can be used for delivery of different perfusates into the region dialyzed. With practice, this can also be done manually. Each perfusate switch can result in the introduction of air bubbles into the probe (resulting in a decrease in the active membrane area). Therefore, care must be taken when performing this step.
12. Interestingly, this method can be employed also to study neuroregenerative strategies, since Santiago and colleagues (25) investigated whether a short-lasting infusion of MPP^+ into the SN can be used as a chronic model of parkinsonism, by allowing a 1-month delay after intranigral MPP^+ administration, and it was shown that infusion of MPP^+ into the SN may cause chronic damage to the nigrostriatal pathway. Therefore, using various DAergic toxins (e.g., 6-OHDA in SNc) and DA releasing agents such as high K^+ or amphetamine, besides MPP^+, for the test challenge of the second day or after 1–2 weeks after the lesion, it is possible to employ this method for both the study of neuroprotective or restorative treatments for PD.
13. Typically, the mobile phase is recycled and replaced one to two times a week. If background noise is detected, however, the recycled mobile phase should be degassed once again. Mobile phase composition is critical for HPLC/EC detection of dopamine and its metabolites. Retention times are affected by alterations in room temperature as well as by pH, the concentration of ion-pairing reagent, and the inorganic solvent employed. These can be varied to modify analyte resolution. Increasing pH decreases the retention time of acidic metabolites (e.g., DOPAC and HVA). Increasing the concentration of ion-pairing reagents (e.g., octanesulfonate, octylsulfate, pentanesulfonate, or hexanesulfonate) delays retention of amines, enabling better separation from charged molecules. Increasing solvent concentration results in a decreased retention time of amines and their metabolites. Each of these parameters can be modified to increase analyte resolution.
14. Filtration and complete degassing is essential for sensitive electrochemical detection of catecholamines. Particulate matter in the mobile phase can clog the HPLC/EC detector system, creating high pressure and baseline disturbances. Air bubbles

from improper degassing are the most common problem in EC detection and can wreak havoc on an HPLC/EC system.

15. Under these conditions, the sensitivity for DA is about 0.35 pg/20 μl with a signal to noise ratio of 3:1.
16. Samples are injected manually into the HPLC and detection of DA, DOPAC, or 3,4-DHBA is carried out with a coulometric detector (Coulochem II, ESA, Bedford, MA, USA) coupled to a dual electrode analytic cell (model 5014) (Fig. 7).

Acknowledgments

Authors are indebted to the EU COST Action CM1103 "Structure-based drug design for diagnosis and treatment of neurological diseases: dissecting and modulating complex function in the monoaminergic systems of the brain" for supporting our international collaboration.

References

1. Blandini F, Nappi G, Tassorelli C, Martignoni E (2000) Functional changes of the basal ganglia circuitry in Parkinson's disease. Prog Neurobiol 62:63–88
2. Di Giovanni G (2008) Will it ever become possible to prevent dopaminergic neuronal degeneration? CNS Neurol Disord Drug Targets 7:28–44
3. Schober A (2004) Classic toxin-induced animal models of Parkinson's disease: 6-OHDA and MPTP. Cell Tissue Res 318:215–224
4. Beal MF (2001) Experimental models of Parkinson's disease. Nat Rev Neurosci 2: 325–334
5. Ungerstedt U (1991) Microdialysis: principles and applications for studies in animals and man. J Intern Med 230:365–373
6. Hammarlund-Udenaes M (2000) The use of microdialysis in CNS drug delivery studies: pharmacokinetic perspectives and results with analgesics and antiepileptics. Adv Drug Deliv Rev 45:283–294
7. Di Giovanni G, Esposito E, Di Matteo V (2009) In vivo microdialysis in Parkinson's research. J Neural Transm Suppl 73:223–243
8. Giovanni A, Sonsalla PK, Heikkila RE (1994) Studies on species sensitivity to the dopaminergic neurotoxin 1-methyl-4-phenyl-1,2,3,6-tetrahydropyridine. 2. Central administration of 1- methyl-4-phenylpyridinium. J Pharmacol Exp Ther 270:1008–1014
9. Rollema H, Kuhr WG, Kranenborg G, De Vries J, Van den Berg C (1988) MPP^+-induced efflux of dopamine and lactate from rat striatum have similar time courses as shown by in vivo brain dialysis. J Pharmacol Exp Ther 245:858–866
10. Tobón-Velasco JC, Silva-Adaya D, Carmona-Aparicio L, García E, Galván-Arzate S, Santamaría A (2010) Early toxic effect of 6-hydroxydopamine on extracellular concentrations of neurotransmitters in the rat striatum: an in vivo microdialysis study. Neurotoxicology 31:715–723
11. Benveniste H, Huttemeier PC (1990) Microdialysis: theory and application. Prog Neurobiol 35:195–215
12. Paxinos G, Watson C (1986) The rat brain in stereotaxic coordinates. Academic, New York
13. Di Matteo V, Pierucci M, Di Giovanni G, Di Santo A, Poggi A, Benigno A, Esposito E (2006) Aspirin protects striatal dopaminergic neurons from neurotoxin-induced degeneration: an in vivo microdialysis study. Brain Res 1095:167–177
14. Di Matteo V, Benigno A, Pierucci M, Giuliano DA, Crescimanno G, Esposito E, Di Giovanni G (2006) 7-nitroindazole protects striatal dopaminergic neurons against MPP+-induced degeneration: an in vivo microdialysis study. Ann N Y Acad Sci 1089:462–471
15. Chiueh CC, Krishna G, Tulsi P, Obata T, Lang K, Huang S-J, Murphy DL (1992) Intracranial

microdialysis of salicylic acid to detect hydroxyl radical generation through dopamine autooxidation in the caudate nucleus: effects of MPP^+. Free Radic Biol Med 13:581–583

16. Teismann P, Schwaninger M, Weih F, Ferger B (2001) Nuclear factor-kappaB activation is not involved in a MPTP model of Parkinson's disease. Neuroreport 12:1049–1053
17. Obata T (2006) Nitric oxide and MPP^+-induced hydroxyl radical generation. J Neural Transm 113:1131–1144
18. Halliwell B, Kaur H, Ingelman-Sundberg M (1991) Hydroxylation of salicylate as an assay for hydroxyl radicals: a cautionary note. Free Radic Biol Med 10:439–441
19. Halliwell B, Kaur H (1997) Hydroxylation of salicylate and phenylalanine as assays for hydroxyl radicals: a cautionary note visited for the third time. Free Radic Res 27:239–244
20. Montgomery J, Ste-Marie L, Boismenu D, Vachon L (1995) Hydroxylation of aromatic compounds as indices of hydroxyl radical production: a cautionary note revisited. Free Radic Biol Med 19:927–933
21. Ste-Marie L, Boismenu D, Vachon L, Montgomery J (1996) Evaluation of sodium 4-hydroxybenzoate as an hydroxyl radical trap using gas chromatography-mass spectrometry and high-performance liquid chromatography with electochemical detection. Anal Biochem 241:67–74
22. Amin AR, Vyas P, Attur M, Leszczynska-Piziak J, Patel IR, Weissmann G, Abramson SB (1995) The mode of action of aspirin-like drugs: effect on inducible nitric oxide synthase. Proc Natl Acad Sci U S A 92(17):7926–7930
23. Liu M, Liu S, Peterson SL, Miyake M, Liu KJ (2002) On the application of 4-hydroxybenzoic acid as a trapping agent to study hydroxyl radical generation during cerebral ischemia and reperfusion. Mol Cell Biochem 234(235): 379–385
24. Bogdanov MB, Ferrante RJ, Mueller G, Ramos LE, Martinou JC, Beal MF (1999) Oxidative stress is attenuated in mice overexpressing BCL-2. Neurosci Lett 262:33–36
25. Santiago M, Westerink BHC, Rollema H (1991) Responsiveness of striatal dopamine release in awake animals after chronic 1-methyl-4-phenylpyridinium ion-induced lesions of the substantia nigra. J Neurochem 56:1336–1134

Chapter 3

Application of Reverse Microdialysis in Neuropharmacological Studies

Christian Höcht, Facundo Martín Bertera, and Carlos Alberto Taira

Abstract

A recently developed application of microdialysis is the introduction of a substance into the extracellular space via the microdialysis probe. The inclusion of a higher amount of a drug in the perfusate allows the drug to diffuse through the microdialysis membrane to the tissue. This technique, reverse microdialysis, not only allows the local administration of a substance but also permits the simultaneous sampling of the extracellular levels of endogenous compounds. In neuropharmacological studies, reverse microdialysis has been extensively used for the study of the effects of diverse pharmacological and toxicological agents on neurotransmission at different central nuclei, such as antidepressants, antipsychotics, antiparkinsonians, hallucinogens, drugs of abuse, and experimental drugs. In addition, reverse microdialysis allows the simultaneous monitoring of effects of locally applied drugs on physiological parameters, including blood pressure, heart rate, thermoregulation, neuronal activity, and behavior of the experimental animal.

Key words: Reverse microdialysis, Neurotransmission, Physiological parameters, Microinfusion, Stereotaxic, Analytical methods, In vitro recovery

1. Introduction

Microdialysis sampling is an attractive research technology for the evaluation of pharmacokinetic and pharmacodynamic behavior of therapeutic and toxicological agents in both laboratory animals and human beings (1, 2). Traditionally, microdialysis has been used for the determination of interstitial levels of endogenous compounds and drugs. Nowadays, microdialysis has found important application in the field of blood pharmacokinetics, tissue distribution of drugs, and pharmacokinetic–pharmacodynamic (PK–PD) modeling (2, 3).

More recently, microdialysis sampling has been applied for the local administration of compounds into the extracellular space via the microdialysis probe (4). The inclusion of a higher amount of a

Giuseppe Di Giovanni and Vincenzo Di Matteo (eds.), *Microdialysis Techniques in Neuroscience*, Neuromethods, vol. 75,
DOI 10.1007/978-1-62703-173-8_3, © Springer Science+Business Media, LLC 2013

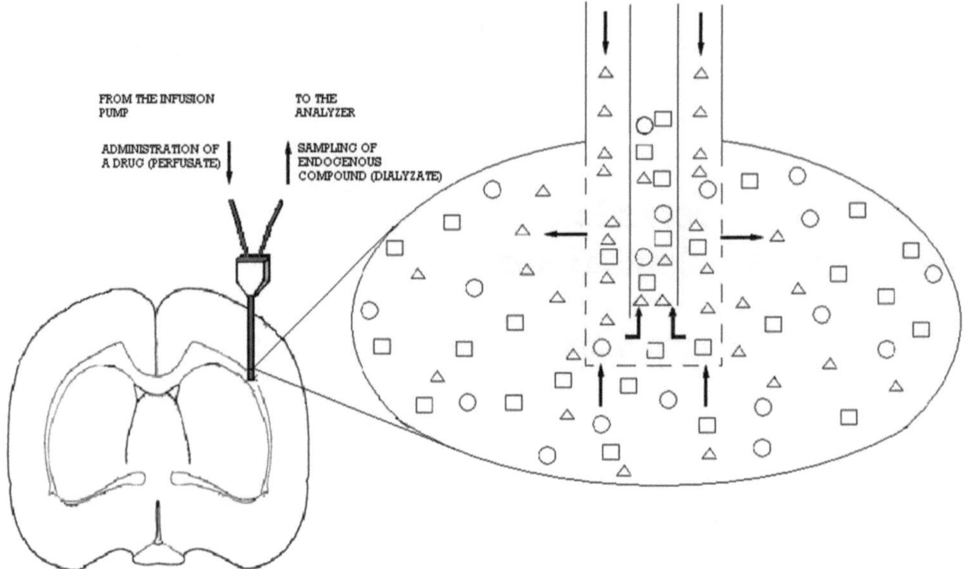

Fig. 1. Reverse microdialysis principle. A concentric microdialysis probe inserted in the extracellular space of a tissue is perfused with a solution containing the exogenous compound (*triangles*) to administer. The solution is pumped through the inner cannula allowing the diffusion of the exogenous compounds into the surrounding tissue. Simultaneously, neurotransmitters (*circles*) and their metabolites (*squares*) diffuse from the extracellular space into the microdialysis probe and are collected in the dialysate for further analysis.

drug in the perfusate with regard to the perfused tissue allows the drug to diffuse through the microdialysis membrane into the tissue. This technique is known as reverse microdialysis (5) (Fig. 1).

Based on the dialysis principle, reverse microdialysis is a powerful tool in neurochemical research considering that it not only allows the local administration of a substance but also permits the simultaneous sampling of the extracellular levels of endogenous compounds (Fig. 1). In other words, reverse microdialysis serves both for administration and as a sampling technique.

Reverse microdialysis is an interesting alternative to other methods for local drug delivery in the central nervous system, such as microinfusion through an implanted intracranial cannula (Table 1). Drug administration by microinfusion has several drawbacks compared with reverse microdialysis. In microinfusion studies the compound is administered by the introduction of a determined volume of fluid into the extracellular space of the tissue under investigation. As volume expansion occurs, this technique is rather useful for chronic drug administration, especially for chronic infusion into brain parenchyma (6). In addition, microinfusion may produce nonspecific tissue damage due to the volume of fluid introduced. Conversely, reverse microdialysis delivers the drug without net gain of fluid making this technique suitable for chronic drug administration. It is important only to ensure that the recovery of the microdialysis probe did not change during the

Table 1
Advantages of reverse microdialysis for local drug administration at the central nervous system

No volume expansion of central nuclei
Suitable for chronic administration for long periods
Maintenance of constant drug concentration at the surrounding tissue
Applied drug targets larger volume of the tissue
Able to monitor the proper performance of the microdialysis probe
Suitable for dose–response studies in the same animal

experiment because of alterations in the surrounding tissue such as gliosis or inflammation.

Moreover, microinfusion provides only a specific amount of a drug, targeting a limited area of the tissue under investigation. Once a drug has been given, it is diffused rapidly out of the brain into the peripheral bloodstream making interpretation of results extremely difficult (7). During reverse microdialysis experiments, the drug is chronically perfused into the target site maintaining more constant drug concentration at the surrounding tissue. In addition, the drug targets a larger volume of the tissue because the drug diffuses to the interstitial space from the whole microdialysis membrane. Therefore, an important issue in reverse microdialysis experiments is the selection of the adequate membrane size in order to ensure drug administration in a great volume of the target tissue.

Solutions injected by microinfusion through an intracranial implanted cannula are assumed to enter the brain as a spherical drop, causing a variation in drop size, resulting from pressure injection delivery, which may produce variability in the diffusion area (8). Another advantage of reverse microdialysis drug delivery is the potential to monitor the proper performance of the microdialysis probe throughout the period of chronic drug administration. In intracranial injection studies, incomplete expulsion due to cannula obstruction has been reported (9). On the other hand, reverse microdialysis is more suitable for dose–response studies in the same animal, because there is no need to reimplant the microdialysis probe for further dose administration. Perhaps the only drawback of reverse microdialysis with regard to other delivery methods is the greater cost of the equipment.

As commented previously, reverse microdialysis not only allows the local perfusion of a drug into the target tissue but also serves as a sampling technique by recovery of endogenous compounds for further quantification (Table 2). Push–pull perfusion and in vivo voltammetry are alternative sampling methods to reverse

Table 2
Advantages of reverse microdialysis as a sampling technique at the central nervous system

Advantages
Allows simultaneous local drug administration and evaluation of the effects on neurotransmitters and their metabolites
High spatial and temporal resolution
Permits simultaneous monitoring of multiple analytes
The actual animal serves as its own control

microdialysis. However, these techniques do not allow drug administration. Push–pull probes consist of two concentric tubes so that the perfusion fluid is introduced by means of a pump through the inner tube and removed by a second pump through the outer tube (10). One advantage of this method with regard to microdialysis is that the recovery of endogenous compounds is greater, allowing a higher output of the recovered endogenous substance. The absence of a membrane barrier in push–pull probes makes this method more adequate for the recovery of compounds which adsorb to a dialysis membrane or which hardly diffuse through the membrane such as peptides. Nevertheless, push–pull perfusion has several drawbacks with regard to microdialysis sampling. In push–pull studies, high flow rates are required to prevent blockage of fluid flow generating turbulence and consequently significant tissue damage. This sampling technique does not prevent enzymatic degradation of the endogenous compounds and needs cleanup procedures for further analysis. Moreover, push–pull perfusion is technically more complicated than microdialysis perfusion, considering that two exactly calibrated pumps are necessary to transport both the push and the pull fluids (10).

In vivo voltammetry represents an alternative sampling technique of neurocompounds which consists of the measurement of electroactive substances at the surface of an implanted carbon electrode (11). This technique involves oxidation or reduction of an analyte as the result of an applied potential. Although each compound has a characteristic oxidation potential, the differences between oxidation potentials are so small that specificity is a serious challenge. In voltammetry using enzyme-based biosensors, the specificity of the enzyme establishes the identity of the endogenous compounds. A great limitation of in vivo voltammetry is the fact that only electroactive compounds or enzyme-specific substances can be determined with this technique. Nevertheless, compared with microdialysis sampling, in vivo voltammetry has a higher time resolution and is therefore the most adequate sampling method for behavioral studies (11).

2. Materials

The experimental setting of a reverse microdialysis trial is based on the insertion of a probe that mimics the function of a capillary blood vessel into the target tissue (Fig. 1). The probe is composed of a hollow fiber permeable to water and small molecules that allows the exchange of molecules through the dialysis membrane by diffusion in both directions in favor of a gradient of concentration during perfusion of the microdialysis catheter with a physiological fluid. As the perfusion media is pumped continuously through the microdialysis probe, dialysate samples are obtained periodically and the content of compounds is quantified using highly sensitive techniques. More specifically, microdialysis is able to sample endogenous compounds, because levels of neurotransmitters and metabolites are higher in the extracellular space than in the perfusion fluid.

The inclusion of a higher concentration of a drug or an endogenous compound in the perfusate allows the substance to diffuse through the dialysis membrane to the tissue and therefore serves for the local application of xenobiotics in the target tissue (1). The reverse microdialysis approach is highly attractive for neurochemical research since it allows the local application of a known amount of drugs into a specific brain nucleus without inducing major changes in tissue physiology (1).

During a reverse microdialysis experiment, the researcher will need the following materials.

2.1. Perfusion Pump

There are different commercially available perfusion pumps that vary in perfusion flow rate ranges, inclusion of user programmable timed perfusions, and flow directions (CMA Microdialysis AB, Skmedica, Meql). During reverse microdialysis experiments, adequate flow rate ranges from 0.1 to 2.0 μl/min approximately.

2.2. Microdialysis Probe

The microdialysis probe is perhaps the nucleus of the microdialysis experiment and consists of a tubular dialysis fiber that is connected with an inlet and an outlet tube. The inlet and outlet tubes are connected with thin and flexible tubing with a perfusion pump and with a fraction collector, respectively. Microdialysis probes vary in design, type, and size of dialysis membrane.

2.2.1. Probe Design

Although there are commercially available microdialysis probes (CMA Microdialysis AB, Harvard Apparatus) many researchers have designed their own probe as its construction lasts not longer than several minutes. Different probe designs are available for microdialysis experiments:

- Concentric cannula probe.
- Linear probe.
- Shunt probe.

However the brain is a solid tissue that requires the implantation of rigid concentric microdialysis probes. In addition, a loop dialysis catheter may be suitable for microdialysis experiments at the spinal cord (12).

2.2.2. Type of Dialysis Membrane

Dialysis membranes (Hospal, Sorin Biomedica) are composed of different materials and vary in their cutoff size. The membranes are traditionally made of (2):

- Cuproamonic rayon.
- Cellulose.
- Polycarbonate.
- Polyethersulfone.
- Cuprophan.

Conventional microdialysis probes are constructed with 5–20 kDa molecular weight cutoff membranes enabling the measurement or perfusion of small molecules, including carbohydrates, amino acid, neurotransmitters, and different metabolites. In order to allow exchange of larger molecules such as neuropeptides and cytokines, a 100 kDa molecular weight cutoff microdialysis membrane is available (13). The researcher must select the most suitable dialysis membrane bearing in mind that only compounds with a molar mass lower than the weight cutoff of the membrane are capable of diffusing into the surrounding tissue or into the probe. For instance, if the drug to be perfused through the probe during a reverse microdialysis experiments weighs 5,000 Da, a 20 kDa cutoff dialysis membrane is needed.

2.2.3. Size of the Microdialysis Membrane

Specifically in neuromethods, the size of the microdialysis membrane is limited by the fact that the brain is highly heterogeneous. The microdialysis membrane must not exceed the size of the central nuclei to be studied. Therefore, reverse microdialysis has been used for the study of drug effects in relatively large central nuclei. The researcher should select a suitable size of dialysis membrane taking into account the size of the central tissue and that most of the compounds have slow infusion rates and do not diffuse more than 1 mm from the membrane into the tissue (14).

2.3. Stereotaxic

In order to ensure adequate implantation of the microdialysis probe into the target central nuclei, a stereotaxic frame equipped with carrier for guide cannula or dialysis probe/guide assembly is needed. Stereotaxic frames vary according to laboratory animal species (Stoelting, David Kopf Instruments).

2.4. Perfusion Fluid

In order to prevent changes in the volume of the dialysate, ideally the composition, ion strength, osmotic value, and pH of the perfusion solution should be as close as possible to those of the extracellular fluid of the target tissue. Therefore, the perfusate is an aqueous solution of sodium and potassium salts and other ions in a minor proportion, without proteins or with a very small concentration of them. In some cases, proteins should be added to the perfusion medium to prevent drugs sticking to the microdialysis probe and tubing connections. Although different perfusion fluids are available for reverse microdialysis experiments (for review see (2)), the most adequate media is the artificial cerebrospinal fluid (Tocris Bioscience) composed by (in mM):

- NaCl 125.
- KCl 2.
- Na_2HPO_4 1.25.
- $NaHCO_3$ 26.
- $CaCl_2$ 3.0.
- $MgCl_2$ 3.0.
- Glucose 10.
- pH 7.3–7.4.

In addition, it is important to ensure that the compound to be perfused via the microdialysis probe does not change pH or tonicity of the perfusion medium, because these alterations could affect transmitter release and consequently the results of the study (15). However, a small concentration of the compound added to the perfusion solution is unlikely to change the perfusion fluid properties.

2.5. Equipment for Analytical Quantification of Microdialysis Samples

Reverse microdialysis experiments are mainly designed in order to assess the effect of perfused drugs on extracellular levels of neurotransmitters and metabolites at the target nuclei. Microdialysis provides small-volume samples (1–10 μl) because of the need of slow perfusion rates (0.1–2 μl) to obtain high recoveries of the endogenous compounds maintaining an adequate temporal resolution (2). Also, endogenous compounds, especially neurotransmitters, are often contained at very low concentration in the interstitial space (pM–fM range). For these reasons, the detection limit of the analytical method must be lower than the lowest in vivo expected concentration of the analyte.

Non-separation-based methods allow the detection of one analyte at a time, in contrast with separation-based methods that can be used for the detection of multiple analytes in each sample.

Non-separation-based methods include:

- Biosensors (Biosensors International, Azosensors).
- Immunoassays (Invitrogen, Medcompare).
- Mass spectrometry (Waters, Interlink Scientific Services).

Separation-based methods are the most adequate for reverse microdialysis studies, because the investigator can monitor not only the extracellular levels of the drug but also the concentration of different chemically related endogenous compounds. These methods include:

- Liquid chromatography (LC).
- Microbore LC.
- Capillary LC.
- Capillary electrophoresis (CE).

LC systems are coupled commonly to the following detectors:

- Immunoassay (Invitrogen, Medcompare).
- Ultraviolet (UV) absorbance detector (Konik, Waters).
- Electrochemical (EC) detector (Shiseido, Bioanalytical Systems).
- Fluorescence detector (Thermo, Shimadzu).
- Mass spectrometric detection (Waters, Interlink Scientific Services).

On the other hand, for capillary electrophoresis, the most used are:

- Electrochemical detector (Shiseido, Bioanalytical Systems).
- Laser-induced fluorescence (LIF) detector (Unimicro Technologies, Picometrics).

Analytical methods used for detection of neurotransmitters and metabolites in dialysate samples are highlighted in Table 3 (16).

2.6. Additional Equipment

Although reverse microdialysis has been largely used to study the effects of diverse pharmacological and toxicological agents on neurotransmission at different central nuclei, this technique also allows the evaluation of locally applied drugs on other parameters, including effect on physiological parameters (e.g., blood pressure, locomotor activity, among others) and toxicity on neurons and oligodendrocytes (Table 4).

3. Methods

A successful reverse microdialysis experiment requires an adequate design (Fig. 2). Before the experiment, the researcher needs to evaluate the following points:

3.1. Solubility of the Drug to Be Perfused Through the Microdialysis Probe

It is important to establish if the drug (Fluka) to be perfused into the target nuclei via the microdialysis cannula is soluble in the perfusion media. If the aim of the reverse microdialysis experiment is the perfusion of a high concentration of endogenous compounds

Table 3
Analytical techniques for detection of neurocompounds in dialysate samples

Endogenous compounds	Example	Analytical technique	Sensitivity range (nM)
Acetylcholine		LC-EC, LC-MS/MS	0.1–0.3
Catecholamines	DA, 5HT, NA, A, MOPEG, DOPAC, 5-HIAA, HVA, DOPEG	LC-FL, LC-EC, LC-MS/MS	0.2–0.50
Neuroactive amino acids	GABA, taurine, glycine, serine, glutamate	CE-LIF, LC-FLD	5–100
Neuropeptides	Methionine-enkephalin	cLC-EC	0.02
	Dynorphins, β-Endorphin	LC-MS	0.04
	Somatostatin	RIA, ELISA	0.013–0.024
	Neurotensin	RIA	0.0019
	Angiotensin	Nano LC-MS	0.05
Nucleotides	Adenosine, Guanosine, Cytosine	LC-MS/MS	0.5

CE capillary electrophoresis, *EC* electrochemical detection, *LC* liquid chromatography, *LIF* laser-induced fluorescence detection, *FL* fluorometric detection, *RIA* radio immunoassay, *MS* mass spectrometry, *MS/MS* tandem mass spectrometry, *DA* dopamine, *5-HT* serotonine, *NA* noradrenaline, *A* adrenaline, *MOPEG* 3-methoxy-4-hydroxy-phenylglycol, *DOPAC* 3,4-dihydroxyphenylacetic acid, *5-HIAA* 5-hydroxyindoleacetic acid, *HVA* homovanillic acid, *DOPEG* dihydroxyphenylglycol, *GABA* gamma amino butyric acid
Adapted from (2, 16)

Table 4
Additional equipment for simultaneous monitoring of effect of drugs applied by reverse microdialysis on physiological parameters

Physiological function or system	Physiological parameters	Equipment	References
Behavior	Locomotion, resting/sleeping and alertness	Direct observation	(17)
Cardiovascular	Blood pressure, heart rate, blood pressure variability	Pressure transducer coupled to polygraph	(18, 19)
Depression	Immobility time in the forced swimming test	Glass cylinders	(20)
Feeding	Alcohol intake, eating, and drinking	Direct observation	(17, 21)
Neuronal activity	Electrophysiological characteristics	Intracellular recording electrodes	(22)
Neuronal toxicity	Densities of surviving neurons or oligodendrocytes	Cell counts	(23)
Sleep	Wake–sleep states	Polygraph	(24)
Thermoregulation	Corporal temperature	Telemetry device	(25)

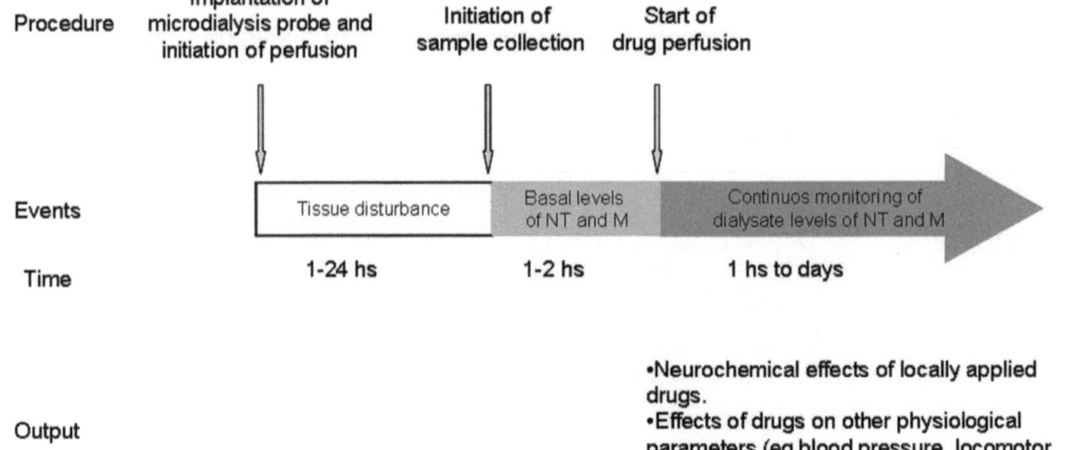

Fig. 2. Time course of a reverse microdialysis experiment. *NT* neurotransmitters, *M* metabolites.

(e.g., neurotransmitter, metabolites) solubility will probably not be an important issue considering that endogenous compounds are water soluble. Conversely, most central acting drugs (e.g., benzodiazepines, antidepressant, and antipsychotics) are lipophylic; therefore the researcher needs to establish if the drug achieves complete dissolution in the perfusion media at the desired concentration. If the drug is not completely soluble in the perfusion media, the researcher can first dissolve the drug in another solvent and then dilute with the perfusion media. For instance, pindolol (100 μM) was first dissolved in perchloric acid (1%) and then diluted in artificial cerebrospinal fluid (20).

3.2. Cleaning and Storing a New Microdialysis Probe

If a new microdialysis probe is used it is recommended to clean the cannula with a solution of ethanol 10% v/v in distilled water. After a microdialysis experiment, the probe must be cleaned by perfusion distilled water in order to wash out salts. Microdialysis probes must be stored in vials containing distilled water and be placed at 4°C during storage. It is important to keep the membrane wet to prevent it from shrinking.

3.3. Dialysance of Drug to Be Perfused Through the Microdialysis Probe

A key part of a reverse microdialysis experiment is to ensure that the drug to be perfused through the microdialysis probe is able to cross the dialysis membrane to achieve the desired concentration in the target central nuclei. In addition, it is necessary to rule out sticking of the drug to the tubing of the microdialysis probe. For this purpose, the researcher needs to estimate the in vitro recovery of the drug by loss and by gain.

During determination of the in vitro recovery by loss, the microdialysis probe, placed in a vial containing the perfusion media, must be perfused with the perfusion media containing the target drug at the desired concentration (2). Before starting the collection of samples, the probe must be perfused at the selected flow rate (e.g., 2 μl/min) for 30 min in order to allow equilibration of the dialysance process. Then, approximately four samples are taken and analyzed for the content of the perfused drug. In vitro recovery of the drug by loss is estimated as follows:

$$R_{\text{loss}} = \frac{C_{\text{per}} - C_{\text{dial}}}{C_{\text{per}}},$$

where R_{loss} is the in vitro recovery by loss, C_{per} is the drug concentration in the perfusate, and C_{dial} is the drug concentration in the dialysate.

For estimation of the in vitro recovery by gain, the microdialysis probe is placed in a vial containing the target drug at the desired concentration dissolved in the perfusion media, and is perfused with the perfusion media without the drug. Before starting the collection of samples, the probe must be perfused at the selected flow rate (e.g., 2 μl/min) for 30 min allowing equilibration of the dialysance process. Then, approximately four samples are taken and analyzed for the content of the dialyzed drug. In vitro recovery of the drug by gain is estimated as follows:

$$R_{\text{gain}} = \frac{C_{\text{dial}}}{C_{\text{vial}}},$$

where R_{gain} is the in vitro recovery by gain, C_{dial} is the drug concentration in the dialysate, and C_{vial} is the drug concentration in the vial.

In vitro recovery of the drug by gain or by loss must be similar considering that the dialysis process is not affected by direction of exchange. If the in vitro recovery by loss is greater than the in vitro recovery by gain the researcher must suspect the existence of sticking of the drug to the microdialysis probe.

In vitro recovery represents an estimation of the amount of drug that will achieve the target central nuclei during a reverse microdialysis experiment. It will be interesting to estimate in vitro recovery at different perfusion flow rates in order to establish the most appropriate rate during the reverse microdialysis experiment.

3.4. Selection of Stereotaxic Coordinates

Using a stereotaxic atlas (Academic Press), the researcher needs to determine stereotaxic coordinates for the target brain region by using the bony landmarks on the skull (bregma) as the reference point. Stereotaxic atlases are available for different laboratory animal species, including the "Mouse brains" atlas software (26) and the Paxinos and Watson "Rat Brain in Stereotaxic Coordinates" (27). In the case of the Paxinos Atlas, it is important to consider that the atlas can be successfully used with male or female rats weighing from

250 to 350 g (27). Frequently, preliminary studies should be conducted to validate or adjust the coordinates if necessary.

3.5. Selection of Animal Condition

A reverse microdialysis experiment can be performed in anesthetized or freely moving animals. When using anesthetized animals, the effect of the anesthetic preparation on the pharmacological response of the applied drug and its effect on specific transmitter systems must be taken into account (28). Anesthetized animals are more suitable for experiments with more than one microdialysis probe, local injection cannula, and in experiments requiring dissections (15). In reverse microdialysis experiments conducted in awake animals, it should be considered that stress could introduce artifacts in the measurements and therefore, the impact of environmental stress must be established (29).

3.6. Duration of Local Drug Administration

Reverse microdialysis allows the continuous local application of drugs and/or endogenous compounds chronically. The researcher needs to take into account that duration of locally applied drugs at a specific central nucleus determines the experimental design. In fact, chronic application of drugs through the microdialysis probe can only be performed in awake animals and requires special equipment, such as the Raturn interactive cage to house the experimental animal. In addition, chronic implantation of the microdialysis probe induces histological changes, such as edema, minor hemorrhages, and leukocyte accumulation in the surrounding tissue of the probe (30), which can change probe performance and need to be considered during the analysis of the results.

During reverse microdialysis experiments, the following steps are needed:

3.7. Anesthetize the Laboratory Animal

In order to place the microdialysis probe in the target central nuclei using the stereotaxic frame, anesthesia of the animal is needed. Different anesthetics are available for microdialysis experiments, including chloral hydrate, pentobarbital, urethane, ketamine, propofol, halothane, and isoflurane (for review see (31)). Selection of the anesthesia depends on whether the reverse microdialysis experiment will be performed in anesthetized or in awake animals. A long-acting anesthesia is preferable for experiments performed in anesthetized animals. In our experience, combination of urethane and chloralose is a highly attractive option.

3.8. Placement of Microdialysis Probe

The microdialysis probe can be implanted directly at the target site or after placement of a guide cannula above the region of interest (32). If the reverse microdialysis experiment is performed in a conscious animal, use of guide cannula is preferable. For implantation of the microdialysis probe, the animal must be mounted in the stereotaxic apparatus and the skull must be exposed. Thereafter, a needle must be placed at the bregma (midline intersection of the coronal and sagittal sutures) and the antero-posterior (AP), lateral (L), and dorsoventral (DV) coordinates must be read. To determine the working coordinates, the researcher needs to add or subtract the coordinates of the target nuclei obtained from the stereotaxic atlas. Then, the needle must be moved to the calculated AP and L coordinates and the skull must be drilled until the dura is punctured. Then, the needle must be introduced into brain parenchyma until the working DV coordinate is achieved. Then the needle must be removed and the same procedure must be performed with the microdialysis probe or the guide cannula. However, the microdialysis probe needs to be slowly introduced in brain parenchyma (0.5 mm/min) in order to prevent membrane damage.

3.8.1. Equilibration Period

As microdialysis probe implantation is an invasive technique, an equilibration period is needed to allow the recovery of tissue function. It is a well-known fact that probe insertion initially induces the release of neurotransmitters in the surrounding area, but their levels rapidly return to baseline. In addition, a disturbance in tissue function lasting from 30 min to 24 h also has been described and is characterized by increased glucose metabolism and decreased blood flow (15). On the other hand, microdialysis probe implantation has only minor effects on blood–brain barrier integrity (30).

Commonly, reverse microdialysis in anesthetized preparations needs an equilibration period of 2 h before beginning sample collection (32, 33). In a conscious animal, the standard equilibration time ranges typically 6–10 h (32).

3.8.2. Estimation of Basal Parameters

An attractive aspect of reverse microdialysis experiments is the fact that the actual animal serves as its own control requiring fewer subjects than do some other methods. Therefore, in reverse microdialysis experiments the desirable information is the relative change of the neurotransmitter and/or metabolite concentration induced by drug local administration. An advantage of this approach is that accurate in vivo calibration of the microdialysis probe in pharmacodynamic studies is not necessary.

The microdialysis probe must be perfused at the working flow rate for approximately 1–2 h in order to obtain approximately four microdialysis samples to assess basal parameters (neurotransmitter and metabolite concentration and basal levels of physiological parameters). During collection of microdialysis samples, it is important to store samples (preferably at –70°C) immediately following their collection to help prevent sample degradation or evaporation (32). Although microdialysis sampling prevents enzymatic degradation of compounds, several neurotransmitters, especially catecholamines, are highly sensitive to chemical and photodegradation.

The content of neurotransmitters and metabolites in microdialysis samples must be quantified using an analytical method of choice. Previous to drug perfusion, it is important to evaluate the data in order to demonstrate that baseline samples are at steady state by determining that basal dialysate levels remain constant over several consecutive microdialysis samples.

3.8.3. Perfusion of the Drug Through the Microdialysis Probe

After assessment of basal conditions, perfusion media must be replaced by the same fluid containing the drug to be perfused in order to initiate the local application of the drug at the target central nuclei. If the dead volume of the inlet of the microdialysis probe is large, the researcher can consider initially increasing the perfusion flow rate for a few minutes in order to reduce the delay in arrival of the drug to the dialysis membrane. The perfusion rate used during the basal period and during the drug perfusion must be the same considering that recovery of the endogenous compounds depends on the flow rate.

Microdialysis samples obtained during perfusion of the drug can be fractionated in order to estimate the amount of drug administered by reverse microdialysis and to determine changes in dialysate concentrations of endogenous compounds induced by drug perfusion. In fact, reverse microdialysis allows the possibility to manipulate and estimate the amount of drug that diffuses into the extracellular space. As only a fraction of the drug included in the perfusion media is diffused to the surrounding tissue, the amount of the added substance that is delivered to the interstitial space could be calculated by the determination of the concentration of the compound in the perfusate and the dialysate. The difference of these concentrations represents the amount of drug delivered into the tissue during the sampling time. Therefore, the sum of the difference of concentrations between perfusate and dialysate of the drug for each sampling interval represents the total amount of drug delivered during the reverse microdialysis experiments (5). The total amount of the applied drug can be determined with the following equation:

$$D_{\mathrm{A}} = \sum_{i=1}^{n} (C_{\mathrm{per}} - C_{\mathrm{dial}}) \times R \times \Delta t,$$

where n is the succession of sampling, $i = 1$ to n, D_{A} is the total amount of the drug A administered via reverse microdialysis, C_{per} is the concentration of A in the perfusate, C_{dial} is the concentration of A in the dialysate, R is the flow rate of perfusion, and Δt is the time of the sampling interval. In the reverse microdialysis technique, drug administration into the site of perfusion is of zero order similar to an intravenous constant rate infusion. Therefore, constant levels at the site of perfusion are attained when the diffusion rate of the drug from the tissue surrounding the microdialysis probe due to distribution in other tissues is similar to the diffusion rate of the drug from the probe to the extracellular space.

In addition, the content of neurotransmitters and metabolites in the second fraction of the microdialysis samples must be analyzed

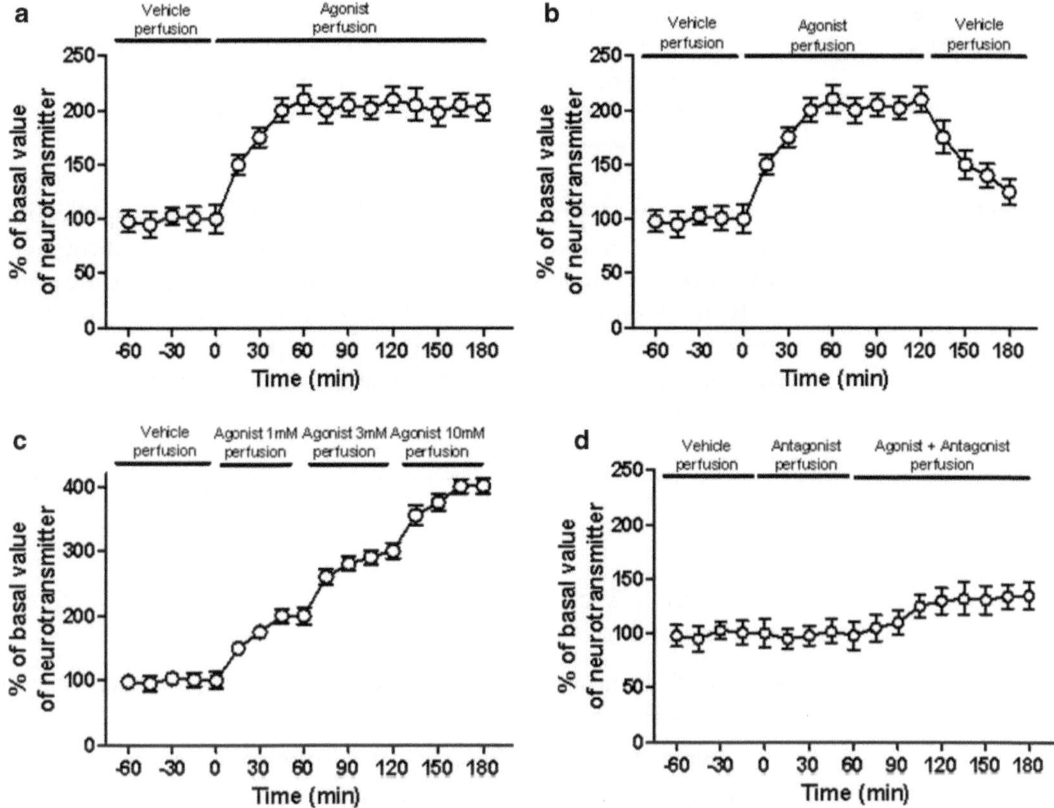

Fig. 3. Different protocol designs of reverse microdialysis studies. (**a**) Perfusion of a single drug concentration of the experimental drug. (**b**) Perfusion of a single drug concentration of the experimental drug followed by vehicle perfusion. (**c**) Perfusion of increasing drug concentration of the experimental drug. (**d**) Coperfusion of antagonist with the experimental drug.

using standard analytical methods (Table 3) in order to establish the relative change in neurotransmission induced by drug perfusion. Simultaneously to reverse microdialysis of the drug, the researcher can also assess changes in different physiological parameters induced by local drug application (Table 4).

Time of perfusion of the drug can vary from few hours to several days and depends on the design of the reverse microdialysis experiment.

As control experiment, the researcher needs to consider the inclusion of an experimental group perfused with perfusion media alone.

Reverse microdialysis is highly versatile considering that it gives the opportunity to apply different treatments in the same experiment. In fact, after estimation of basal conditions, it is possible to (1) only perfuse the experimental drug during the complete assay (Fig 3a), (2) allow the recovery of basal conditions after drug exposure (Fig 3b), (3) perfuse the experimental drug at different concentrations (Fig 3c), or (4) test the effect of antagonist coperfusion on experimental drug action (Fig 3d).

3.8.4. Localization of the Microdialysis Probe

Accurate implantation into the area to be sampled is crucial for reverse microdialysis experiments, especially when small central nuclei are perfused. Therefore, after finishing the reverse microdialysis experiment, verification of microdialysis probe placement by histology is necessary.

3.9. Analysis of the Data

As commented previously, the effects of drug local application through the microdialysis probe are analyzed by relating neurotransmitters and/or metabolite concentrations at each time point with the basal levels assessed prior to drug perfusion. Therefore, the outcome of a reverse microdialysis experiment is expressed as the percentage of change with regard to basal conditions.

4. Notes

To ensure a successful reverse microdialysis trial and adequate interpretation of the results, the researcher needs to:

- Confirm the ability of the target drug to cross the microdialysis membrane by means of in vitro assays.
- Rule out sticking of the drug to probe components.
- Store adequately microdialysis samples prior to quantification.
- Exclude interference by anesthesia and tissue disturbance.
- Select the suitable membrane size with regard to the target central nucleus.
- Include a control group only perfused with blank media.
- Check localization of the microdialysis probe after the experiment.

Appendix: List of Main Suppliers and Web Page Address

Supplier	Product	Web page address
CMA Microdialysis AB	Microdialysis probes, perfusion pumps	http//www.microdialysis.se
Skmedica	Perfusion pumps	http//www.Skmedica.com
Meql	Perfusion pumps	http//www.meql.com
David Kopf Instruments	Stereotaxic	http//www.kopfinstruments.com
Stoelting	Stereotaxic instrumentation	http//www.stoeltingco.com

(continued)

Supplier	Product	Web page address
Bioanalytical systems	Analytical equipment	http//www.basinc.com
Harvard apparatus	Microinfusion pump, microdialysis probes	http//www.harvardapparatus.com
Academic press	Brain atlas	http//www.elsevierdirect.com
Eppendorf	Tubes	http//www.eppendorf.com
Fluka	Analyte standards	http//www.sigmaaldrich.com
Hospal	Microdialysis fibers	http//www.hospal.it
Tocris bioscience	Artificial cerebrospinal fluid	http//www.tocris.com
Microbiotech	FEP tubing	http//www.microbiotech.se
Sorin biomedica	Microdialysis fibers	http//www.sorin.com
Spark Holland	Autosamplers	http//www.sparkholland.com
Univentor	Fraction collectors	http//www.univentor.com
Waters	Mass detectors	http//www.waters.com
Interlink scientific services	Mass detectors	http//www.iss-store.co.uk
Biosensors International	Biosensors	http//www.biosensors.com
Azosensors	Biosensors	http//www.azosensors.com
Invitrogen	Immunoassay equipment	http//www.invitrogen.com
Medcompare	Immunoassay equipment	http//www.medcompare.com
Konik	UV detectors	http//www.konik-group.com
Waters	UV detectors	http//www.waters.com
Shiseido	Electrochemical detectors	http//www.shiseido.co.jp
Bioanalytical systems	Electrochemical detectors	http//www.basinc.com
Thermo	Fluorescence detectors	http//www.thermoscientific.com
Shimadzu	Fluorescence detectors	http//www.shimadzu.com

(continued)

Supplier	Product	Web page address
Unimicro technologies	LIF detectors	http//www.unimicrotech.com
Picometrics	LIF detectors	http//www.picometrics.com
Microbore	Microbore LC	http//www.microbore.de
SGE	Capillary HPLC equipment	http//www.sge.com
Agilent technologies	Capillary HPLC equipment	http//www.chem.agilent.com

References

1. Höcht C, Opezzo JA, Taira CA (2007) Applicability of reverse microdialysis in pharmacological and toxicological studies. J Pharmacol Toxicol Methods 55(1):3–15
2. Höcht C, Opezzo JA, Taira CA (2004) Microdialysis in drug discovery. Curr Drug Discov Technol 1(4):269–285
3. Höcht C, Opezzo JAW, Bramuglia GF, Taira CA (2006) Application of microdialysis for pharmacokinetic-pharmacodynamic (PK–PD) modelling. Exp Opin Drug Discov 1(4):289–301
4. Galvan A, Smith Y, Wichmann T (2003) Continuous monitoring of intracerebral glutamate levels in awake monkeys using microdialysis and enzyme fluorometric detection. J Neurosci Methods 126(2):175–185
5. Chan SHH, Chan JYH (1999) Application of reverse microdialysis in the evaluation of neural regulation of cardiovascular functions. Anal Chim Acta 379(3):275–279
6. Bazzett TJ, Becker JB, Albin RL (1991) A novel device for chronic intracranial drug delivery via microdialysis. J Neurosci Methods 40(1):1–8
7. Evans BK, Armstrong S, Singer G, Cook RD, Burnstock G (1975) Intracranial injection of drugs: comparison of diffusion of 6-OHDA and guanethidine. Pharmacol Biochem Behav 3(2):205–228
8. Rice ME, Gerhardt GA, Hierl PM, Nagy G, Adams RN (1985) Diffusion coefficients of neurotransmitters and their metabolites in brain extracellular fluid space. Neuroscience 15(3):891–902
9. Hargraves R, Freed WJ (1987) Chronic intrastriatal dopamine infusions in rats with unilateral lesions of the substantia nigra. Life Sci 40(10):959–966
10. Westerink BHC, Justice JB Jr (1991) Microdialysis compared with other in vivo release models. In: Robinson TE, Justice JB Jr (eds) Microdialysis in the neurosciences, techniques in the behavioral and neural sciences, vol 7. Elsevier, New York, pp 23–46
11. Fillenz M (2005) In vivo neurochemical monitoring and the study of behaviour. Neurosci Biobehav Rev 29(6):949–962
12. Marsala M, Malmberg AB, Yaksh TL (1995) The spinal loop dialysis catheter: characterization of use in the unanesthetized rat. J Neurosci Methods 62(1–2):43–53
13. Hutchinson PJ, O'Connell MT, Nortje J, Smith P, Al-Rawi PG, Gupta AK, Menon DK, Pickard JD (2005) Cerebral microdialysis methodology–evaluation of 20 kDa and 100 kDa catheters. Physiol Meas 26(4): 423–428
14. Westerink BH, De Vries JB (2001) A method to evaluate the diffusion rate of drugs from a microdialysis probe through brain tissue. J Neurosci Methods 109(1):53–58
15. Ungerstedt U (1991) Introduction to intracerebral microdialysis. In: Robinson TE, Justice JB Jr (eds) Microdialysis in the neurosciences. Elsevier, Amsterdam, pp 3–22
16. Perry M, Li Q, Kennedy RT (2009) Review of recent advances in analytical techniques for the determination of neurotransmitters. Anal Chim Acta 653(1):1–22
17. Duva MA, Tomkins EM, Moranda LM, Kaplan R, Sukhaseum A, Jimenez A, Stanley BG (2001) Reverse microdialysis of N-methyl-D-aspartic acid into the lateral hypothalamus of rats: effects on feeding and other behaviors. Brain Res 921(1–2):122–132

18. Di Verniero CA, Bertera F, Buontempo F, Bernabeu E, Chiappetta D, Mayer MA, Bramuglia GF, Taira CA, Höcht C (2010) Enantioselective pharmacokinetic-pharmacodynamic modelling of carvedilol in a N-nitro-l-arginine methyl ester rat model of secondary hypertension. J Pharm Pharmacol 62(7):890–900
19. Höcht C, Opezzo JA, Taira CA (2005) Hypothalamic antihypertensive effect of metoprolol in chronic aortic coarctated rats. Clin Exp Pharmacol Physiol 32(8):681–686
20. Guilloux JP, David DJ, Guiard BP, Chenu F, Repérant C, Toth M, Bourin M, Gardier AM (2006) Blockade of 5-HT1A receptors by (+/-)-pindolol potentiates cortical 5-HT outflow, but not antidepressant-like activity of paroxetine: microdialysis and behavioral approaches in 5-HT1A receptor knockout mice. Neuropsychopharmacology 31(10):2162–2172
21. Engleman EA, McBride WJ, Wilber AA, Shaikh SR, Eha RD, Lumeng L, Li TK, Murphy JM (2000) Reverse microdialysis of a dopamine uptake inhibitor in the nucleus accumbens of alcohol-preferring rats: effects on dialysate dopamine levels and ethanol intake. Alcohol Clin Exp Res 24(6):795–801
22. West AR, Grace AA (2002) Opposite influences of endogenous dopamine D1 and D2 receptor activation on activity states and electrophysiological properties of striatal neurons: studies combining in vivo intracellular recordings and reverse microdialysis. J Neurosci 22(1):294–304
23. McAdoo DJ, Wu P (2008) Microdialysis in central nervous system disorders and their treatment. Pharmacol Biochem Behav 90(2):282–296
24. Crochet S, Onoe H, Sakai K (2006) A potent non-monoaminergic paradoxical sleep inhibitory system: a reverse microdialysis and single-unit recording study. Eur J Neurosci 24(5): 1404–1412
25. Ishiwata T, Saito T, Hasegawa H, Yazawa T, Otokawa M, Aihara Y (2006) Changes of body temperature and extracellular serotonin level in the preoptic area and anterior hypothalamus after thermal or serotonergic pharmacological stimulation of freely moving rats. Life Sci 75(22):2665–2675
26. Hof PR, Young WG, Bloom FE, Belichenko PV, Celio MR (2000) Comparative cytoarchitectonic atlas of the C57BL/6 and 129/Sv mouse brains with CD ROM. Elsevier, San Diego, CA
27. Paxinos G, Watson CH (2007) The rat brain in stereotaxic coordinates, 6th edn. Academic, New York
28. Claassen V (1994) Anaesthesia. In: Claaseen V (ed) Neglected factors in pharmacology and neuroscience research. Elsevier, Amsterdam, pp 382–421
29. Bourne JA (2003) Intracerebral microdialysis: 30 years as a tool for the neuroscientist. Clin Exp Pharmacol Physiol 30(1–2):16–24
30. Benveniste H, Hansen AJ (1991) Practical aspects of using microdialysis for determination of brain interstitial concentrations. In: Robinson TE, Justice JB Jr (eds) Microdialysis in the neurosciences. Elsevier, Amsterdam, pp 23–43
31. Souza Silva MA, Müller CP, Huston JP (2008) Microdialysis in the brain of anesthetized vs. freely moving animals. In: Westerink BHC, Cremers TIFH (eds) Handbook of Microdialysis. Elsevier, Amsterdam, pp 71–91
32. Zapata A, Chefer VI, Shippenberg TS (2009) Microdialysis in rodents. Curr Protoc Neurosci 7.2.1–7.2.29
33. Parsons LH, Justice JB Jr (1994) Quantitative approaches to in vivo brain microdialysis. Crit Rev Neurobiol 8(3):189–220
34. Gaddum JH (1961) Push-pull cannulae. J Physiol 155:1P–2P

Chapter 4

Microdialysis and Advances for Sampling Synaptic and Extrasynaptic Pools

Cheng-fu Chen, Brian T. Rasley, Benjamin P.E. Warlick, Tom K. Green, Kristian E. Swearingen, and Kelly L. Drew

Abstract

Intercellular communication plays a key role in information processing in the nervous system, to immune response, to cellular growth and differentiation, and to most processes fundamental to life for multicellular organisms. As a sampling technique microdialysis enables in vivo studies of brain and other tissues and has advanced our understanding of intercellular signal processing. The first part of this chapter is an overview of microdialysis in the context of a how-to-guide with reference to sampling extrasynaptic glutamate and GABA using conventional methodology. The limitations and challenges associated with sampling the synaptic pool of fast neurotransmitters are then addressed. The last part of this chapter presents ideas of advancing the microdialysis technique that may bring the microdialysis membrane closer to the synapse.

Key words: Microdialysis, Membrane, Diffusion, Microfluidics, Droplet-based microdialysis

1. Introduction

Surveillance of intercellular communication is essential to advance our understanding of information processing in the nervous system, immune response, cellular growth and differentiation, and most processes fundamental to life for multicellular organisms. Understanding intercellular communication requires a capacity to sample and detect the interstitial milieu. In this regard, microdialysis has opened a window into the brain and other tissues that has advanced our understanding of intercellular signal processing (1–5). Nonetheless, the microdialysis probe design has barely changed since 1966 (except for the use of shorter cannulas for mice). Here we provide an overview of microdialysis in the context of a how-to-guide. We then discuss specific challenges associated

Giuseppe Di Giovanni and Vincenzo Di Matteo (eds.), *Microdialysis Techniques in Neuroscience*, Neuromethods, vol. 75,
DOI 10.1007/978-1-62703-173-8_4,

with sampling the synaptic pool of fast neurotransmitters such as glutamate and GABA. Finally we present ideas for advancing the microdialysis technique that may bring the microdialysis membrane closer to the synapse.

2. Overview of Microdialysis

Microdialysis is an invasive sampling technique whereby a probe with a porous or a semipermeable membrane is inserted into tissue such that one side of the membrane is in contact with extracellular fluid and the other side is in contact with a dialysis fluid (e.g., Ringer's solution or artificial cerebrospinal fluid). By diffusion (per Fick's law), molecules that are smaller than the molecular weight cutoff of the microdialysis membrane move across the membrane from high concentration to low concentration. In this way intercellular signaling molecules flow from the extracellular space into the perfusate; or in the case of "reverse dialysis" drugs dissolved in the dialysis fluid flow from the dialysis fluid into the extracellular space. Microdialysis becomes a biosensor when sampling is coupled with analytical techniques to separate and detect the molecules collected in the analyte-laden perfusate.

Since its inception (6), microdialysis has become one of the most frequently employed methods for sampling the extracellular fluid compartment and has a wide variety of neuroscience and other research and clinical applications (7–12). Microdialysis offers some advantages over other in vivo sensor techniques. For example, microdialysis samples can be divided between two or more analytical techniques for concurrent detection of several types of analytes (13). Separative analytical techniques coupled with microdialysis sampling enhance confidence of identity and quantification of a given analyte compared with in vivo sensor technologies. Moreover chemical analysis achieved on a laboratory bench in a controlled environment is not vulnerable to movement or other sources of artifact encountered in freely moving animals. Nonetheless with regard to sampling synaptic levels of fast neurotransmitters microdialysis lacks temporal and spatial resolution (14, 15).

2.1. What to Consider Before Using Microdialysis

2.1.1. Spatial and Temporal Resolution

Conventional microdialysis typically provides spatial and temporal resolution of 0.1 mm^3 and 1–10 min or longer depending on the analyte (12). This rate of sampling targets extrasynaptic pools of neurotransmitters and neuromodulators. Due to limited spatial and temporal resolution, conventional microdialysis is best suited for sampling extrasynaptic pools of neurotransmitters and neuromodulators (15). Spatial resolution refers to the ability to resolve changes in analyte levels within a given volume of tissue. Temporal resolution refers to the ability to resolve changes in analyte levels within a given period of time. Synaptic and extrasynaptic pools of neurotransmitters serve distinct functions in the brain (4, 14, 16, 17), and the spatial and temporal resolution of the sampling technique determines which pool of neurotransmitter is monitored.

2.1.2. Choice of Probes and Accessories

Custom-made probes can increase temporal resolution. The smaller the internal volume of the dialyzing tip the better the temporal resolution will be. This is because smaller volumes equilibrate faster with the extracellular fluid and have less internal fluid to mix with and dilute analyte diffusing into the lumen of the dialyzing tip. Microdialysis probes with internal volumes smaller than commercially available probes can be constructed in house (see Appendix for probe assembly protocol). Nonetheless, commercial probes are well suited for many applications and users may benefit from working with a commercial supplier that will provide probes, accessories, and detailed instructions on how to work with their particular products (Table 1). Related accessories such as pumps, syringes, tubing, tubing adapters, probe holders, and stands can also be purchased from companies that supply microdialysis probes. In general continuous drive syringe pumps are favored over peristaltic pumps because they provide pulse-free flow. Airtight glass syringes are preferred over plastic syringes, but this preference may depend on your desire for sterile, disposable plastic verses gastight glass. Sterile conditions are essential when sampling glu for more than a few hours since bacteria release glutamate (18). Fluorinated Ethylene Propylene (*FEP*) tubing is flexible and chemically inert; however fused glass capillary tubing offers smaller internal diameters. Both types can be autoclaved. The internal volume of tubing that connects the syringe to the probe (i.e., the inlet tubing) and the volume of the probe inlet will determine the time it takes for a given solution to travel from the syringe to the dialyzing tip. The internal volume, also referred to as dead volume, is the volume of the probe tip and outlet cannula. The internal volume and the length of outlet tubing will determine the transit time for an analyte from the probe to the collection vial. When working with outlet tubing with small internal diameters keep in mind that the smaller the internal diameter and the longer the outlet length, the greater the backpressure at the probe tip. High backpressures may cause the probe to "sweat" or to break and leak. Backpressure will therefore limit the flow rate that can be used for microdialysis sampling or to flush your system. Probes can be cleaned and flushed at whatever flow rate the system will handle without compromising the probe. Narrow bore and lengthy outlets will require lower flow rates as indicated by the manufacturer.

2.1.3. Choice of Flow Rate

Flow rates during sample collection generally vary between 1 and 2 μL/min. As flow rate is increased concentration of analyte in the dialysate will decrease while total volume collected per sampling

Table 1
Some suppliers of microdialysis probes and accessories for neuroscience applications

Vendor	Product	Membrane dimensions	Membrane and MW cut-off	Dead volume	Sterilized for surgery
CMA http://www.microdialysis.se	CMA 12	0.5 mm OD 1,2,3 or 4 mm lengths	Polyarylethersulfone (PAES) 20 kD and Polyethersulfone (PES) 100 kD	3 μL	No
	CMA 11	0.24 mm OD 1,2,3 or 4 mm lengths	Cuprophane (regenerated cellulose) 6 kD	1.0 μL	No
	CMA 7	0.24 mm OD 1 or 2 mm lengths	Cuprophane (regenerated cellulose) 6 kD	0.3 μL	May be purchased β-irradiated
BASi http://www.basinc.com	BR-2	0.32 mm OD 2 mm lengths	Polyacrylonitrile (PAN) 30 kD	2.1 μL	No
	BR-4	0.32 mm OD 4 mm lengths	PAN 30 kD	2.3 μL	No
PlasticsOne http://www.plastics1.com	HMD-1	2–5 mm lengths	13 kD	Not available	No
	HMD-2	2–5 mm lengths	18 kD	Not available	No
Noncommercial "low-volume" Appendix		Custom	Regenerated cellulose 13 kD	0.1 μL	No

BASi "BR Brain Probes User's Guide": http://www.basinc.com/mans/brbp.pdf
CMA applications notes: http://www.microdialysis.se/basic-research/application-notes
CMA "how to design your microdialysis experiments": http://www.microdialysis.se/public/file.php?REF=9a1158154dfa42caddbd0694a4e9bdc8&art=42&FILE_ID=20070626112805_1_1.pdf

period will increase. Higher flow rates are an advantage in that they provide ample volume for handling and for sharing between two or more analytical techniques. Unfortunately, higher flow rates may not concentrate your analyte of interest sufficiently to reach your limit of detection. Moreover, higher flow rates may drain components from the extracellular space (19). Slower flow rates are advantageous for many reasons. Slow flow rates do not drain components from the extracellular space and slow flow rates yield highly concentrated analyte. Very slow flow rates such as 0.05–0.1 μL/min and slower allow for the perfusion fluid to equilibrate with the extracellular space such that concentration of analyte in the perfusate approaches the concentration of analyte in the extracellular fluid (20–22). Very slow flow rate is thus one approach to quantitative microdialysis (23, 24). The disadvantage of slow flow rates, however, is that they yield low sample volume. Necessary sampling frequency, subsequent sample volume, analyte concentration, and desire for quantitative or relative measures of extracellular concentration will ultimately influence your choice of flow rate.

2.1.4. Choice of Analytical Technique

Many of these choices will also depend on the sensitivity and optimal injection volume of your analytical technique. Thus, choice and optimization of a suitable analytical technique are paramount to successful microdialysis. The technique must be sensitive enough to detect analyte collected at a sampling frequency appropriate for the question at hand and accommodate the volume of available sample. HPLC is a common analytical method (25) with advances being made in capillary LC (26). Capillary electrophoresis (CE) is used less frequently but offers several advantages in terms of sensitivity, "injection" volume, and liquid handling (27–30). Liquid chromatography tandem mass spectrometry (LC/MS/MS) offers promise for enhanced sensitivity for both small molecules and proteins (31, 32). It is best to optimize the analytical technique prior to conducting any in vivo experiments. As a first step, establish the limit of detection, reproducibility, linear range, and stability of your analyte(s). Confirm identity and purity of peaks by spiking a sample with known analyte and removing analyte from a biological sample by treating the sample with an appropriate enzyme such as glutamate oxidase to remove glutamate.

2.1.5. Option to Perform In Vitro Tests

In vitro tests provide an opportunity to assess probe performance. In vitro tests are highly recommended for homemade probes or used commercial probes and are also an excellent means to train someone new to the technique and to test equipment and other aspects of the approach prior to working with live animals. In vitro tests are conducted as an in vivo experiment except that the dialyzing tip is immersed in a solution of known quantity of analyte rather than into living tissue. The closer the details are to the in vivo test, the better the in vitro reproduction will prepare a first time user for the in vivo experiment. Many suppliers of microdialysis probes provide stands and probe holders that will facilitate in vitro testing.

Probe performance is assessed from relative recovery of analyte. Relative recovery represents a percentage of analyte recovered from a given solution of known analyte concentration, and can be addressed as the ratio of concentration of analyte in the perfusate to the concentration of analyte in the solution in which the probe is immersed. Relative recovery may be less than 10% or as great as 100% and will depend on several factors including the molecular weight of analyte, molecular cutoff of the membrane, length and diameter of the dialyzing tip, perfusion flow rate, and integrity of the probe. Relative recovery of endogenous neurotransmitter determined in vitro will *not* predict relative recovery of neurotransmitter in vivo and *cannot* be used to calculate true in vivo concentrations of endogenous neurotransmitter (33) (this may not be the case for studies of exogenously applied drugs that are not metabolized or cleared from the extracellular space as are endogenous molecules). High-purity water is often used as a perfusion fluid for in vitro tests when the purpose is simply to assess probe performance.

2.1.6. Choice of Perfusion Fluid

Once sampling begins in vivo, however, choice of perfusion fluid may impact the physiological relevance of your results. Perfusion fluid flows from the syringe, through the inlet tubing, and into the lumen of the probe where it is in contact with the extracellular fluid of your target tissue. Molecules and gases in your perfusion fluid will diffuse into or out of the tissue according to concentration gradient. The perfusion fluid travels from the probe tip through the outlet tubing and into a collection vial or an analytical instrument. Thus, the perfusion fluid must be free of any contaminant that will interfere with analysis or will influence the physiological integrity of your target tissue. The contents of the perfusion fluid should match the composition of the extracellular milieu as closely as feasible because deviations will influence the composition of the extracellular milieu and the target tissue.

Perfusion fluids such as artificial cerebrospinal fluid (aCSF) will, at a minimum, consist of electrolyes (Table 2). aCSF may also contain glucose and a buffer. PO_2 and PCO_2 may be adjusted to within a range of normal values (34). The primary electrolyes are sodium, chloride, potassium, calcium, and magnesium (35). Sodium concentration is generally around 148 mM. Decreasing sodium in the aCSF will decrease sodium in the extracellular space and can reverse sodium-dependent transport of neurotransmitters. Na^+ and Cl^- contribute significantly to osmotic pressure in the extracellular space. If sodium concentration is decreased, for example to facilitate reversal of transporters, sodium may be replaced by choline to avoid significant decreases in osmotic pressure. Extracellular potassium influences the resting membrane potential, so raising it will

Table 2
Contents of artificial cerebrospinal fluid (aCSF), a common perfusion fluid

	Electrolytes (mM)	Nutrients	Gases and pH[a]
Unbuffered aCSF (81)	146 mM NaCl 2.7 mM KCl 1.2 mM $CaCl_2$ 0.8–1.0 mM $MgCl_2$	None	Unaltered
Bicarbonate-buffered aCSF (22)	124 mM NaCl 2.7 mM KCl 1.2 mM $CaCl_2$ 0.85 mM $MgCl_2$ 24 mM $NaHCO_3$ 309 mOsmol/L	1.4 mM glucose	pH 7.4 Po_2 70–80 mmHg Pco_2 30–40 mmHg

[a]Adjusted by bubbling with 95% N_2/5% CO_2

cause neurons to depolarize. Doubling potassium concentrations will increase basal levels of some neurotransmitters. It is expected that raising potassium to 40 mM or higher will produce a rapid and significant increase in neurotransmitter release and overflow into the extrasynaptic space. Extracellular calcium facilitates exocytosis so that increasing calcium concentrations increases extracellular concentrations of most neurotransmitters. Magnesium is a calcium channel blocker and will counteract effects of calcium. Magnesium also gates glutamate-activated NMDA receptors, so raising magnesium concentrations will influence glutamatergic signaling and attenuate excitotoxicity. The brain consumes large amounts of glucose as its preferred energy substrate and glucose is a precursor for glutamate and GABA during periods of high neuronal activity (36). Including glucose in the aCSF minimizes loss of glucose from the extracellular space. Addition of glucose, however, also promotes bacterial growth; so glucose should be added on the day of the experiment. To avoid bacterial contamination that can affect glutamate concentrations (18) all solutions should be sterilized by filtering through 0.2 μm filters using sterile technique. aCSF should be prepared from high-purity water to minimize chemical contaminants that may interfere with analysis. Water samples should be analyzed periodically to confirm purity and to avoid contaminating peaks. To avoid movement of water into the perfusate the osmolarity of aCSF should be around 300 mOsmol/L. Affinity agents may be added to enhance recovery of peptides (37). Antioxidants such as ascorbate (0.2 mM) may be added to the perfusion fluid. Samples may also be collected into acidic solutions containing antioxidants to stabilize catecholamines. For example, samples may be collected into a volume of 1.1 N perchloric acid solution containing 50 mg/L Na_2EDTA and 50 mg/L sodium metabisulfite.

2.1.7. Choice of an Awake or Anesthetized Preparation

Acute, anesthetized preparations provide slightly higher throughput than awake, freely moving animals. Anesthetized preparations may, however, not allow for full recovery after mild blunt trauma caused by inserting the probe into brain tissue (38). Moreover, a strength of microdialysis is that it is well suited for sampling from awake animals and awake preparations are preferred for any study that could be confounded by the influence of anesthesia. If you have chosen an awake preparation does your experimental design require acute sampling within one 24-h period or repeated sampling over days? Repeated sampling from an individual over days can be compromised by gliosis around the implantation site. Gliosis can influence diffusion and uptake of neurotransmitters as well as delivery of substances by reverse dialysis. If repeated sampling is desired a probe can remain in place for several days or be removed and reinserted at desired intervals. In one study evidence favored the removable probe system (39, 40). Sterile technique can greatly improve long-term sampling.

2.1.8. Sterile Technique

Sterile technique for surgery, probe insertion into tissue, and preparation of aCSF is advised, and recommended when monitoring amino acid neurotransmitters. Bacteria release glutamate and can confound interpretation of neurotransmitter levels (18). Probes should be autoclaved (if recommended by the manufacturer) or gas (ethylene oxide) sterilized by your institution's veterinary services or other clinical department. aCSF should be filtered through a 0.2 μm filter (such as sterile Acrodisc® syringe filters) into a sterile container. Copolite Copalite®, an antimicrobial varnish (Cooley and Cooley, Houston, TX), applied immediately after exposing the skull during surgery enhances long-term maintenance of head stages when using chronically implanted guide cannula.

Prior to placing the probe into tissue, ensure that your system components and perfusion fluid are free of contaminants by analyzing samples of perfusion fluid that are collected before and after the fluid has passed from inlet to outlet in vitro. This is especially true for glu, GABA, and other amino acids because these analytes are ubiquitous in all biological systems and system components are easily contaminated. While flushing the system in vitro search for leaks or blocked tubing and insure that the volume delivered per unit time is consistent with your flow rate. Air bubbles can be minimized by warming the solution in a water bath set to a tepid temperature 1–2°C warmer than the ambient temperature. Fill the syringes and flush the tubing and probe while the solution is warm. To preserve the probes and tubing once the experiment is complete, rinse syringes and flush tubing and probes with copious amounts of water to wash out all residual salts.

3. Challenges Associated with Sampling Fast Neurotransmitters Such as Glutamate and GABA

3.1. Overview of Microdialysis for the Study of Synaptic and Extrasynaptic Pools of Neurotransmitters

Traditional concepts of neurotransmission focus on the chemical synapse with associated vesicular release and rapid clearance of neurotransmitters. Historically, microdialysis unveiled an extrasynaptic pool of glutamate and GABA that failed to respond to manipulation of vesicular release (41, 42). These extrasynaptic pools have since been shown to be physiologically and pathologically significant (17, 43, 44). Evidence suggests that reversal of GABA transport regulates tonic inhibition and can contribute to inhibition (45). Cystine–glutamate exchange regulates an extrasynaptic pool of glutamate (43, 46) that underlies cocaine relapse (47). Activation of extrasynaptic NMDA receptors alone accounts for Ca^{2+} overload and excitotoxic cell death (48) and it is this pool that is sampled with microdialysis. Clinical monitoring of the extrasynaptic pool reflects dysfunction of cerebral energy metabolism and is an early indicator of delayed neurological deterioration (49, 50). Microdialysis studies furthermore demonstrate that synaptic pools of some neurotransmitters spill into the extrasynaptic space. For this reason, microdialysis has provided a window into the extra synaptic pool that has significantly advanced the understanding of neurotransmitter control of behavior. Nonetheless, the ability to sample synaptic pools of neurotransmitters with conventional microdialysis is limited.

Extracellular glu originates from heterogenous synaptic and extrasynaptic pools. Whole cell recordings of NMDA-mediated excitatory postsynaptic current (EPSC) suggest that ambient (resting) synaptic concentrations of glu are near 25 nM (51). These ambient, synaptic concentrations are significantly lower than the 1–4 μM ambient concentrations measured outside of the synapse with conventional microdialysis (18, 43) or direct sampling of the extracellular compartment (21). In vivo amperometry using 15 × 333 μm electrodes with a temporal resolution of <1 s detects synaptic spillover that responds to manipulations of vesicular release. This pool of glu reaches concentrations in the 25 μM range (14) albeit most likely outside of the range of synaptic influence (52). Taken together measurements acquired with these varied techniques depict three pools of extracellular glu. Ambient concentrations within the synaptic pool are low so that glu transients produce phasic, postsynaptic responses. After a synaptic transient, the synapse is rapidly cleared of glu to concentrations too low to cause significant receptor activation or desensitization (51). Immediately outside of the synapse ambient glu appears to reach the highest concentrations that are ultimately diluted as glu diffuses into the extracellular mileu sampled most frequently by microdialysis. Given the versatility of microdialysis it would be an advantage if a microdialysis device could gain access to the area of synaptic spillover in order to sample glutamatergic signaling derived from vesicular release. Interestingly limited clearance of DA and other monoamines makes the synaptic pool of these neurotransmitters more accessible than glu and GABA using conventional microdialysis techniques. Sensitivity to manipulations of vesicular release as well as models of synaptic dynamics support this interpretation (52–54).

3.2. Spatial Resolution Is Limited by Conventional Probe Size

Even the smallest of conventional microdialysis probes are orders of magnitude greater in volume than a synapse. Spatial resolution is limited largely by the distance from synapse to membrane. Spatial resolution is also affected by the constitution of the tissue. Because extracellular space is full of interstitial milieu, molecules travel a tortuous diffusion path (which is longer than the straight distance). The tortuosity factor (55, 56) is a measure of the influence of the tortuous paths in porous media, such as the extracellular space and semipermeable membrane, on the reduction of the coefficient of diffusion.

In addition to the interstitial milieu that separates the implanted microdialysis probe from the sampling target, the damaged tissue around the probe also forms a diffusion barrier reducing the sampling performance. Injury caused when the probe is inserted into the tissue produces a layer of damaged cells that separates the membrane from the most representative extracellular milieu (57, 58). Probes with large cross-sectional area can cause significant tissue damage and limit spatial resolution (58–60). They also can initiate a prolonged inflammatory reaction, which will adversely affect in vivo sampling. Mechanically, the damaged tissue forms a passive layer surrounding the implanted probe and creates an unwanted tortuous path between the source and the probe (61). Tissue damage caused by probe implantation limits the ability to sample the most representative tissue in human and animal studies (8, 58). Examination of these criticisms in microdialysis (60, 62) suggests that tissue injury surrounding the implanted probe is a significant limitation of conventional microdialysis technique.

Tissue damage has been one major hurdle in applying microdialysis to sample neurotransmitters. Recent work with small carbon fibers (63) suggests that smaller probes may minimize traumatic tissue damage resulting from probe implantation. Decreasing the dimensions of conventional microdialysis probes described almost 30 years ago (64) is therefore expected to minimize perturbation of the surrounding tissue and improve spatial resolution. For brain microdialysis spatial resolution may be improved by shortening the diffusion distance and thinning the diffusion barrier remembering that the shorter the diffusion path, the quicker the particle migrates.

3.3. Temporal Resolution Is Limited by Dead Volume, Liquid Handling, and Limit of Detection

Temporal resolution with microdialysis remains limited by mass transport effects within the dialysis probe. Temporal resolution is affected by the sampling rate and mixing of the advancing perfusion fluid with the analyte filled perfusate within the lumen of the microdialysis probe. The internal or "dead" volume creates a liquid compartment where the analyte mixes with perfusion fluid. This mixing compromises temporal resolution. Minimizing dead volume will improve temporal resolution. Probe design determines the dead volume of a particular probe. Long outlet tubing sometimes used for sample collection in awake animal preparations also allows for mixing of analyte and disturbs temporal resolution. Finally, the rate of sample collection and detection will also limit temporal resolution. Sampling rate is restricted by the sensitivity of the analytical technique and capacity to handle low fluid volumes.

4. Advance in Microdialysis

Advances in probe design, fluid collection and handling, as well as analytical techniques, addressed in more detail below, are making way for breakthrough advances in microdialysis that have potential to move the membrane closer to the synapse.

4.1. Advance in Analyte Handling

On line systems for the analysis of microdialysis samples avoid handling of small-volume droplets and when coupled with sufficiently fast and sensitive analytical techniques can resolve glu and other analytes within seconds (30, 65) and detect changes in synaptic concentrations of fast transmitters such as glutamate (66). More recently a method has been described whereby the dialysis probe is integrated with a poly (dimethylsiloxane), or PDMS chip that merges dialysate with fluorogenic reagent and segments the flow into 8–10 nL plugs at 0.3–0.5 Hz separated by perfluorodecalin. The plugs extracted to an aqueous stream and analyzed by CE with laser-induced fluorescence detection were analyzed every 35 s (67). An off-line in vivo neurochemical monitoring approach was developed based on collecting nanoliter microdialysate fractions as an array of "plugs" segmented by immiscible oil in a piece of Teflon tubing. This approach achieved sampling rates of up to 7 s (67).

4.2. Miniaturization of Microdialysis

Size reduction in microdialysis probes and interconnected cannulae is a promising solution to improve temporal and spatial resolution. One challenge in miniaturizing fluid channels, however, comes from the physical reality that backpressure produced by resistance to fluid flows will be magnified in smaller channels. Backpressure resists continuous flow by dissipating the fluid momentum and energy along the flow stream, which is a natural phenomenon

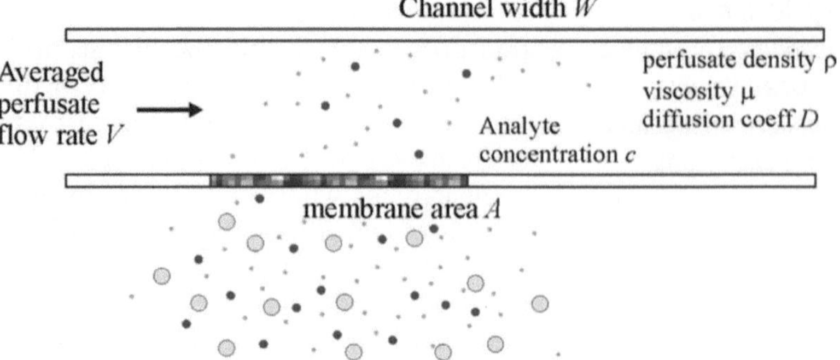

Fig. 1. Schematic of microdialysis sampling.

caused primarily by the viscosity of fluids. Backpressure is also in part attributed to the boundary conditions of flow such as the wall ruggedness. The backpressure effect becomes more significant for a fluid flowing in smaller channels wherein a larger surface-to-volume ratio raises the effect of viscous dragging (68, 69). Therefore, backpressure in microdialysis probes is the primary technical constraint that prevents microdialysis from miniaturization.

Here we introduce preliminary results on a technique under development toward probe miniaturization for microdialysis (70). As sampling devices are miniaturized reduced channel dimensions results in an increase in backpressure that makes continuous flow infeasible. As opposed to the conventional microdialysis which uses continuous perfusate, this new technique produces and operates discrete droplets of perfusate that are separated by air. The use of air as a medium to separate discrete droplets can entirely eliminate backpressure within the lumen of the probe. In the following sections we discuss the principle of air-separated droplet-based microdialysis and how it could be used, in conjunction with CE-LIFD, for chemical analysis.

4.3. The Scaling Law of Backpressure in Quantitative Microdialysis

In this section we elaborate on the adverse scaling effect of the backpressure on miniaturization of conventional microdialysis probes operating with continuous flow of perfusate. The discussion will lead to the idea of using droplets separated by air for microdialysis sampling. For quantitative microdialysis, the perfusate at a constant flow rate is used to reach a constant relative recovery of 100%. Therefore, we express the scaling effect of the probe dimensions on the backpressure subject to a constant relative recovery. The model illustrated in Fig. 1 is used to formulate the process of recovering particles from the extracellular space into a microdialysis probe. Albeit its 2-dimensional configuration, this simple model has included the six most important parameters that have a direct impact on the relative recovery of microdialysis (71, 72). By using c_{∞} as the analyte concentration at equilibrium in the tissue (far away from the membrane), the relative recovery of in vitro microdialysis is a function of three dimensionless parameters (70):

$$\frac{c}{c_\infty} = f\left[\frac{WV}{D}, \frac{\rho WV}{\mu}, \frac{A}{W^2}\right]. \quad (1)$$

The first two parameters are the Péclet number (Pe) and the Reynolds number (*Re*) (which are two basic dimensionless numbers in fluid mechanics), while the last parameter is the area ratio of the membrane to the probe's cross section. This equation addresses that a relative recovery can be maintained at a constant level as long as the values of the three dimensionless parameters remain unchanged regardless of the individual changes in the six individual parameters. For example, the first group can be unchanged if W becomes tenfold smaller while V is tenfold faster.

Equation (1) sheds light into the scaling effect of probes on the backpressure gradient under the condition that the relative recovery is constant. Let us reduce the dimensions A and W while deliberately fixing the non-geometric parameters such as μ, ρ, and D to be constant. In order to sustain a constant relative recovery it requires WV and A/V^2 to be constant in (1). It has been shown (70) that, for a constant relative recovery, the average flow of perfusate must be l-fold faster when a probe dimension W is downsized by l times. In other words, the scaling law in microdialysis under constant relative recovery can be explicitly expressed by

$$\begin{aligned} W &\sim l^{-1} \\ V &\sim l \end{aligned} \quad (2)$$

where for "~" we mean "proportional to" and l (>1) is a scaling factor (e.g., $l = 10$ for downscaling a probe tenfold smaller in one dimension of its cross section).

The scaling law for the backpressure in miniaturized microdialysis can be addressed as follows. The backpressure gradient of a laminar flow in an enclosed channel is often described by the Hagen–Poiseuille equation (73):

$$\frac{\Delta p}{\Delta L} = \frac{64}{Re}\frac{1}{W}\frac{\rho V^2}{2}. \quad (3)$$

By introducing (2) into (3) we find the scaling law for addressing the backpressure for quantitative microdialysis under a constant relative recovery:

$$\frac{\Delta P}{\Delta L} \sim l^3. \quad (4)$$

The increase in the backpressure gradient in a smaller probe, per the scaling ratio $W{\sim}l^{-1}$, is in response to the increase in the flow rate for maintaining the relative recovery (at the steady state) per the scaling rule $V{\sim}l$. Equation (4) predicts that, for example, to reach the same relative recovery, a 10×-downscaled probe (with the same membrane-to-probe area ratio as its pre-scaled counterpart) will experience a 1,000× larger backpressure gradient. The scaling law of the backpressure in quantitative microdialysis under continuous perfusate flow is

$$\Delta p \sim l^2. \quad (5)$$

Equivalently, the backpressure will be 100-fold larger in a 10-fold downscaled probe. Apparently, backpressure will be raised as a new technical issue when pumping perfusate solution into a miniaturized microdialysis probe. Backpressure will be raised further by long, narrow bore outlet tubing. Although supplying a larger pressure source could overcome the backpressure, an elevated pressure in the channel would produce ultrafiltration—fluid exchange from the probe chamber to the tissue—an unwanted phenomenon in microdialysis. Backpressure will also retard fluid transportation, which in turn will enhance the Taylor dispersion phenomenon (74) and increase the chance of unwanted sample mixing (75).

The escalated backpressure in miniaturized microdialysis calls for a game-changing operation methodology for running microdialysis. Miniaturization of the microdialysis technique is not only dedicated to downscaling dialysis devices but also involves developing and verifying new operation principles that accommodate the scaling effects and the handling of the increased backpressure. We introduce droplets-in-air-based microdialysis designed to eliminate the backpressure issue for probe miniaturization. We also discuss how this new device can be integrated with capillary electrophoresis/laser-induced fluorescence detection for chemical analysis.

5. Droplets-in-Air-Based Microdialysis

5.1. Principle

The operation principle of the droplet-in-air-based microdialysis is portrayed in Fig. 2, as described previously (70). Droplets, separated by air, march one by one, to the semipermeable membrane where each droplet resides for a measured time while coming to equilibrium with the contents of extracellular fluid. Dynamic equilibrium can be reached quickly with the use of small droplets to improve temporal resolution. After sampling, each analyte-laden droplet is sequentially transported manually or automatically for further chemical analysis. Because droplets are sequentially produced and transported to the membrane site, the train of droplets is loaded with rich, time-series information about the chemical contents in a novel digital manner. In order to eliminate backpressure,

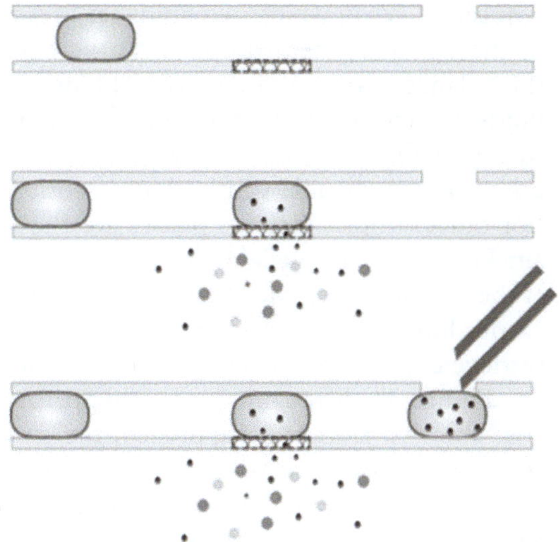

Fig. 2. Operation principle of droplet-based microdialysis. Droplets are sequentially produced and transported to the membrane site, sitting there for a measured time for sampling, and then moved to the outlet for further analysis.

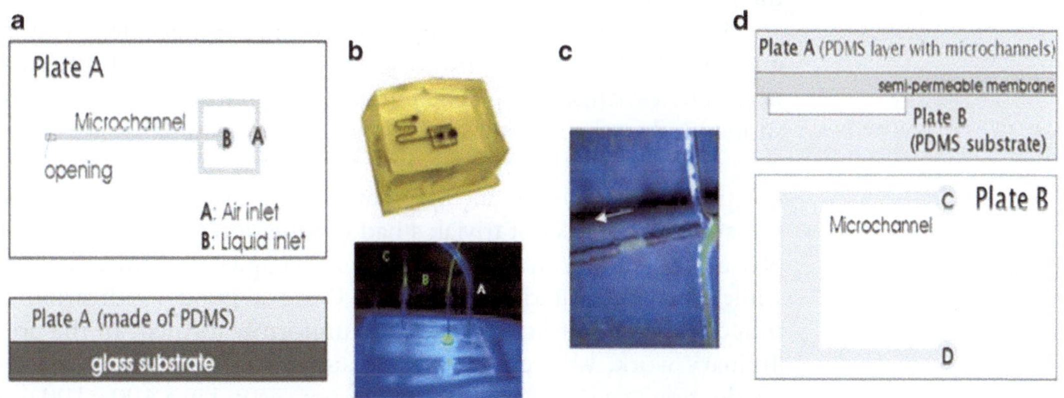

Fig. 3. (**a–c**) A microfluidic prototype for producing aqueous droplets separated by air in the microchannel. (**b**) Droplets (*outlet C*) are produced on a microfluidic chip by compressing air (*inlet A*) to cut off liquid (*inlet B*). (**c**) A droplet is produced by applying an air flow rate = 15.9 mL/h and water flow rate = 0.397 mL/h. (**d**) A stacking design for droplet-based microdialysis in vitro.

the droplets are separated in air. A similar approach to "digital" microdialysis has demonstrated improved temporal resolution using droplets separated in oil (76). Separating droplets in air will allow for miniaturization of the sampling device to improve both temporal and spatial resolution.

5.2. Design for Droplet Production

A preliminary design of microfluidic devices in Fig. 3a–c shows how droplets can be produced. As designated by Plate A in Fig. 3a, this design is capable of producing droplets which are separated by air in microchannels. Figure 3b–c demonstrates the results. The principle of this design is similar to the water-in-oil or oil-in-water emulsion (77) in which two fluids are pressurized to meet at a place in the channel where droplets can be automatically generated by the balance between the interfacial tension and the viscosity of the fluids. In fact, the droplet size and production rate can be quantified by the capillary number (76) which is a dimensionless number governed by the channel width, fluid viscosity, and interfacial tension. For our example, the two phases are the fluorescenced water and air. Producing droplets in air can be beneficial in the later process. If the droplets are generated in another liquid phase—the so-called carrier liquid (77)—then another microfluidic challenge will be to retrieve each droplet from the carrier liquid.

Figure 3b shows one successful prototype made of PDMS by a cost-effective fabrication approach, the so-called shrinky dink process (78), which prepares the master mold by printing channel patterns onto thermoplastic sheets. Any office laser printer can be used for the printing, which makes this special fabrication process popular for its short turnover time for quick proof of concept without the need for cleanroom facilities. After the master mold is printed and baked, replica molding with PDMS is subsequently applied to obtain the prototype (79).

The operation of producing droplets in air is as follows: The air inlet and fluid inlet in the prototype (Fig. 3b) are individually connected to a syringe pump. The pumping flow rate (i.e., the pressure supply) fed to the air and fluid inlets must be modulated appropriately for droplets to form. Because the air has much less viscosity, the pressure fed to the air inlet must be higher. Modulating the pressure feedings is not trivial; a bad modulation may result in one-phase flow—either the liquid or the air occupies the entire channel while the other fluid is stagnated. Once a good modulation is tuned, droplets (or liquid plugs) are uniformly formed. In the preliminary work, we were capable of automating the water droplet production at a rate up to 20 droplets per second in a $400 \times 100\ \mu m$ microchannel. For example, Fig. 3c demonstrates the production of a water droplet sheared off and pushed forward by pressurized air. Preliminary results showed that uniform droplets can be produced at 15.9 and 0.397 mL/h flow rates to the air inlet and the liquid inlet, respectively. Recent experiments (76) show that the droplet size is a nonlinear function of the capillary number, which is a dimensionless number parameterized by the channel size, surface tension of the liquid, and contact angle of the liquid and the channel wall, as well as the flow rates of the air and liquid.

5.3. Design Considerations for In Vitro Microdialysis

Sampling the extracellular fluid from cultured cells and tissues is another application for microdialysis. Design for this type of in vitro microdialysis needs a semipermeable membrane sitting between the microfluidic channel and the culture media or cultured cells. A trilayer, sandwich-like design as portrayed in Fig. 3d, which features a semipermeable membrane is stacked between

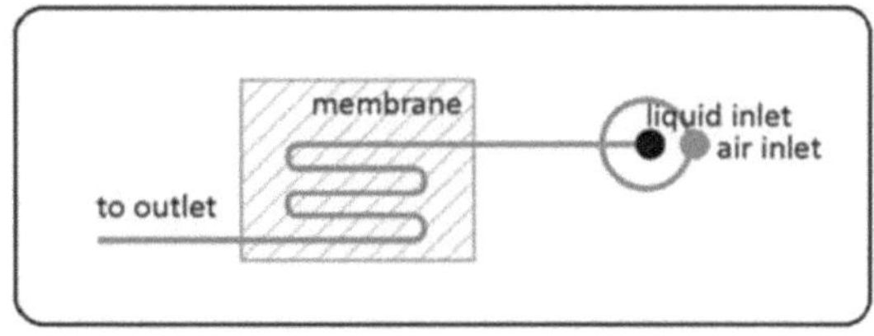

Fig. 4. Designs for keeping droplets in contact with membrane by a measured time.

Plate A (which produces droplets, Fig. 3a) and Plate B (which circulates well-stirred solution, Fig. 3d). The polycarbonate track-etch membrane can be used in the stacking design, because it has various defined pore sizes and can be bonded to PDMS or SU-8 by oxygen plasma treatment (80).

For testing, the culture well or the sample reservoir, as per the design in Fig. 3d, can be filled with a solution that is made by adding amino acids to aCSF in concentrations that simulate real samples (1–25 μM). During sampling, a droplet will be produced and pneumatically transported to where the membrane window contacts the sample reservoir.

Figure 4 illustrates a potential design for keeping a droplet in contact with the semipermeable membrane for a measured time using a serpentine channel. This design allows a train of droplets to be present in the channel at the same time. The time of contact with the membrane will be determined by the times when a droplet enters and leaves the membrane site. In this design, the contact time can be adjusted by the length of the serpentine channel and the motion speed of the droplets which is governed by the air supply.

5.4. How Long Does a Droplet Take for Sampling?

The predictive model in Fig. 5a suggests that the droplet-based sampling will achieve equilibrium in 1 min (70). The predictive results will be experimentally verified using the design in Fig. 4. It is worth noting that the numerical model has considered the influence of porous-medium diffusion through the tortuous extracellular space and the semipermeable membrane.

The time to reach equilibrium is proportional to the linear dimension of the channel squared, $(W3)^2$, as shown in Fig. 5b. For a microchannel in the cross-sectional dimension 50×400 μm with a 50 μm long membrane, the equilibrium time can be achieved at 1 min after sampling starts. The droplet volume corresponding to this example is 1 nL, which can be routinely handled by capillary electrophoresis. Better sampling performance can be achieved by using smaller droplets, as indicated in Fig. 5b.

5.5. Droplet Collection and Storage

We envision that a train of sample-laden droplets can be collected and stored for later analysis without the loss of temporal information sampled. This technique, once successful, would greatly increase the degree of robustness in coupling microdialysis with

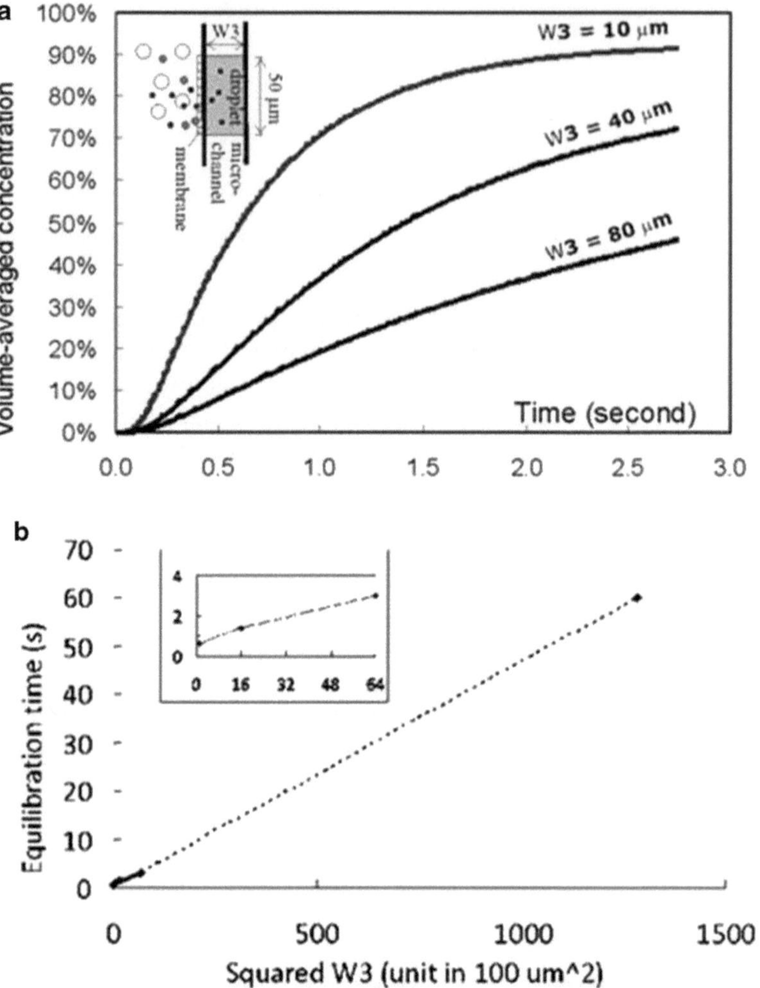

Fig. 5. Predictive modeling for droplet-based microdialysis. (**a**) Equilibration time versus droplet size. The model assumes that a droplet just fully covers the membrane and the channel has an out-of-plane dimension of 50 μm. By varying the channel width (*W3*) to be 10, 40, or 80 μm, the associated droplet volume is 0.025, 0.100, or 0.200 nL, respectively. (**b**) Extrapolation of sampling time. The *inset* shows the data retrieving from the (**a**). The *dashed line* is extrapolated from the modeling result. The *abscissa* is scaled as indicated.

many platforms of chemical analysis. For example, the analyte droplets collected at Laboratory A on the West Coast can be collected, frozen, and then delivered across the continent to Laboratory B on the East Coast for analysis. This mode would enable a much wider span of collaborations than online analysis.

Figure 6 illustrates two conceptual designs for collection and storage of droplets after sampling. The collection and storage may take place either in tubing that is connected to the outlet of the microchannel or in a separate storage chip that has an array of microwells. Both designs would enable a portable storage device, in which an array of droplets carrying time-stamped information is stored for later analysis (82).

Fig. 6. (**a**) In-tubing design for droplet storage. (**b**) A preliminary demonstration for in-tube droplet storage, where tubing is connected to the outlet for collecting droplets. (**c**) On-chip droplet storage design.

6. Summary

Conventional microdialysis sampling of fast neurotransmitters such as glu and GABA is restricted to an extrasynaptic pool. Although this pool has important physiological and pathological significance, use of microdialysis to access spillover of these fast neurotransmitters from the synaptic pool will require a radical shift from conventional probe design. As probes and hence fluid channels are miniaturized backpressure becomes limiting. Producing discrete droplets in air eliminates backpressure and creates opportunity to develop novel, digital sampling devices with improved temporal and spatial resolution.

Appendix: Construction of Low Dead Volume Microdialysis Probes

1. *Prepare capillary*: Using a ceramic capillary tubing cutter, cut up to a 50 cm piece of capillary tubing[1] (this will be the inlet to the probe, so adjust length as needed). Measure and cut another piece for the outlet. For acute, anesthetized preps this may be 10 cm. For awake animal preparations leave the outlet long enough to reach your fraction collector or collection vial. Note that longer outlets will decrease the flow rate that a probe can handle before breaking or seeping fluid (ultrafiltration). Ensure that cuts are smooth; check smoothness using a microscope. Smoothness means that the tube is not shattered and the cut does not leave a point on the tube.
2. *Adjust offset between inlet and outlet that will define the length of dialyzing tip*: On a glass slide with double-sided tape on the topside of the slide, lay the long capillary piece on the slide and next to it lay the short capillary piece. Bring the ends of the two pieces flush with one another, and then make the long capillary tube stick out further than the short capillary tube by 1–4 mm

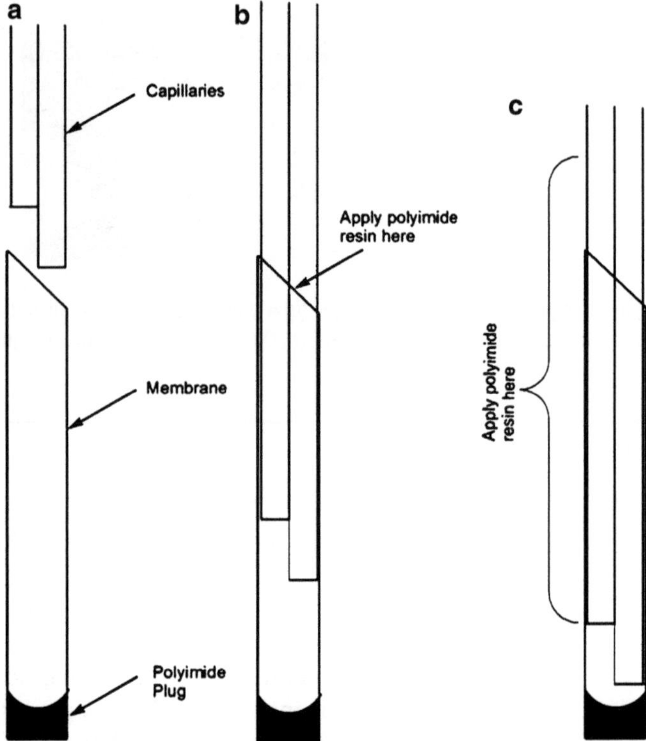

Fig. 7. Offset short and long piece of capillary to define the length of dialyzing tip (**a**). Feed into membrane, sealed on one end with polyimide resin (plug not to exceed 0.5 mm) and trimmed at a 45° angle on the other end. After inserting the capillary tubes into the membrane about 1 mm, apply polyimide resin to the capillaries at the opening of the membrane (**b**). Coat the membrane and capillary tubes with polyimide resin as shown leaving the dialyzing tip free of resin (**c**). These membrane-covered capillaries are now referred to as "the probe".

or longer depending on the desired length of your dialyzing tip. Put this in an area where it will not be disturbed.

3. *Prepare membrane for sliding onto capillaries:* Cut 0.5 cm of cellulose membrane[2] from the roll at a 45° angle. Put the cut membrane on a glass slide with double-sided tape on the top. Using polyimide resin[3], seal the cut membrane end that is not angled. The sealed part should not exceed a thickness of 0.5 mm (Fig. 7a). Leave this to dry for several hours.
4. *Insert capillary tubes into membrane:* Insert offset capillary tubes into a dry, sealed membrane tube. For easier insertion put the short tube on the long side of the angled membrane opening (Fig. 7b).
5. *Coat non-dialyzing surface of membrane and define the final length of dialyzing tip:* After inserting capillary tubes about 1 mm apply polyimide resin onto the capillaries and to the open end of the membrane (the end cut to 45°). Continue feeding the capillaries into the membrane. Adjust the capillaries

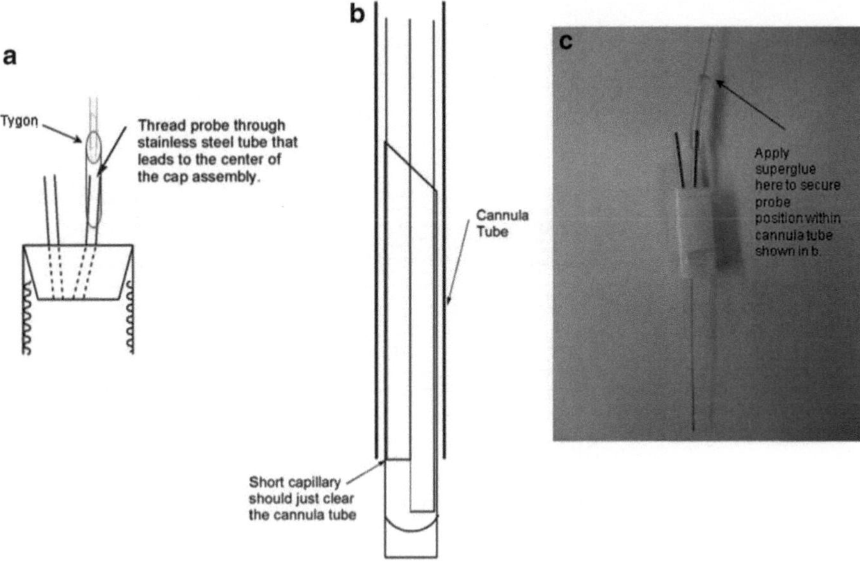

Fig. 8. The probe is threaded through a piece of Tygon tubing affixed to the stainless steel tube that leads to the center of the "cap" that is the push–pull connector-no internal tube (Cat No. C313ICP/NIT) (**a**). The probe that is now connected to the cap is then threaded through the push–pull guide cannula (Cat No. C316GP/0) and the two pieces are screwed together and tightened for a reproducible fit. Adjust the position of the probe within the cannula assembly so that the short capillary just clears the cannula tip and the active portion of the membrane extends beyond the tip (**b**). Apply a dab of super glue where the probe enters the Tygon tubing (**c**) and let dry for at least 20 min.

so that they remain offset as before, but the long capillary is about 0.1 mm from the polyimide plug at the end of the membrane. Spread polyimide resin on the membrane entrance and the membrane itself. Stop spreading resin just before the opening of the furthest back capillary tube. Remove excess polyimide resin and let dry for several hours (Fig. 7c).

6. *Insert the completed probe into a push–pull cannula assembly:* The assembly consists of two pieces, the push–pull connector-no internal tube referred to here as the "cap" (C313ICP/NIT[4] where NIT indicates "no internal tube"), and the push–pull guide cannula (C316GPIO/uncut[4]). Attach a 1 cm piece of Tygon tubing (0.020–0.025" id) to the stainless steel tube on the top of the cap that leads to a hole directly in the middle of the cap (Fig. 8a). Thread the probe through this piece of Tygon and stainless steel tube. Next insert the probe through the push–pull guide cannula and then screw the two pieces together for a tight, reproducible fit.

7. *Adjust and secure probe within cannula assembly:* With the capillary mounted on sticky tack, adjust the push–pull cannula assembly so that active portion of the probe membrane extends just beyond the end of the cannula tip (Fig. 8b). When this is done, apply a little bit of super glue to the entrance of the Tygon to hold the probe in position. Let this dry for 20 min. The fully assembled probe is shown in Fig. 8c and Fig. 9.

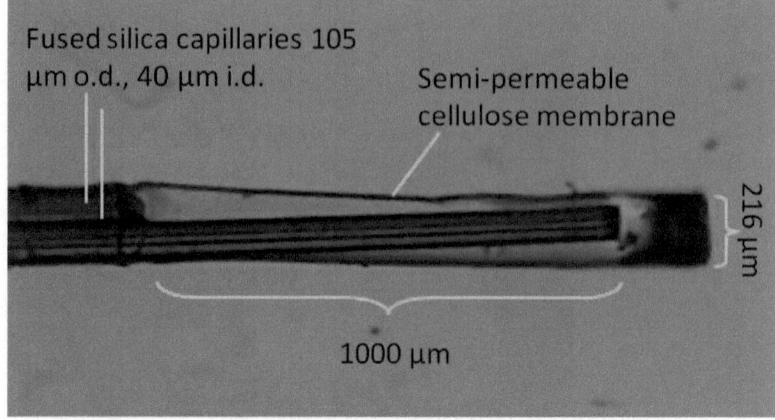

Fig. 9. Photograph of dialyzing tip of an assembled probe.

8. *Prepare capillary inlet to connect to a syringe filled with perfusion fluid:* To fit the capillary tubing onto a luer lock syringe you need to first put the capillary through the luer adapter fitting and then insert the capillary through a red NanoTight®sleeve[5]. After inserting the capillary through the red NanoTight® sleeve, pull the tubing a little further out and cut off about 3–4 cm of tubing to get rid of capillary tubing that might be obstructed, leaving 1 mm projecting from the red NanoTight® sleeve. Place a ferrule on the sleeve and connect using normal HPLC fittings.
9. *Prepare cannula to fit probe holder needed for stereotaxic surgery:* If probes will be used in an acute, anesthetized preparation, unscrew the cap and use a dremel tool to file the edges of the outer threads of the internal cannula to make two parallel, flat surfaces. Use a CMA clip designed to hold the flat body of a CMA 11 or 12 guide cannula[6] for surgery. If probes will be used for an awake preparation that requires chronic guide cannula implantation, use a nut that threads onto the top of the guide cannula. Flatten the sides of the nut so that it fits securely within the CMA clip. To insure accurate depth of placement take care to adjust the length of capillary and dialyzing membrane to an exact length that is consistent for all probes.
10. *Notes on performance:* Due to the low internal diameter of the outlet capillary, these probes will sweat at relatively slow flow rates. Probes with short outlets (15 cm) sweat between 1.0 and 8.0 µL/min (mean of 4.7 µL/min, $n=8$). Probes with long outlets (100 cm) sweat between 0.3 and 0.8 µL/min (mean of 0.6 µL/min, $n=8$). In vitro tests show that when these probes are perfused at 0.1 µL/min the temporal resolution, or time to equilibrate to a new concentration of glu ranging from 0.5 to 2.0 µM, is about 1 min.

A.1. Supply List for Microdialysis Probe Construction

Manufacturer-Specific Items

1. Silica capillary Tubing; 105 μm od, 40 μm id (Polymicro Technologies; http://www.polymicro.com).
   Part No. TSP040105.
2. Regenerated cellulose membrane; 13 kDa cut-off (Spectrum Labs; http://www.spectrumlabs.com).
   Part No.132294.
3. Polyimide sealing resin; part No.5825 (W.R. Grace and Co.; http://www.discoverysciences.com).
4. Push–pull cannula and cannula guides (Plastics 1; http://www.plastics1.com).
   Part No.C313ICP/NIT Push–pull connector-internal cannula.
   Part No. C316GPIO/uncut (maximum length) push–pull guide cannula.
5. HPLC fittings and sleeves (Upchurch Scientific; http://www.upchurch.com).
   Part No. F-237x NanoTight FEP sleeve (0.005"; red).
   Part No. F-142Nx HPLC ferrule.
   Part No. F-331Nx HPLC fitting.
6. CMA 11 and 12 clip (CMA Microdialysis AB; http://www.microdialysis.se).

General Items (Found in Hardware Store or Scientific Supply Catalog)

Small Tygon Tubing (0.020–0.025" id).

Super Glue.

Epoxy Glue.

2-sided tape.

Microscope slides.

Ceramic capillary tubing cutter.

Poster putty.

References

1. Herrera-Marschitz M, Arbuthnott G, Ungerstedt U (2010) The rotational model and microdialysis: significance for dopamine signalling, clinical studies, and beyond. Prog Neurobiol 90:176–189
2. Dupre KB, Ostock CY, Eskow Jaunarajs KL et al (2011) Local modulation of striatal glutamate efflux by serotonin 1A receptor stimulation in dyskinetic, hemiparkinsonian rats. Exp Neurol 229(2):288–299
3. Lönnroth P, Jansson PA, Smith U (1987) A microdialysis method allowing characterization of intercellular water space in humans. Am J Physiol 253:E228–E231
4. Willuhn I, Wanat MJ, Clark JJ et al (2010) Dopamine signaling in the nucleus accumbens of animals self-administering drugs of abuse. Curr Top Behav Neurosci 3:29–71
5. Zetterstrom T, Sharp T, Marsden CA et al (1983) In vivo measurement of dopamine and its metabolites by intracerebral dialysis: changes after d-amphetamine. J Neurochem 41:1769–1773
6. Bito L, Davson H, Levin E et al (1966) The concentrations of free amino acids and other electrolytes in cerebrospinal fluid, in vivo dialysate of brain and blood plasma of the dog. J Neurosci 13:1057–1067

7. Bourne J (2003) Intracerebral microdialysis: 30 Years as a tool for the neuroscientist. Clin Exp Pharmacol Physiol 30:16–24
8. Benjamin RK, Hochberg FH, Fox E et al (2004) Review of microdialysis in brain tumors, from concept to application: first annual carolyn Frye-Halloran symposium. Neuro Oncol 6(1):65–74
9. Johansen MJ, Newman RA, Madden T (1997) The use of microdialysis in pharmacokinetics and pharmacodynamics. Pharmacotherapy 17: 464–481
10. Wang PC, DeVoe DL, Lee CS (2001) Integration of polymeric membranes with microfluidic networks for bioanalytical applications. Electrophoresis 22:3857–3867
11. Páez X, Hemández L (2001) Biomedical applications of capillary electrophoresis with laser-induced fluorescence detection. Biopharm Drug Dispos 22:273–289
12. Watson CJ, Venton BJ, Kennedy RT (2006) In vivo measurements of neurotransmitters by microdialysis sampling. Anal Chem 78(5): 1391–1399
13. Garrison KE, Pasas SA, Cooper JD et al (2002) A review of membrane sampling from biological tissues with applications in pharmacokinetics, metabolism and pharmacodynamics. Eur J Pharm Sci 17:1–12
14. Dash MB, Douglas CL, Vyazovskiy VV et al (2009) Long-term homeostasis of extracellular glutamate in the rat cerebral cortex across sleep and waking states. J Neurosci 29:620–629
15. Drew KL, Pehek EA, Rasley BT et al (2004) Sampling glutamate and GABA with microdialysis: suggestions on how o get the dialysis membrane closer to the synapse. J Neurosci Methods 140:127–131
16. Bordji K, Becerril-Ortega J, Nicole O et al (2010) Activation of extrasynaptic, but not synaptic, NMDA receptors modifies amyloid precursor protein expression pattern and increases amyloid-ss production. J Neurosci 30:15927–15942
17. Okubo Y, Iino M (2011) Visualization of glutamate as a volume transmitter. J Physiol 589: 481–488
18. Zhou F, Braddock JF, Hu Y et al (2002) Microbial origin of glutamate, hibernation and tissue trauma: an in vivo microdialysis study. J Neurosci Methods 119:121–128
19. Sam PM, Justice JB Jr (1996) Effect of general microdialysis-induced depletion on extracellular dopamine. Anal Chem 68:724–728
20. Langemann H, Alessandri B, Mendelowitsch A et al (2001) Extracellular levels of glucose and lactate measured by quantitative microdialysis in the human brain. Neurol Res 23:531–536
21. Kennedy RT, Thompson JE, Vickroy TW (2002) In vivo monitoring of amino acids by direct sampling of brain extracellular fluid at ultralow flow rates and capillary electrophoresis. J Neurosci Methods 114:39–49
22. Osborne PG, Hu Y, Covey DN et al (1999) Determination of striatal extracellular gamma-aminobutyric acid in non-hibernating and hibernating arctic ground squirrels using quantitative microdialysis. Brain Res 839:1–6
23. Justice JB Jr (1993) Quantitative microdialysis of neurotransmitters. J Neurosci Methods 48:263–276
24. Parsons LH, Justice JB Jr (1994) Quantitative approaches to in vivo brain microdialysis. Crit Rev Neurobiol 8(3):189–220
25. Pehek EA, Nocjar C, Roth BL et al (2006) Evidence for the preferential involvement of 5-HT2A serotonin receptors in stress- and drug-induced dopamine release in the rat medial prefrontal cortex. Neuropsychopharmacology 31:265–277
26. Liu Y, Zhang J, Xu X et al (2010) Capillary ultrahigh performance liquid chromatography with elevated temperature for sub-one minute separations of basal serotonin in submicroliter brain microdialysate samples. Anal Chem 82: 9611–9616
27. Kaul S, Faiman MD, Lunte CE (2011) Determination of GABA, glutamate and carbamathione in brain microdialysis samples by capillary electrophoresis with fluorescence detection. Electrophoresis 32:284–291
28. Kirschner D, Jaramillo M, Green TK (2007) Enantio separation and stacking of cyanobenz(f) isoindole-amino acids by reverse polarity capillary electrophoresis and sulfated β-cyclodextrin. Anal Chem 79:736–743
29. Kirschner DL, Green T (2009) Separation and sensitive detection of D-amino acids in biological matrices. J Sep Sci 32:2305–2318
30. Nandi P, Lunte SM (2009) Recent trends in microdialysis sampling integrated with conventional and microanalytical systems for monitoring biological events: a review. Anal Chim Acta 651:1–14
31. Carrozzo MM, Cannazza G, Pinetti D et al (2010) Quantitative analysis of acetylcholine in rat brain microdialysates by liquid chromatography coupled with electrospray ionization tandem mass spectrometry. J Neurosci Methods 194:87–93
32. Dahlin AP, Wetterhall M, Caldwell KD et al (2010) Methodological aspects on microdialysis protein sampling and quantification in biological fluids: an in vitro study on human ventricular CSF. Anal Chem 82: 4376–4385

33. Glick SD, Dong N, Keller RW Jr et al (1994) Estimating extracellular concentrations of dopamine and 3,4-dihydroxyphenylacetic acid in nucleus accumbens and striatum using microdialysis: relationships between in vitro and in vivo recoveries. J Neurochem 62: 2017–2021
34. Wang L, Li Y, Han H et al (2003) Perfusate oxygen and carbon dioxide concentration influence basal microdialysate levels of striatal glucose and lactate in conscious rats. Neurosci Lett 344:91–94
35. Osborne PG, O'Connor WT, Ungerstedt U (1991) Effect of varying the ionic concentration of a microdialysis perfusate on basal striatal dopamine levels in awake rats. J Neurochem 56:452–456
36. Oz G, Berkich DA, Henry PG et al (2004) Neuroglial metabolism in the awake rat brain: CO2 fixation increases with brain activity. J Neurosci 24:11273–11279
37. Duo J, Stenken JA (2011) In vitro and in vivo affinity microdialysis sampling of cytokines using heparin-immobilized microspheres. Anal Bioanal Chem 399:783–793
38. Benveniste H, Drejer J, Schousboe A et al (1987) Regional cerebral glucose phosphorylation and blood flow after insertion of a microdialysis fiber through the dorsal hippocampus in the rat. J Neurochem 49:729–734
39. Fumero B, Guadalupe T, Valladares F et al (1994) Fixed versus removable microdialysis probes for in vivo neurochemical analysis: implications for behavioral studies. J Neurochem 63:1407–1415
40. Osborne PG (1995) Fixed versus removable microdialysis probes for in vivo neurochemical analysis: implications for behavioral studies. J Neurochem 64:1899–1901
41. Drew KL, O'Connor WT, Kehr J et al (1989) Characterization of gamma-aminobutyric acid and dopamine overflow following acute implantation of a microdialysis probe. Life Sci 45:1307–1317
42. Timmerman W, Westerink BH (1997) Brain microdialysis of GABA and glutamate: what does it signify? Synapse 27:242–261
43. Baker DA, Shen H, Kalivas PW (2002) Cystine/glutamate exchange serves as the source for extracellular glutamate: modifications by repeated cocaine administration. Amino Acids 23:161–162
44. Richerson GB (2004) Looking for GABA in all the wrong places: the relevance of extrasynaptic GABA(A) receptors to epilepsy. Epilepsy Curr 4:239–242
45. Wu Y, Wang W, Diez-Sampedro A et al (2007) Nonvesicular inhibitory neurotransmission via reversal of the GABA transporter GAT-1. Neuron 56:851–865
46. Augustin H, Grosjean Y, Chen K et al (2007) Nonvesicular release of glutamate by glial xCT transporters suppresses glutamate receptor clustering in vivo. J Neurosci 27:111–123
47. Amen SL, Piacentine LB, Ahmad ME et al (2011) Repeated N-acetyl cysteine reduces cocaine seeking in rodents and craving in cocaine-dependent humans. Neuropsychopharmacology 36:871–878
48. Stanika RI, Pivovarova NB, Brantner CA et al (2009) Coupling diverse routes of calcium entry to mitochondrial dysfunction and glutamate excitotoxicity. Proc Natl Acad Sci U S A 106:9854–9859
49. Sarrafzadeh A, Haux D, Plotkin M et al (2005) Bedside microdialysis reflects dysfunction of cerebral energy metabolism in patients with aneurysmal subarachnoid hemorrhage as confirmed by 15 O-H2 O-PET and 18 F-FDG-PET. J Neuroradiol 32:348–351
50. Sarrafzadeh AS, Nagel A, Czabanka M et al (2010) Imaging of hypoxic-ischemic penumbra with (18)F-fluoromisonidazole PET/CT and measurement of related cerebral metabolism in aneurysmal subarachnoid hemorrhage. J Cereb Blood Flow Metab 30:36–45
51. Herman MA, Jahr CE (2007) Extracellular glutamate concentration in hippocampal slice. J Neurosci 27:9736–9741
52. Barbour B (2001) An evaluation of synapse independence. J Neurosci 21:7969–7984
53. Cragg SJ, Rice ME (2004) Dancing past the DAT at a DA synapse. Trends Neurosci 27:270–277
54. Schonfuss D, Reum T, Olshausen P et al (2001) Modeling constant potential amperometry for investigations of dopaminergic neurotransmission kinetics in vivo. J Neurosci Methods 112:163–172
55. Nicholson C (2001) Diffusion and related transport mechanisms in brain tissue. Rep Prog Phys 64:815–884
56. Nicholson C (2005) Factors governing diffusing molecular signals in brain extracellular space. J Neural Transm 112:29–44
57. Chen KC (2006) Effects of tissue trauma on the characteristics of microdialysis zero-net-flux method sampling neurotransmitters. J Theor Biol 238:863–881
58. Clapp-Lilly KL, Roberts RC, Duffy LK et al (1999) An ultrastructural analysis of tissue surrounding a microdialysis probe. J Neurosci Methods 90:129–142
59. Connelly CA (1999) Microdialysis update: optimizing the advantages. J Physiol 514(Pt 2):303

60. Bungay PM, Newton-Vinson P, Isele W et al (2003) Microdialysis of dopamine interpreted with quantitative model incorporating probe implantation trauma. J Neurochem 86:932
61. Yang H, Peters JL, Allen C et al (2000) A theoretical description of microdialysis with mass transport coupled chemical events. Anal Chem 72:2042–2049
62. Chen KC (2005) Preferentially impaired neurotransmitter release sites not their discretiness compromise the validoty of microdialysis zero-net-flux method. J Neurochem 92:29–45
63. Allen C, Peters JL, Sesack SR et al (2001) Microelectrodes closely approach intact nerve terminals in vivo, while larger devices do not: a study using electrochemistry and electron microscopy. Monitoring molecules in neuroscience.In: Proceedings of the international conference on in vivo methods, 9th. Dublin, Ireland 89–90
64. Jacobson I, Sandberg M, Hamberger A (1985) Mass transfer in brain dialysis devices–a new method for estimation of extracellular amino acids concentration. J Neurosci Methods 15(3):263–268
65. Lada MW, Vickroy TW, Kennedy RT (1997) High temporal resolution monitoring of glutamate and aspartate in vivo using microdialysis on-line with capillary electrophoresis with laser-induced fluorescence detection. Anal Chem 69:4560–4565
66. Lada MW, Vickroy TW, Kennedy RT (1998) Evidence for neuronal origin and metabotropic receptor-mediated regulation of extracellular glutamate and aspartate in rat striatum in vivo following electrical stimulation of the prefrontal cortex. J Neurochem 70:617–625
67. Wang M, Roman GT, Perry ML et al (2009) Microfluidic chip for high efficiency electrophoretic analysis of segmented flow from a microdialysis probe and in vivo chemical monitoring. Anal Chem 81:9072–9078
68. Croce G, D'Agaro P (2005) Numerical simulation of roughness effect on microchannel heat transfer and pressure drop in laminar flow. J Phys D: Appl Phys 38:1518–1530
69. Nguyen N-T, Huang X, Chuan TK (2002) MEMS-micropumps: a review. J Fluids Eng 124:384–392, ASME
70. Chen C, Drew KL (2008) Droplet-based microdialysis concept, theory, and design consideration. J Chromatogr A 1209:29–38
71. Bungay PM, Morrison PF, Dedrick RL (1990) Steady state theory for quantitative microdialysis of solutes and water in vivo and in vitro. Life Sci 46:105–119
72. Maidment NT, Evans CJ (1991) Measurement of intracellular neuropeptides in the brain: microdialysis linked to solid-phase radioimmunoassays with sub-femtomole limits of detection. In: Robinson TE (ed) Microdialysis in the Neurosciences. Elsevier, Amsterdam
73. Çengel YA, Cimbala JM (2006) Fluid Mechanics fundamentals and applications. McGraw Hill, New York
74. Taylor RI (1953) Dispersion of soluble matter in solvent flowing slowly through a tube. Proc R Soc Lond A 219:186–203
75. Yang H, Nguyen NT, Huang X (2006) Micromixer based on Taylor disperson. J Phys 34:136–141, Conference series
76. Tan J, Li SW, Wang K et al (2009) Gas-liquid flow in T-junction microfluidic devices with a new perpendicular rupturing flow route. Chem Eng J 146:428–433
77. Teh S-Y, Lin R, Hung LH et al (2008) Droplet microfluidics. Lab Chip 8:198–220
78. Chen C-S, Breslauer DN, Luna JI et al (2008) Shrinky-Dink microfluidics: 3D polystyrene chips. Lab Chip 8:622–624
79. Xia Y, Whitesides GM (1998) Soft lithography. Angew Chem Int Ed 37:550–575
80. Hsieh Y-C, Zahn JD (2007) On-chip microdialysis system with flow-through sensing components. Biosens Bioelectron 22: 2422–2428
81. Polston JE, Rubbinaccio HY, Morra JT et al (2011) Music and methamphetamine: conditioned cue-induced increases in locomotor activity and dopamine release in rats. Pharmacol Biochem Behav 98:54–61
82. Shim, J.-U., G. Cristobal, D. Link, T. Thorsen, and S. Fraden (2007). Using microfluidics to decouple nucleation and growth of protein crystals. Cryst Growth Des 7, no. 11 2192–94.

Chapter 5

In Vivo Microdialysis for Measurement of Oxidative Metabolism in the Living Brain

H. Ronald Zielke, Carol L. Zielke, and Peter J. Baab

Abstract

Microdialysis allows for measurement of multiple compounds dissolved in the interstitial fluid of the rat brain, including carbon dioxide, if a radioactive tracer is included. Upon introduction of a ^{14}C-labeled compound into the interstitial fluid of the brain, the compound is transported into cells and metabolized into various products including CO_2. The ^{14}C-metabolites, including $^{14}CO_2$ released to the interstitial fluid, equilibrate with fluid in the microdialysis probe and can be collected in the microdialysis effluent. The described methodology allows determination of the rate of oxidative metabolism of potential substrates for brain energy metabolism. Inclusion of nonradioactive competitive substrates provides additional information on oxidative pathways. Our laboratory has utilized microdialysis to measure changes in oxidative metabolism in the brains of adult rats and of 8-day-old rat pups subjected to ischemia/hypoxia.

Key words: Energy metabolism, Brain energy substrates, Oxidative pathways, Microdialysis with rat pups, CO_2 formation in brain, Fluorocitrate, Acetyl-L-carnitine

1. Introduction

Oxidative metabolism in the brain is a measure of energy metabolism. The most direct measurements of energy metabolism have been performed with cultured cells (1), slices, and extracts (2) which are incubated with ^{14}C-labeled substrates yielding $^{14}CO_2$. The latter is trapped in a basic solution and quantitated using a liquid scintillation counter. Another approach used is the measurement of oxygen consumption (3) and measurement of metabolic products, other than CO_2, formed during the oxidative process (4–6). We describe a method of measuring $^{14}CO_2$ formation in the intact living rat brain and the effect of inclusion of metabolic inhibitors on $^{14}CO_2$ formation (7–11). The method allows labeling of the metabolic pool of a potential substrate followed by determination of oxidative metabolic rates, determination of competition

Giuseppe Di Giovanni and Vincenzo Di Matteo (eds.), *Microdialysis Techniques in Neuroscience*, Neuromethods, vol. 75, DOI 10.1007/978-1-62703-173-8_5,

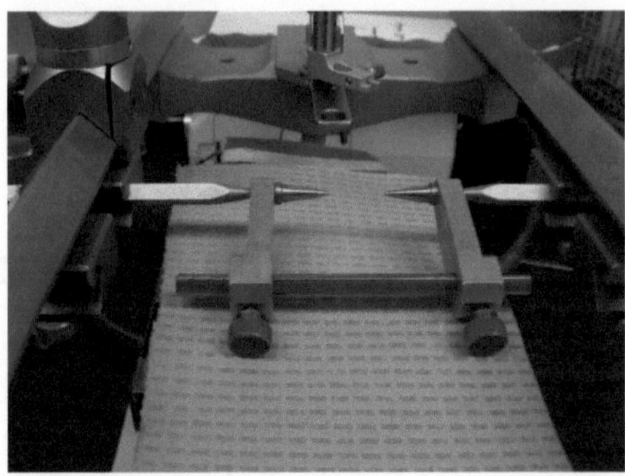

Fig. 1. The fabricated ear bars (which serve as a set of ear bars within the ear bars of the stereotaxic frame) are inserted into the rat's ears and the set screw is tightened and then inserted into the stereotaxic frame.

between more than one substrate, and identification of metabolic pathways. Limitations are that without the determination of substrate concentrations in the interstitial space, the absolute rate of oxidative metabolism cannot be determined. Furthermore, time resolution of less than 60 min is difficult to achieve because of the time required to achieve equilibrium of the specific activity of the radiolabeled substrate in the cell and dilution of radioactivity in the multistep metabolic process in CO_2 formation.

2. Materials

2.1. Microdialysis: Adult Rats

1. 190 g male Sprague Dawley rats are purchased from Zivic Miller Laboratories (Zelienople, PA, USA) and used when they weighed approximately 240 g.
2. A Kopf stereotaxic frame assembly, model 1430, equipped with a 1210 Table Mount Base Plate, a 920 Rat Adapter, and 1770 Electrode Holder was purchased from David Kopf Instruments (Tujunga, CA, USA).
3. Custom-made ear bars were fabricated (Fig. 1) to facilitate the insertion of ear bars into the rat.
4. A foot-operated handheld drill, model 1471, was purchased from Kopf.
5. Stereotaxic trephine drill bits (0307431058) were purchased from CMA (North Chelmsford, MA, USA).
6. Drill bits (8 J60 Drill HSS or D#60 drill bit high speed, steel, 1 mm) for drilling holes for anchoring screws were purchased from Plastic One (Roanoke, VA, USA).

7. Screwdriver, model SD-80, from Plastics One has features to hold small screws during insertion process.
8. Gelfoam sterile sponges, 3 mm thick (NDC 0009-0301-01), were purchased from Upjohn (Kalamazoo, MI, USA).
9. A guide cannula holder consisting of a clamp (MD1520) and a rod (MD1521) was purchased from BAS (Lafayette, IN, USA).
10. The guide cannula (MD2251) was purchased from BAS.
11. Microdialysis probes (MD2200, 2 mm) from BAS are sturdy and are reusable multiple times.
12. Dental cement: Ortho Jet Powder and Lang's Jet Liquid Acrylic were purchased from Lang Dental Manufacturing Co. (Wheeling, IL, USA).
13. Glass shell vials, 15×45 mm, (03-339-30C), were purchased from Fisher Scientific (Pittsburgh, PA, USA).
14. A DigiSense Temperature Controller, heating pad, and a general-purpose temperature probe were purchased from Cole Palmer (Vernon Hills, IL, USA).
15. Sodium pentobarbital (50 mg/ml) was purchased from Southern Anesthesia and Surgical (West Columbia, SC, USA).
16. Providone-Iodine swabsticks (MDS093901) were purchased from Medline Industries (Havre de Grace, MD, USA) and Rimadyl (carprofen), 50 mg/ml, was purchased from Pfizer Animal Health.
17. Stainless steel swivels (53622) with tether were purchased from Stoelting Co. (Wood Dale, IL, USA).
18. Peek tubing, 0.65 mm OD×0.12 mm ID (blue, MD1511, and tan, MF5366), was purchased from BAS.
19. Connectors (3409500) for connecting Peek tubing to swivel and microdialysis probe were purchased from CMA.
20. Dennison Bar-Lok Cable ties, 7 in., #08470, Avery-Dennison, were purchased from Cole Parmer. The ties have the appearance of a ladder and are used as collars for the rats.
21. Jewelry C-clasp for attaching the spring tether to the plastic collar was purchased from a local hobby shop.
22. A plastic bowl, CMA 120 System for freely moving animals, with counter balance arm assembly (8309032) was purchased from CMA.
23. Four Baby Bee syringe drivers (MD1001) with 3 ft Bee Controller cables (MD1000N) and Bee Hive controller for four syringes (MD1020) were purchased from BAS.

24. The Harvard syringe pump, model 22, was purchased from Harvard Apparatus (Holliston, MA, USA). Any syringe pump with capacity for a 5 ml syringe is suitable.
25. One ml Hamilton syringes (model 80965) were purchased from Hamilton Co. (Reno, NV, USA).
26. Kontes stoppers (882320-0000) and Kontes center wells (882310-0000) are obtained from Kontes Scientific (Vineland, NJ, USA).
27. Syringes and needles were obtained from Fisher Scientific.
28. Gentamycin sulfate solution (50 mg/ml, 1,000×) was purchased from Mediatech, Inc. (Herndon, VA, USA).
29. Benzethonium hydroxide and 10% trichloroacetic acid were purchased from Sigma Chemical Co. (St. Louis, MO, USA).
30. Radioactive ^{14}C-labeled substrates were purchased from GE Healthcare Bio-Science Corp. (Piscataway, NJ, USA) and Perkin Elmer NEN Radiochemicals (Waltham, MA, USA).
31. Whatman 20 mm wide chromatography paper, cat. No. 3030-614, was purchased from Fisher Scientific.
32. Sterile artificial CSF (aCSF) was prepared: The aCSF was composed of 150 mM Na, 3.0 mM K, 1.4 mM Ca, 0.8 mM Mg, 1.0 mM P, and 155 mM Cl and adjusted to pH 7.4 (12).
33. EcoLite Scintillation fluor was purchased from MP Biomedicals (Irvine, CA, USA).

2.2. Microdialysis: 7–10-Day-Old Rats

1. Time pregnant Sprague Dawley rats are purchased from Zivic Miller Laboratories.
2. Microdialysis guide cannulae (P000138) and microdialysis probes (P000083) were purchased from CMA.
3. Screws for anchoring dental cement cap (7431021) were purchased from CMA.
4. Mouse adapter, #1226 with soft tissue cups, was obtained from Kopf.
5. Model 907 mouse anesthesia mask was obtained from Kopf.
6. Super glue in gel form was purchased from a local vendor.
7. Carbide burr, 0.8 mm, RS-6280C-1, was purchased from Roboz (Gaithersburg, MD, USA).
8. SurgiVet Model 100 isoflurane vaporizer was purchased from Smith Medical (Dublin, OH, USA).
9. Isoflurane, USP, was purchased from Halocarbon Products Corp. (River Edge, NJ, USA).
10. Two 1 cm holes are drilled through the lids of 500 ml glass jars purchased from Fisher.

3. Methods

3.1. Advance Preparation

1. The stereotaxic frame is prepared to include the BAS guide cannula holder.
2. A mechanical jack is positioned to support the rat in the stereotaxic unit. A heating unit, covered with absorbent paper to provide separation between the heated surface and the skin, is placed on the jack.
3. Modify the spring that will tether the rat to the swivel. Create a kink in the spring tether about 5 cm from each end. To avoid confusing the inflow and effluent lines thread two sections of plastic tubing of different colors through the kinks. The two lengths of tubing protrude 6–8 cm through each kink. Threading the tubing through the kinks rather than the end of the spring keeps the tubing in place and allows greater freedom of manipulating the tubing when making the necessary connections. The Peek tubing leading from the syringe pump to the probe is one color (tan, BAS, 0.65 mm OD, 0.12 mm ID, peek tubing, MF5377) while the tubing from the probe to the collection tube is a second color (blue, BAS, peek tubing MD1511). These sections of tubing will go from the swivel to the microdialysis probe and from the effluent side of the probe to a 500 μl collection tube. Do not yet attach tubing to swivel.
4. Connect swivel to spring that turns the swivel with the movement of the rat (Stoelting Co. PT53622, unit of all three components). The other end of the spring has an attached clasp from a hobby shop that clips to a collar around the rat's neck. More expensive systems were less functional.
5. In an autoclave bag, place the swivel as configured above and two additional 20 cm sections of tubing that will be used to flush the microdialysis probe and to connect the syringe pump to the swivel. In a second autoclave bag place a 1 ml Hamilton glass syringe plus its plunger and needle removed from the barrel. Steam sterilize for 15 min.
6. Prepare Kontes center wells by inserting a fluted piece of filter paper (0.5 × 2 cm) into the center well.
7. Insert each center well into a Kontes stopper.
8. Pierce the top of 500 μl microfuge tubes to be used for collection of the effluent from the microdialysis probe.
9. Prepare sterile aCSF.

3.2. Preparations on Day Before Oxidation Studies

1. Steps 1–4 are completed in a sterile laminar flow hood. A total of 6–10 plastic connectors (CMA, 3409500) are sterilized by soaking in 70% ethanol for 10 min. Immediately slip connectors on all tubing ends of the sterilized tubing and swivel as

described above. (*Leaving the connectors longer in ethanol more than 10 min softens the plastic so that the connectors do not provide tight junction with the tubing or swivel. Unused connectors may be dried in a sterile plastic petri dish for use at another time at which time the ethanol treatment is repeated.*)

2. Partially open the sterile autoclave bag. Remove ethanol-sterilized connectors with sterile forceps and slip connector onto each end of the three sections of sterile tubing except the end of the effluent line that will go to the collection tube.
3. Connect tubing to each end of the swivel. Do not yet assemble the 1 ml Hamilton syringes.
4. Determine that the swivel and tubing connections are free of leakage, not blocked, and free of ethanol. For this a Harvard syringe pump is used since it has a larger capacity than the BAS pump used for microdialysis. Attach swivel and tubing to a 5 ml sterile syringe with a sterile connector and pump sterile water through all the tubing at 50 μl/min. Once connections to the swivel are shown to be tight, the tubing open and free of air bubbles, disconnect the pump and reinsert swivel assembly into its autoclave bag. Store bagged assembly in sterile hood until the next day. If two rats are being used, prepare two assembled swivels and tubing.
5. Preparation of radioactive solutions: The total volume (μl) equals its infusion volume plus 300 μl to be used for blanks and counting. The infusion volume (μl) equals the length of the experiment (in minutes) times the pump rate (2 μl/min). The total microcuries equal the total volume (μl) of the radioactive solution times 0.02 ± 0.005 μcuries per μl. The appropriate amount of ^{14}C-labeled substrate is blown dry with a gentle stream of nitrogen in a chemical hood and resuspended in sterile aCSF. If a change in solutions (addition of a nonradioactive effector) is made during the experiment, reduce the volume of aCSF to compensate for any additional volume added. Immediately sterilize by filtration through a 0.22 μm sterile syringe filter of appropriate size.

3.3. Implantation of Microdialysis Probe Prior to In Vivo Oxidation Studies in Adult Rats

1. Anesthetize rat by IP injection of sodium pentobarbital (60 mg/kg) using a 1 ml TB Syringe with 25 gauge needle. If required, an additional one-third the initial dose is given.
2. The fabricated ear bars (which will be placed within the ear bars of the stereotaxic frame) are inserted into the rat's ears and the set screw is tightened (Fig. 1).
3. The combination of rat and ear bars is inserted into the stereotaxic unit using the ear bars of the unit.
4. The tooth bar is set at desired angle, i.e., a setting of 5 below the horizontal.

Table 1
Typical setting for performing microdialysis of the hippocampus of an adult rat

Position	Rostral	Lateral	Vertical
Coordinates for the surface of brain at bregma	81.7	34.3	15.5
Stereotaxic coordinates for the hippocampus (13)	-4.1	+3.7	-2.5
Desired location (sum)	77.6	38.0	13.0

5. The rat's teeth are hooked on the tooth bar and gently extended.
6. The top of the head is scrubbed with a Providone-Iodine Swabstick followed by 100% ethanol. The hair is parted down the midline using a small spatula.
7. A 2 cm incision is made down the midline. 0.05 ml of Rimadyl (carprofen, 50 mg/ml) is infused with a 25 gauge needle around the incision site. If necessary, alligator clips are used to hold the skin apart so that the lambda and bregma are visible. The bregma is the zero point for all measurements.
8. The incision may have to be made larger depending on where the microdialysis probe is to be placed. The loose skin is not cut off, but will be under the dental cement cap, which leaves no free ends for the rat to reach and damage.
9. Insert guide cannula into the stereotaxic arm and tighten.
10. Position the stereotaxic arm such that the guide cannula is over the bregma.
11. Rostral and lateral readings for the desired brain location are taken from the stereotaxic frame at the bregma. Ascertain setting for desired brain location from a stereotaxic atlas (13). These values are added to the readings at the bregma (Table 1).
12. Reposition the stereotaxic arm with the guide cannula to the calculated settings.
13. Lower guide cannula, lift up, and mark skull using pencil.
14. Swing arm out of the way.
15. Drill a hole through the skull bone using the trephine drill bit. One can feel as the skull is penetrated.
16. Change to different drill bit (8J60 drill HSS). Drill two holes 4–5 mm from the mark, 90° to each other. Drill until it penetrates the skull bone.
17. Using SS screws (080 × 1/8) with screw driver model SD-80 that holds the screws, insert the two screws to the depth of two threads into the bone. Do not penetrate the bone completely.

18. Reposition the guide cannula to the probe coordinates and lower it to the dura.
19. Read the depth of the guide cannula. Record and add the desired vertical setting.
20. Lift and move the guide cannula out of the way.
21. Using a 25 gauge sterile syringe needle, nick the dura. This will allow the guide cannula to penetrate the dura and brain.
22. Lower guide cannula to the calculated depth.
23. Blot any fluid or blood with a sterile cotton pad to prevent wetting of the Gelfoam.
24. Cut a 4–5 mm square of Gelfoam. From one side make a cut to the center of the Gelfoam to allow easy placement of Gelfoam around guide cannula. Tap down with forceps.
25. Add solid dental cement to two glass vials. Add powder to a depth of 5–6 mm and use approximately1 ml of liquid, i.e., up to 3 cm on barrel of Pasteur pipet. Add liquid cement to one vial so that it has a light consistently. Triturate with a glass Pasteur pipet that has been cut so that the thinnest portion of the pipet is no more than 1.5 cm long (otherwise the gel will solidify in the thin part of the pipet). It is easier to add the liquid to the powder and mix. If too runny, add more powder or triturate which will allow it time to become more firm. Suction all the mixed dental cement into the pipet and use to distribute. If it is done at the correct speed, most of the material will be expelled before it hardens.
26. The cap of dental cement around the guide cannula is constructed in stages. First layer a base around the guide cannula, the Gelfoam, and the screws. Then, using the same pipet, carefully coat the guide cannula on all sides. By this time any remaining dental cement in the Pasteur pipet becomes too hard to use.
27. A second vial of dental cement is prepared using a fresh pipet. The dental cement cap is built up so that it forms a cone around the guide cannula. The cap will harden in 3–5 min. Tapping the cap with a metal spatula yields a sharp sound.
28. Loosen clamp that holds the guide cannula and carefully move it out of the way to avoid jarring the dental cement cap.
29. Remove the tooth bar from the rat's mouth. With the custom-fabricated ear bars still inserted in the rat, remove the rat from the stereotaxic unit. Then remove the fabricated ear bars. (If done in a regular rat stereotaxic unit, remove the ear bars after removing the tooth bar.)
30. Place rat in open topped plastic cage with bedding, food, and water. Warm the animals with a 60 W bulb placed 8 in. from the rat. Check temperature in cage with a thermometer. Animal

will awaken 1.5–2 h after surgery and move about the cage. Rarely the rat will not awaken and move freely. These rats are euthanized.

31. The rat is placed in a cage and returned to the animal facility to recover from the anesthesia and surgery. Only one rat is placed in a cage because the animals will chew on the caps if more than one is housed in one cage. The animals are used for microdialysis the next day about 18 h after implantation of the guide cannula.

3.4. In Vivo Oxidation Studies in Adult Rats

We will describe this methodology by use of an example for ^{14}C-lactate oxidation and the inhibitory effect of fluorocitrate.

1. Steps 1–4 are performed in a sterile laminar flow hood. Have available 4–10 connectors soaked in 70% ethanol for 10 min in case the ones on tubing do not hold properly. Position a sterile 5 ml syringe filled with water in the Harvard syringe pump with a 20 cm segment of sterile tubing connected to a microdialysis probe. Submerge the microdialysis probe in sterile water through a hole in the top of a glass vial and pump sterile water through the probe at 10 μl/min. Sterility is critical for all components. Bacteria in tubing, swivel, or solutions will oxidize ^{14}C-metabolites to $^{14}CO_2$.
2. Replace the water-filled syringe with a 5 ml syringe containing aCSF, flush out all tubing, probe, and swivel, and leave all tubes filled with aCSF.
3. Assemble and fill Hamilton syringe with radioactive solution(s) and leave in sterile hood.
4. Attach tubing to probe. Loosen the microdialysis probe from its storage vial so that it can be easily removed using one hand.
5. Take a cable tie and make a loose loop. Slip the loop around the head of the rat taking care not to disturb the cap. Tighten gently so that it does not pinch the rat. Cut off excess using small cutting pliers, leaving about 5 cm behind.
6. While gently holding the rat, remove the dummy plug from the guide cannula implanted in the rat brain.
7. Remove the BAS microdialysis probe from its vial and insert into the guide cannula. Twist the C-ring portion of the microdialysis probe to lock in place. Place rat back in experimental bowl that will be used for microdialysis.
8. Clip the stainless steel swivels and tether (Stoelting 53622) to the counter balance arm (CMA 8309032).
9. Place the 1 ml Hamilton syringe in the BAS syringe pump. Connect the tubing from the swivel to the BAS syringe pump. At this point the probe is inside the guide cannula and all sections of tubing are connected.

10. Pump aCSF at 10 μl/min with the BAS pump. The effluent should come out only from the effluent tube end. If no effluent is observed, it is probable that the microdialysis membrane is broken and has to be replaced.
11. Reduce pump rate to 2 μl/min, and place the end of the effluent tubing into an empty 500 μl microfuge tube held by its clip to the tether. Perfuse the microdialysis probe with aCSF for 1–1.5 h to let amino acid levels attain an equilibrium following the insertion of probe. Collect 20-min fractions for amino acid determination, if desired.
12. During the equilibration time, number and add 5 μl of 100 mM KOH to a series of microfuge tubes with pierced tops that allow insertion of tubing from probe.
13. A 1 ml Hamilton syringe is filled with the ^{14}C-labeled compound prepared in aCSF and positioned in a second BAS syringe pump. To switch between infusion solutions, the tubing is moved from one Hamilton syringe to the second. Continue pumping at 2 μl/min and collect a minimum of 3, and up to 6, 1-h fractions.
14. Experimental manipulations may be undertaken such as adding inhibitors, nonradioactive substrate competitive compounds to the ^{14}C-labeled substrate, etc. These manipulations require the use of separate 1 ml syringes containing the ^{14}C-labeled substrate and a known amount of inhibitor, etc. If a change of solution is made, collect one 30-min fraction before resuming 1-h collection times. Do not use the 30-min fraction for calculations.
15. At the end of the study remove the microdialysis probe from the rat brain, remove tubing, and insert probe through a hole in a plastic cap into a glass vial with sterile water. Attach a short piece of tubing, flush the probe with 500 μl of sterile water pumped at a rate of 50 μl/min, and collect the effluent. Discard the effluent and the glass vial in radioactive waste containers.
16. Storage of microdialysis probe—store in gentamycin (50 μg/ml) in 5 ml glass vials with plastic cap into which holes had been cut to hold the microdialysis probe. Probe can be used up to five times if it is not physically damaged.
17. The rat is euthanized and decapitated. Skin is removed from the skull and the head is placed in 10% formalin for 1 week. The skull is removed from the brain. The brain may be sectioned to verify the placement of the probe and to examine if a cystic area developed around the guide cannula.
18. To recover $^{14}CO_2$ from the collected effluent cut off the cap of the 500 μl microfuge tube containing the 100 mM KOH and effluent. Place microfuge tube into a 16 × 100 mm glass test tube.
19. Add 300 μl benzethonium hydroxide to a Kontes center well.

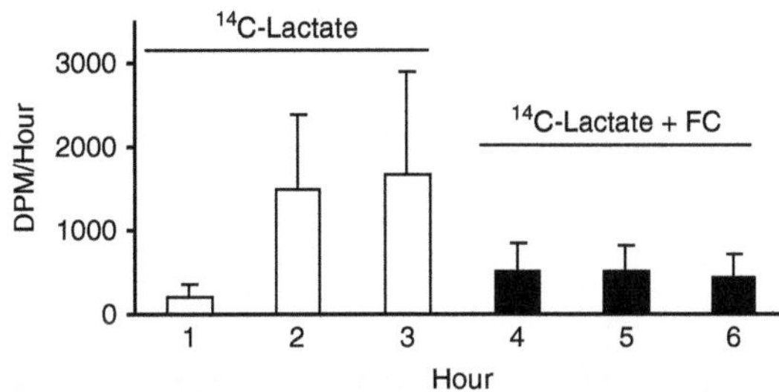

Fig. 2. Effect of 100 μmol/l flurocitrate on the time course of radioactivity recovered as $^{14}CO_2$ following perfusion of (U-14C) lactate into the interstitial space by microdialysis (10), with permission.

20. Seal the test tube with a Kontes stopper and its Kontes center well.
21. Inject 1 ml 10% trichloroacetic acid into the microfuge tube with a 20 gauge × 90 mm spinal needle. Place the test tubes at 4°C for 1 h.
22. Remove the stopper and cut the center well with a small pair of cutting pliers and place in a 5 ml scintillation vial with scintillation fluor. If chemoluminescence is observed due to the benzethonium hydroxide, add 117 ml of concentrated acetic acid to each gallon of fluor.
23. Prepare duplicate blank tubes containing 120 μl of the starting material in separate 500 μl microfuge tubes. Repeat steps 18–22 to recover $^{14}CO_2$ released.
24. Quantitate radioactivity in samples with a scintillation counter.
25. Calculate the DPM/hour minus blank. The first sample after initiation of perfusion with a radioactive solution is not used for calculation because the interstitial specific activity has not yet reached equilibrium (Fig. 2).

3.5. Implantation of Microdialysis Probe Prior to In Vivo Oxidation Studies in 8-Day-Old Rats

We will describe this methodology by use of examples in which ^{14}C-glucose, ^{14}C-lactate, ^{14}C-glutamate, and ^{14}C-glutamine oxidation were measured in the brain of 8-day-old rat pups. The pups are subjected to hypoxia/ischemia to one hemisphere on day 7 (14, 15). The protective effect of acetyl-L-carnitine (ALCAR) on ^{14}C-glutamate and ^{14}C-glutamine oxidative metabolism was also determined on day 8. The method can readily be adapted to measuring oxidative metabolism in one hemisphere in rat pups without hypoxic/ischemic pretreatment. The oxidative measurements allow comparison of oxidative metabolism of different compounds and demonstrate the protective effect of ALCAR (Fig. 3).

1. The rat pups are weighed and each pup is assigned a unique number that is written on its back with a marking pen.
2. The rat pup is anesthetized in a small plastic box with 4.5% isoflurane and a mixture of 70% nitrous oxide plus 30% oxygen

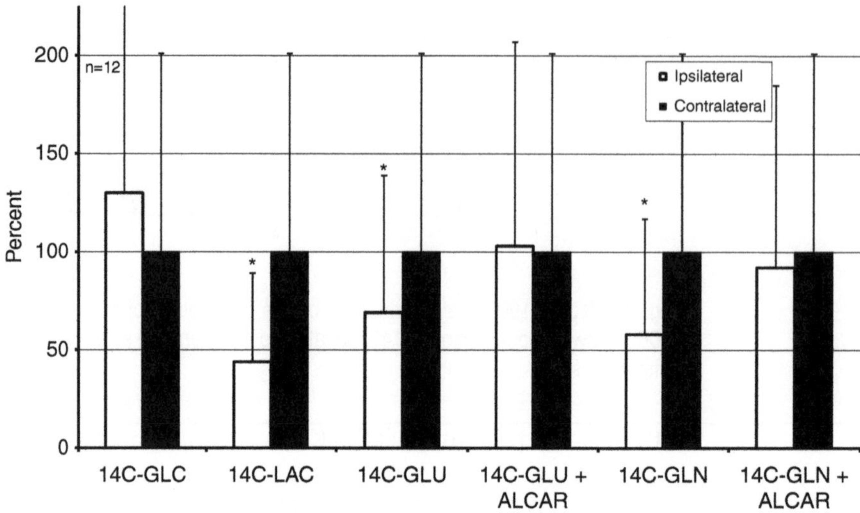

Fig. 3. In vivo oxidative brain metabolism in 8-day-old rat pups 24 h after hypoxia/ischemia that only affected one hemisphere. Oxidative metabolism of lactate, glutamate, and glutamine is reduced in the ipsilateral side while glucose oxidation was more variable. Acetyl-L-carnitine (ALCAR) was injected subscapularly at a dose of 100 mg/kg immediately after the 90-min hypoxia–ischemia period and again immediately prior to microdialysis, the decrease in oxidative metabolism was prevented using labeled glutamate and glutamine. Subscapular injection of ALCAR at a dose of 100 mg/kg immediately after the 90-min hypoxia–ischemia period and again immediately prior to microdialysis prevented the decrease in oxidative metabolism of labeled glutamate and glutamine.

and placed in the Kopf stereotaxic frame using the Kopf mouse adapter #1226 and Model 907 mouse anesthesia mask.

3. The bregma is exposed with a sterile scalpel, the loose tissue above the bregma is cut off, and stereotaxic measurements are recorded.
4. The guide cannula is positioned to the predetermined stereotaxic coordinates for the developing rat (16).
5. The location of the desired position for the guide cannula with respect to the bregma is marked in pencil.
6. If desired, a guide cannula may be inserted into both hemispheres.
7. A drill bit (8J60 drill HSS) is used to drill through the skull for the guide cannula.
8. Two anchoring screw holes are made with a 0.8 mm carbide burr, RS-6280C-1.
9. Gently insert screws (CMA 7431021) so that the thread is engaged to maintain an upright position. Use super glue in gel form to hold screws in place.
10. The CMA P000138 guide cannula is inserted to the desired depth from the dura.
11. Gel foam is not used because the guide cannula fits tightly if the above drill bit is used.

12. Use dental cement to anchor the guide cannula.
13. Make a dental cement cap as described for the adult rat. After the dental cement has hardened, the pup is removed from the stereotaxic frame and allowed to recover under a heat lamp. Monitor temperature with a thermometer inside the cage.
14. When the pup has recovered, return the pup to the dam to nurse for 60 min.
15. The pup is removed from the dam and immediately used for the microdialysis study. Do not leave the pup with its dam overnight since the dam will chew on the cap.

3.6. In Vivo Oxidation Studies in 8-Day-Old Rats

The procedures are the same as for the adult rats with minor changes as noted.

1. The microdialysis probe (CMA P000083) with attached tubing is inserted into the guide cannula. The lines attached to the probe are sufficiently long so that extra tubing is not required.
2. No swivel is used because the pups have limited mobility and there is sufficient flexibility in the tubing to account for this mobility.
3. The rat pup is placed in a 500 ml glass jar and the tubing is passed through the holes in the lids.
4. The jar is partially submerged in a 37°C water bath to prevent hypothcrmia.
5. The remainder of the procedure is identical to the adult protocol.

References

1. Peng L, Zhang X, Hertz L (1994) High extracellular potassium concentration stimulates oxidative metabolism in a glutamatergic neuronal culture and glycolysis in cultured astrocytes but have no stimulatory effect in a GABAergic neuronal culture. Brain Res 663:168–172
2. Tildon JT, Merrill S, Roeder LM (1983) Differential substrate oxidation by dissociated brain cells and homogenates during development. Biochem J 216:21–25
3. Sokoloff L (1960) The metabolism of the central nervous system in vivo. In: Field J, Magoun HW, Hall VE (eds) Handbook of physiology-neurophysiology, vol 3. American Physiological Soc, Washington, DC, pp 1843–1864
4. Bak LK, Walls AB, Schousboe A, Ring A, Sonnewald U, Waagepetersen HS (2009) Neuronal glucose but not lactate utilization is positively correlated with NMDA-induced neurotransmission and fluctuations in cytosolic levels. J Neurochem 109(Suppl 1):87–93
5. Garseth M, Sonnewald U, White LR, Rod M, Zwart JA, Nygaard O, Aasly J (2000) Proton magnetic resonance spectroscopy of cerebrospinal fluid in neurodegenerative diseases: indication of glial energy impairment in Huntington chorea, but not Parkinson disease. J Neurosci Res 60:779–782
6. Nordström C-H (2010) Cerebral energy metabolism and microdialysis in neurocritical care. Childs Nerv Syst 26:465–472
7. Huang Y, Zielke CL, Tildon Zielke HR (1993) Monitoring in vivo oxidation of ^{14}C-labelled substrates to $^{14}CO_2$ by brain microdialysis. Dev Neurosci 15:233–239
8. Zielke HR, Huang Y, Baab PJ, Collins RM Jr, Zielke CL, Tildon JT (1997) Effect of α-ketoisocaproate and leucine on the in vivo oxidation of glutamate and glutamine in the rat brain. Neurochem Res 22:1159–1164
9. Zielke HR, Zielke CL, Baab PJ (2007) Oxidation of ^{14}C-labeled compounds perfused

by microdialysis in the brains of free-moving rats. J Neurosci Res 85:3145–3149
10. Zielke HR, Zielke CL, Baab PJ, Tildon JT (2007) Effect of fluorocitrate on the oxidation of lactate and glucose in the brain of free moving rats. J Neurochem 101:9–16
11. Zielke HR, Zielke CL, Baab PJ (2009) Direct measurement of oxidative metabolism in the living brain by microdialysis: a review. J Neurochem 109(Suppl 1):24–29
12. Sendelbeck L (1987) Recipe for preparation of artificial CSF. In: Ray N (ed) Special Delivery, vol 8(3). ALZA Corp., Palo Alto, CA, p 4
13. Pellegrino LJ, Pellegrino AS, Cushman AJ (1979) A stereotaxic atlas of the rat brain, 2nd edn. Plenum, New York
14. Rice JE, Vannucci RC, Brierley JB (1983) The influence of immaturity on hypoxic-ischemic brain damage in the rat. Ann Neurol 9:131–141
15. Vannucci RC, Brucklacher RM, Vannucci SJ (1996) The effect of hyperglycemia on cerebral metabolism during hypoxia-ischemia in the immature rat. J Cereb Blood Flow 16:1026–1033
16. Sherwood NM, Timiras PS (1970) A stereotaxic atlas of the developing rat brain. University of California Press, Berkeley

Chapter 6

Quantitative In Vivo Microdialysis in Pharmacokinetic Studies

Teodoro Zornoza, María José Cano-Cebrián, Ana Polache, and Luis Granero

Abstract

Recent theoretical studies have yielded a more profound knowledge of the properties of recovery (the key parameter in quantitative microdialysis) and have put in evidence important limitations of the usual in vivo calibration methods used in quantitative microdialysis for pharmacokinetic studies. Recovery values obtained by using the more classical methods of calibration (the variation of flow rate perfusion method, the delivery and retrodialysis methods, and the no net flux method) can only be used to accurately convert dialysate drug concentrations into extracellular concentrations, when the drug of interest is in the body under steady-state conditions. Therefore, these in vivo calibration procedures must not be used when the drug studied has to be administered using modalities of administration which do not provide steady-state concentrations (for example, intragastric, subcutaneous, intraperitoneal, or intravenous bolus injections). The dynamic no net flux (DNNF) method, however, can be considered the only in vivo calibration method useful in PK experiments developed under transient conditions, although this calibration procedure has several serious disadvantages. The new modified version of the ultraslow microdialysis (the MetaQuant technique) overcomes many of the limitations of both the classical calibration and the DNNF methods and, therefore, it could be considered a promising tool in pharmacokinetics.

Key words: Quantitative microdialysis, Recovery, In vivo calibration, Variation of perfusion flow rate method, Ultraslow microdialysis, Delivery method, Retrodialysis method, NNF method, DNNF method

Abbreviations

C_d	Drug concentration in the dialysate
C_{ecf}	Drug concentration in the extracellular fluid
C_{in}	Drug concentration in perfusate
CSF	Cerebrospinal fluid
DNNF	Dynamic no net flux method
E	Extraction fraction, extraction efficiency or recovery
ECF	Extracellular fluid

Giuseppe Di Giovanni and Vincenzo Di Matteo (eds.), *Microdialysis Techniques in Neuroscience*, Neuromethods, vol. 75, DOI 10.1007/978-1-62703-173-8_6, © Springer Science+Business Media, LLC 2013

MD Microdialysis
NNF No net flux
PK Pharmacokinetics
Q Perfusion flow rate

1. Introduction

The mathematical description of the time course of a drug (or its metabolite/s) level in the biological media, after single or repeated administration to a living organism, is probably the most important objective of Pharmacokinetics (PK). This mathematical description has the fundamental goal to allow us to make accurate predictions of the pharmacological response.

To develop their mathematical models, PK has classically used serum or plasma samples, in spite of the knowledge that most drugs exert their effects not within the blood compartment, but in defined tissue sub-compartments in which drug receptors are located. This practice is based on the assumption that changes in serum or plasma drug concentrations are a true reflection of those occurring in the biophase. Although this methodology has provided, in general, satisfactory results for a large majority of drugs, there are notable exceptions. For example, it is now well established that drug penetration into certain types of tumors is variable and poor (1). Consequently, studies on the clinical efficacy of certain anticancer drugs have revealed a lack of correlation between serum drug concentrations and tumor exposure to the drug. Another paradigmatic example comes from the field of anti-infective drugs. Several studies have reported a high inter-tissue and inter-subject variability of antibiotic tissue distribution, with antibiotic tissue concentrations varying considerably from the corresponding plasma levels (2, 3). These, and other examples, clearly suggest that direct concentration measurements at biophase might be more relevant in predicting clinical response than the estimation of response from plasma drug concentrations.

The use of drug concentrations at the level of the drug target to predict the pharmacological response has been a long-standing goal for the PK research. Microdialysis (MD) has allowed researchers to realize this old dream of pharmacologists.

To monitor drug concentrations in target tissues in animals and humans, several techniques (e.g., tissue biopsies, saliva and skin blister fluid sampling, imaging techniques) have been employed. But, in contrast to MD, which allows sequential sampling over time, traditional concentration measurements in bodily secretions or biopsy samples usually only yield a limited number of time points. Moreover, taking biopsies presents ethical limitations, and more important, the resulting total concentrations in homogenized tissue usually lead to over- or underestimation of actual drug tissue concentrations (4).

In vivo microdialysis can be considered the only method useful, at present, to accurately quantify the free extracellular levels of drugs in tissues of living animals. MD has become a common technique for monitoring changes in free concentrations of drugs and metabolites in various tissues. The use of MD in PK studies has increased considerably due, in part, to improvement of our theoretical knowledge of quantitative processes involved in MD and advances in the sensitivity and specificity of the analytical methodology. Numerous research studies incorporating MD have been published in the last few years. The applications of MD to PK research are presently numerous and include blood and tissue pharmacokinetics, in vivo protein binding studies, drug and/or metabolite profiles in peripheral tissues, pharmacokinetics in discrete regions of the brain, studies of transport phenomena in the central nervous system, and so on.

The present chapter deals with the application of MD in PK studies. Since in this scientific field, knowledge of the free drug extracellular concentrations is crucial, it becomes imperative to have adequate calibration protocols that allow one to convert dialysate concentrations into extracellular concentrations. In the following, we describe and analyze the existing methods of probe calibration and the variables and factors affecting them, in the light of recent advances in our theoretical knowledge of mass transport processes implicated in MD.

1.1. Calibration of Probes: A Necessity in MD-PK Studies

The purpose of PK is to study the time course of drug and/or metabolite concentrations in different tissues and excreta and to construct mathematical models suitable for interpreting such data. As commented above, without accurate measures of drug concentrations in tissues of interest it is not possible to carry out adequate predictions for clinical response. In this sense, the major concern in applying MD to PK studies is to accurately establish the absolute drug concentrations in the extracellular fluid (C_{ecf}) of the tissues analyzed. This problem arises from the fact that MD is a dynamic sampling technique where drug molecules diffuse across a semipermeable membrane in the presence of a concentration gradient (Fig. 1).

Diffusing molecules are then swept away by a perfusion medium that is continuously pumped at a constant flow rate (normally between 1 and 5 μl/min) through the MD probe. In this situation, the drug concentration in the microdialysate (C_d) is not that found in the extracellular fluid (ECF). Therefore, direct quantification of the drug in the dialysate samples yields drug concentration values clearly lower than those existing in the ECF surrounding the MD probe. The ratio between the drug concentration in the dialysate and the concentration of the same substance in the ECF distant to the probe is defined as relative recovery or, simply, "recovery" (E), and is expressed either as a ratio or as a percentage.

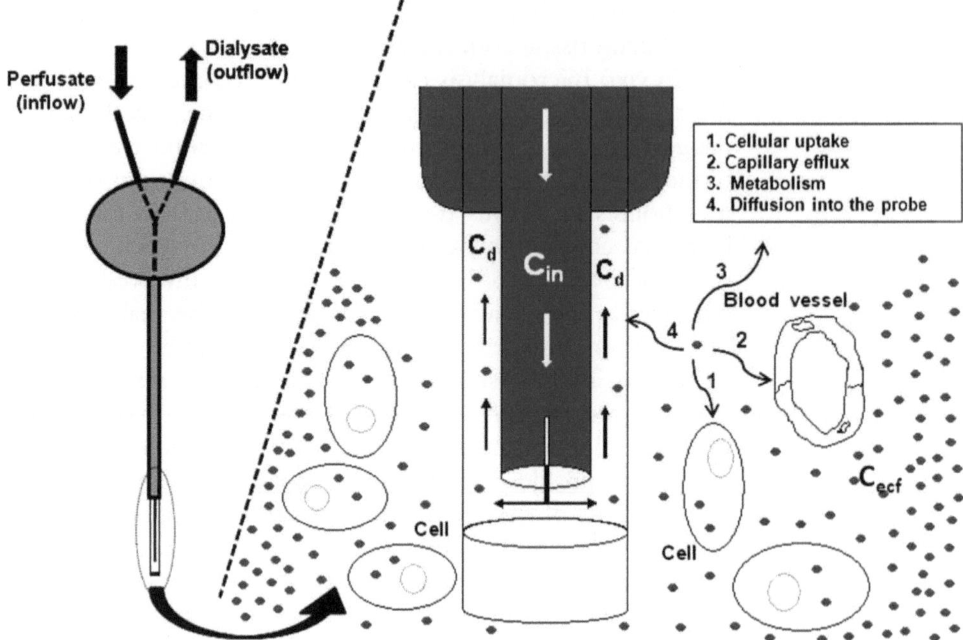

Fig. 1. Schematic representation of a microdialysis concentric probe implanted in a solid tissue, sampling a hypothetic drug systemically administered. The diffusion paths and routes of elimination for a hypothetical drug molecule are also shown. Abbreviations: *ECF* extracellular fluid of the tissue, C_d drug concentration in dialysate, C_{ecf} drug concentration in the ECF at far field distance from the probe, C_{in} drug concentration in the perfusate, *Q* flow rate.

Recovery is dependent upon several factors (5) which are summarized in Table 1; the most important among them are the perfusion flow rate (*Q*) and the properties of the external medium (tissue properties).

Early studies of MD processes assumed that the membrane used to construct the probe constituted the greatest resistance (and therefore the limiting factor) to drug movements. This led to the use of the results of simple in vitro calibration experiments in a quiescent aqueous medium to calculate the corresponding *E* value. The in vitro *E* value thus calculated was then used to transform in vivo C_d into C_{ecf}, assuming that the membrane properties, probe geometry, and flow rate were not different in the in vitro and in vivo situations (27, 28).

It is now clear that this procedure is not valid, however. Several investigations have shown the existence of factors, in biological tissues, that serve to either impede or enhance mass transport of the drug from the ECF to the probe (19, 23, 24, 29).

In a simple medium (for example an aqueous solution), the flux of diffusion, *J*, measured in relation to the total area is given by Fick's law as

$$J = -D\left(\frac{\partial C}{\partial r}\right) \tag{1}$$

Table 1
Factors exerting an influence upon recovery (*E*)

Membrane properties: • Composition and surface charge (6, 7) • Molecular weight cutoff (MWC) (8) • Membrane surface area (S): Outer diameter (OD), length, thickness (9) For example, Polyethersulfone (OD: 0.50 mm; MWC 100 kDa) Cuprophan®(OD: 0.24 mm; MWC 6 kDa) Polyarylethersulfone (OD: 0.50 mm; MWC 20 kDa) HOSPAL AN69® Polyacrylonitrile/sodium methallyl sulfonate copolymer (OD: 0.31 mm; MWC 40 kDa)	Perfusate: • Perfusion flow rate: ↑flow rate implies ↓recovery (9, 11) • Ionic composition (12–16) • Osmolality (17) For example, Saline solution, ringers solution, buffered ringers solution, artificial-cerebrospinal fluids, Krebs Ringer solution
Drug properties: • Molecular weight (8) • Charge (10) • Shape	Tissue properties: • Tissue diffusivity: Diffusion coefficients in tissue are directly dependent on effective volume fraction of the distribution space (α) (18–22) and inversely dependent on tortuosity (λ^2) • Tissue clearance processes: Capillary exchange, cellular uptake processes, local metabolism (23–26)

where D is the diffusion coefficient in the free solution, C is the drug concentration in a unit volume, and r is the distance measured perpendicularly to the area considered. However, in a solid tissue, flux must be calculated by the following equation:

$$J = -\alpha D^{*}\left(\frac{\partial C}{\partial r}\right) \qquad (2)$$

where α is the effective volume fraction and D^* the diffusion coefficient in the tissue ECF (it is assumed that the drug does not penetrate into the cells). As can be seen, flux of diffusion in a solid tissue is directly dependent on the α value, which is usually lower than 1 (for example, in most brain regions $\alpha \approx 0.2$) (21). On the other hand, the diffusion path length in vivo is clearly increased compared to that existing in a free solution because of the tortuosity (λ) of the route. In this context, λ, which describes the increased diffusion pathway in vivo due to the presence of impermeable cell membranes, is equal to $(D/D^*)^{1/2}$ (in brain $\lambda \approx 1.6$). Therefore, the diffusion coefficients in a free liquid medium (D) are usually clearly higher than those found in a porous matrix such as a solid tissue (20, 22). The above theoretical considerations clearly suggest that flux in vivo is always smaller than in vitro. Because MD is essentially a diffusion process, recovery in vitro is expected to be greater than that obtained in vivo at steady state. In fact, this has been empirically demonstrated for many substances. Thus, for example, the ratios for the recoveries of mannitol and sucrose in vitro versus in vivo (determined in brain) were found to be 2.93 (30) and 3.50 (31), respectively.

Nevertheless, the above assertions must not be taken as a rule of thumb; several authors have also reported in vivo recoveries exceeding those obtained in vitro. For example, the ratios for the recoveries were 0.82 for cocaine (32) and 0.72 for dopamine (29, 33). In these latter examples, the discrepancies between in vitro and in vivo recoveries are largely due to the existence of potent drug clearance processes as it will be explained later.

It is clear that in vitro calibration has scant application in MD in vivo. However, in vitro experiments can be very useful for detecting drug–probe membrane interactions and for establishing the absence or existence of probe-to-probe differences—particularly in the case of in-house manufactured probes (6, 7). In vitro experiments in combination with in vivo studies can also yield relevant information on the tissue parameters involved in mass transfer phenomena in MD (34, 35).

At present it is well established that tissue is the most important factor in determining the mass transfer in the majority of MD-PK experiments, and that the calculation of C_{ecf} requires in vivo probe calibration. In order to perform in vivo calibration of the probes, it is customary to calculate, in the tissue of interest, the E value. Originally, E was defined as the ratio between dialysate concentration (C_d) and ECF concentration at far field distance from the probe (C_{ecf}) (1):

$$E = \frac{C_d}{C_{ecf}} \tag{3}$$

A more generalized form of the above equation is that corresponding to the so-called extraction fraction, extraction ratio, or extraction efficiency, which is designated with the same symbol (E) in the remaining discussion below:

$$E = \frac{C_{in} - C_d}{C_{in} - C_{ecf}} \tag{4}$$

where C_{in} is the inflowing drug concentration. When no drug is added to the perfusate (i.e., when we perform a classical recovery experiment, $C_{in} = 0$), diffusion of the drug is from the external medium into the dialysate, and (4) is reduced to (3).

E may also be calculated by including a known concentration of the drug in the perfusion fluid while C_{ecf} is zero (i.e., the so-called delivery experiment). In this case, drug diffusion occurs from the perfusate into the ECF, and (4) is reduced to

$$E = \frac{C_{in} - C_d}{C_{in}} \tag{5}$$

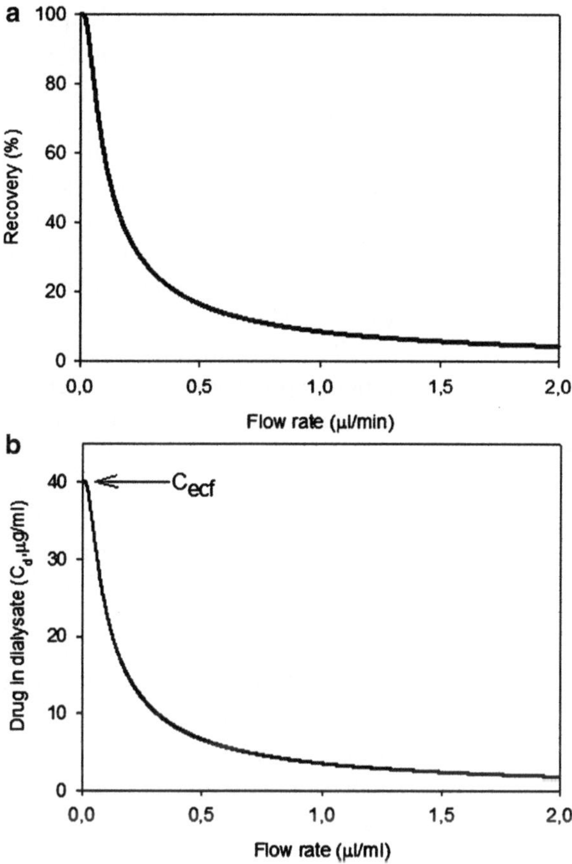

Fig. 2. (**a**) Effect of flow rate (*Q*) on relative recovery (*E*) and (**b**) concentration in dialysate samples (C_d) for a hypothetic drug.

Theoretically, under identical experimental conditions for a given MD probe and drug, *E* values calculated from a recovery experiment or from a delivery experiment are identical.

1.2. Properties of Recovery

A basic and important property of *E* is its dependence on perfusion flow rate (*Q*) (Fig. 2). As indicated in Table 1, *E* is inversely dependent on *Q*. The greater the *Q*, the lower the concentration in dialysate. As can be seen in Fig. 2a, *E* attains its maximum value (i.e., 100%) at zero flow. At this point C_d equals C_{ecf} (Fig. 2b). As we comment below, the dependence of *E* on *Q* was mathematically formulated by Jacobson et al. in 1985 (11), and constitutes the basis for the calculation of *E* used in one of the empirical methods more commonly used in quantitative MD.

Another important property of *E* is related to the time dependency of this parameter. MD is essentially a diffusion process. The insertion of a MD probe, perfused at a flow rate Q, in a tissue in which C_{ecf} is at steady state (the common situation in a classical in vivo calibration experiment), induces the appearance of a depletion zone around the probe. In other words, a concentration gradient is formed around the probe as the drug is removed by the perfusion fluid. For any given diffusion coefficient, the *E* value is determined by the magnitude of this concentration gradient. The latter will be more or less steeper depending on the existence of additional tissue factors such as irreversible metabolism, microvasculature transport processes, and so on. The faster the drug is metabolized in the tissue or is removed by microvascular transport, the steeper the gradient formed, and, consequently, the greater the value of E.

The formation of the concentration gradient is not an instantaneous phenomenon. In a classical recovery experiment, drug diffusion occurs first through the tissue ECF space, then through the membrane, and finally into the dialysate. Attending to the usual geometry of MD probes (cylindrical), the progression to steady-state concentration profile can be very slow, principally if there are no other mechanisms for removing the drug other than extraction through the probe. As *E* depends on the concentration gradient around the probe, the time dependency of *E* is, therefore, an inherent characteristic and is related to the temporal development of spatial concentration gradients in the probe and in the milieu adjacent to the probe. This phenomenon is known as mass transfer transient (36). The time dependency of *E* has been confirmed by experimental data proceeding from both in vitro (31) and in vivo experiments (15).

Theoretical predictions suggest very significant changes in *E* value over time. So, predictions indicate that *E* will decrease progressively over time, starting with the maximum value at time = 0, and, finally, falling to its steady-state value, E^{ss}, as can be appreciated in Fig. 3.

The magnitude and rapidity of the time-dependent variation of *E* are mainly dependent on the drug clearance processes—globally represented by k_{clr} in Fig. 3—taking place in the tissue interstitium (for example, efflux transport and local tissue metabolism). As observed in Fig. 3a, a decrease in k_{clr} value could imply (i) a decrease in the magnitude of E^{ss} and (ii) an increase in the time to reach the E^{ss} value, in some circumstances, this second trend being the most prominent: for example a change in two orders of magnitude in the k_{clr} value is able to induce a 60-fold increase in the time to approach E^{ss}, whereas the same change in k_{clr} induces only a threefold change in the E^{ss} value. Since this property of *E* is very relevant, it should be adequately considered by the users of MD in PK studies.

The scenario is clearly more complicated when we consider other normal situations in MD-PK studies. In many cases, drug administration can involve different modalities (intragastric administrations, intraperitoneal injections, intravenous administration (bolus or infusions), etc.), in which C_{ecf} is not at steady state in the course of the experiment. In these cases, the changes in *E* over time are due to not only the above-mentioned mass transfer transients

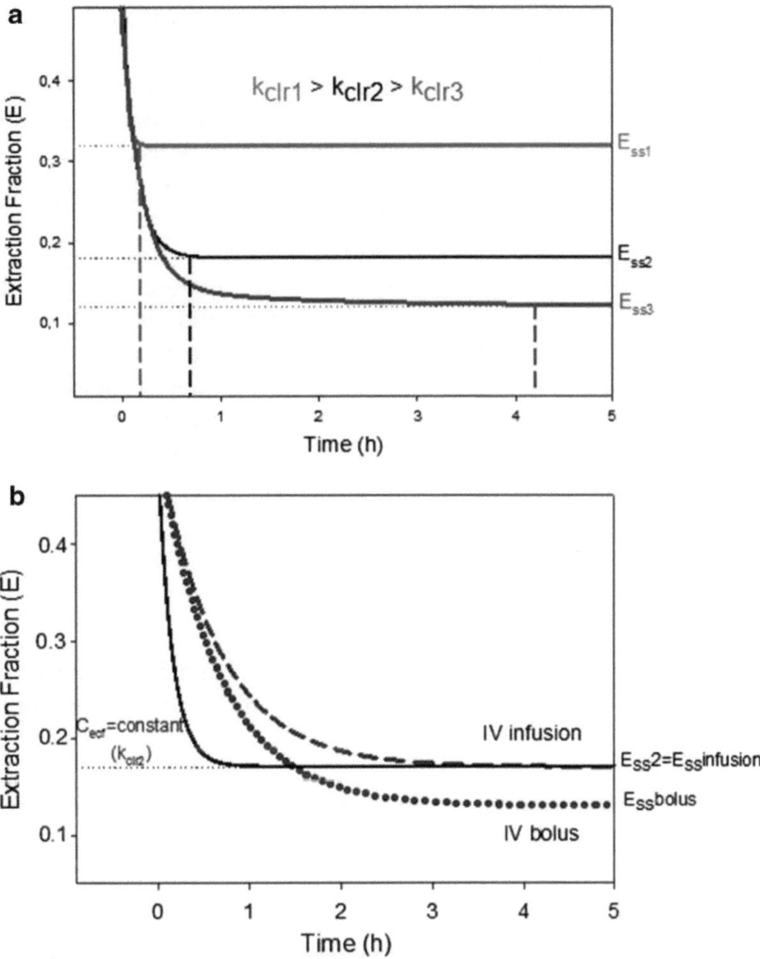

Fig. 3. (**a**) Time course of *E* for a hypothetical microdialysis experiment in which C_{ecf} = constant. The right E^{ss} values indicate the respective steady-state recoveries obtained for three hypothetical drugs that only differed in their overall rate constant for clearance from the tissue (k_{clr}). (**b**) Time course of *E* for a hypothetical drug administered by constant iv infusion (*dashed line*) or iv bolus (*dotted line*). At *right* are indicated the corresponding E^{ss} values. For comparison, the time variation of *E* for the same drug is also shown, with a rate constant equal to k_{clr2}, in the case of C_{ecf} = constant shown in the upper panel (*solid line*). Adapted from (36).

(inherent to MD) but also the time variation of the C_{ecf} values: the so-called pharmacokinetic transients (36). The theoretical limitations derived from this *E* property in determining C_{ecf} in MD-PK studies had not been studied in depth, until some simulations using two kinds of in vivo MD-PK experiments involving either a loading intravenous dose followed by an intravenous constant infusion or intravenous bolus drug administration were performed (36). The most relevant conclusion drawn from these simulations is that the time variation of *E* depends on the nature of the PK experiment. As can be appreciated in the bottom panel of Fig. 3, E^{ss} tended towards the same value obtained in the simulation of the pure mass transient in the case of constant intravenous infusion, though this situation was reached over longer times. The E^{ss} value in bolus administration was also achieved over longer times, though, as can be observed, its value is lower than that obtained in the pure mass transfer transient simulations. In other words, the time variation of E and its steady-state values can be clearly different depending on the mode of administration; this suggests that some methods of probe calibration in vivo could not be correctly used in some types of MD-PK experiments—specifically those in which C_{ecf} is not under steady-state conditions. Thus, some of the usual in vivo calibration methods commented below (in particular, the variation in flow rate method, the delivery and retrodialysis methods, and the no net flux (NNF) method) may be severely limited depending on the mode of administration assayed.

2. Materials

The use of MD in PK studies requires no specific materials other than those commonly used in other applications of MD.

3. Methods

3.1. Methods for Probe Calibration Under In Vivo Conditions

As we have commented above, since the vast majority of tissues in which MD probes are implanted do not constitute a simple and passive environment, and different kinetic processes can take place, it is usual for quantitative work in PK studies to calibrate MD probes directly in the tissue/s of interest. Several in vivo calibration methods have been developed and studied in the last years. In general, the in vivo MD calibration procedures that we present below require C_{ecf} to be zero or a constant value (i.e., they require C_{ecf} to be at steady state). This requirement is very important because it will preclude the use of the E value calculated to correctly convert C_d into C_{ecf} when the tissue drug concentrations are not at steady state.

3.1.1. Variation of Perfusion Flow Rate (Extrapolation to Zero Flow Rate) Method

The exponential dependence of recovery on flow rate (Q) was mathematically formulated by Jacobson et al. in 1985 (11):

$$E = \frac{C_d}{C_{ecf}} = 1 - \exp^{\left(-\frac{K_m \cdot S}{Q}\right)} \qquad (6)$$

where K_m is the average mass transfer coefficient, unique to the drug analyzed and the membrane properties, and S is the membrane surface. This model assumes that K_m is constant and does not vary with Q. By extrapolating the curve to zero flow, E will attain a value of 100%, i.e., at zero flow C_d is equal to C_{ecf} (Fig. 2b).

From a practical point of view, calculations using this method imply measuring C_d in the tissue of interest at several Q values. Using a nonlinear regression analysis program, the parameters of (6.6), $K_m \cdot S$ and C_{ecf}, can be obtained.

The variation of perfusion flow rate method has been extensively used in the MD-PK literature. Numerous reports have utilized this method for their calculations. In general, it can be considered a good method for estimating C_{ecf} concentrations, although some authors (37, 38) have shown that when used in in vitro experiments, it yields smaller estimates of C_{ecf} and higher estimates of E.

An alternative method that overcomes the above problems is based on the fact that at very slow perfusion rates, the residence time of the perfusion fluid through and in the membrane lumen is greatly increased. In such cases, C_d is near to the in vivo drug concentration in ECF. While theoretically it would require zero flow for concentrations to equilibrate, in practice E values higher than 0.9 (90%) are obtained at slow perfusion rates (of around 50 nl/min or lower). In this situation, quantification of C_d in dialysate samples directly yields the C_{ecf} value. The validity of this slow perfusion rate method has been ascertained by several authors both in vitro and in vivo under steady-state C_{ecf} conditions (32, 39). To our knowledge, this method has not been applied to MD-PK studies using non-steady-state conditions, despite its great advantage. Obviously, the method is only applicable when the analytical technique employed to quantify the drug in dialysates is highly sensitive and capable of accurately manipulating sample volumes of less than 1 μl. Other problems also derived from the very slow flow rate used are the following: (i) experiments are time consuming as for example on removing dead volumes from the MD system; (ii) sample evaporation can occur during the long collection times; and (iii) the lack of stability and accuracy of the perfusion flow rates used. At present, technology exists to manipulate and analyze low-volume samples and to overcome some of the problems mentioned above.

In fact, a modified version of the ultraslow microdialysis has been recently described (40). In this new method, the conventional microdialysis probe setup has been modified by introducing an additional flow line (exclusively acting as a carrier flow) that enters into the probe through a different channel in order to merge with the ultraslow dialysate immediately downstream from the microdialysis membrane. The carrier fluid is delivered at a higher flow rate (typical 0.9 μl/min). This shortens the lag time (making possible the in vivo application of the technique) and increases the final sample volume for easier handling. This new type of probe, named MetaQuant by the authors, has been able to generate drug concentration–time curves useful for PK analysis in medial prefrontal cortex of rats, with the use of a small number of rats. It seems that this in vivo MD methodology will become more common in application to MD-PK studies in the near future.

3.1.2. Delivery Method

Lindefors et al. (18) presented this method in 1989. In this approach to calibrating an MD probe in vivo, the amount of drug that diffuses out of the membrane relative to the amount in the perfusate is determined, and it is assumed that the same proportion exists for the amount of drug that will diffuse into the probe relative to the concentration in the ECF. In other words, in an in vivo delivery experiment, the drug is perfused through the MD probe while C_{ecf} is zero, and it is assumed that E is independent of drug mass transport direction.

As indicated above, the validity of the in vivo delivery method as a calibration protocol depends on the assumption that the E value calculated for diffusion out of the probe is the same as E for diffusion into the probe. Several papers support the above assertion for a wide range of compounds (41–43), although some exceptions have been reported (18, 44).

3.1.3. Retrodialysis Method

The retrodialysis method constitutes an alternative to the above-commented in vivo delivery experiment. In this method, a calibrant or internal standard is included in the perfusion fluid.

This approach uses close structural analogs of the drug with a similar molecular size, degree of ionization, lipophilicity, and interaction with the probe membrane and dialysis components. In this case, it is crucial not only to check the diffusivity of the calibrant in vivo but also to consider the potential differences in microvasculature transport, metabolism in tissue, and intra/extracellular space exchange. In addition, the calibrant must not kinetically interact with the drug. For example, it has been demonstrated that zidovudine is able to decrease the E value of aluvodine from 10.6 to 8.7% in brain, but not in adipose tissue (45). This phenomenon was attributed to competition for an active transport site in brain, and demonstrates the importance of appropriate selection of the calibrant.

A retrodialysis experiment can be performed simultaneously with the MD-PK experiment. This is in fact its main feature, and also the most common practice. Retrodialysis can offer several advantages over other in vivo calibration methods. The first advantage is a shortening of the study duration. Methods such as NNF method can only be used when C_{ecf} is at steady state—a fact that may lengthen the experiment. A second advantage is that retrodialysis can be used to continuously monitor and, consequently, to correct for changes in E among and within MD probes during in vivo microdialysis (46–49).

Despite the advantages of retrodialysis, it may be severely limited for in vivo probe calibration when the drug of interest is not under steady-state conditions (36). It must be remembered that transient C_{ecf} conditions will prevail unless the drug is administered according to zero order kinetics. Thus, if the calibrant is neither initially present in the tissue ECF nor systemically administered, the calibrant will exhibit a pure mass transfer transient (see above), and the E^{ss} value obtained for the calibrant by using (6.5) might be clearly different to E^{ss} exhibited for the study drug. In this case, E^{ss} obtained for the pure mass transfer transient (i.e., E^{ss} obtained for a calibrant in a retrodialysis experiment) does not correctly convert the individual C_d values to the corresponding C_{ecf} values.

As we commented above, although the MD parameters of the calibrant agree with those of the drug and, theoretically, it is possible to achieve equality in E^{ss} for the two solutes, the time courses of drug and calibrant recoveries will agree only if, additionally, both solutes exhibit the same plasma kinetics and are administered systemically in an identical fashion. However, if so administered, then C_{ecf} for the calibrant would be nonzero during the course of the experiment and, therefore, E could not be calculated by using (6.5). This suggests that the retrodialysis method for in vivo calibration in PK transient experiments is severely limited.

3.1.4. No Net Flux and Dynamic No Net Flux Methods

Another commonly used calibration method employed under in vivo conditions is the so-called NNF or zero net flux method (50). In this method the drug is added to the perfusion fluid at different concentrations (C_{in}) (higher and lower than the expected C_{ecf}), and the different C_d are measured in the dialysates. C_{in} and C_d are presumed to be equal if there is no net exchange of drug between ECF and the perfusate in the probe. However, when C_{in} is higher than C_{ecf}, the drug will diffuse out into the ECF, resulting in a decrease in C_d in relation to C_{in}. On the other hand, when C_{in} is lower than C_{ecf}, drug will diffuse into the probe from the ECF and, therefore, C_d will increase.

For assessment of this C_{ecf}, the difference between C_{in} and each particular C_d must be calculated. Linear regression of ($C_{in} - C_d$) versus C_{in} is used to determine the intercept with the abscissas axis (the point of NNF), i.e., C_{ecf}. The slope of the regression line provides the in vivo E value (Fig. 4).

For transient conditions, an alternative protocol exists (51). In this method, named Dynamic No Net Flux (DNNF) method, the drug of interest is systemically administered to different groups of animals using both the same dose and route of administration. After dosing, each subject belonging to a concrete group is continuously perfused with one of the perfusion concentrations selected around the expected ECF concentrations and the difference $C_{in} - C_d$ is calculated over time. Different groups are perfused with different C_{in} concentrations and, therefore, different $C_{in} - C_d$

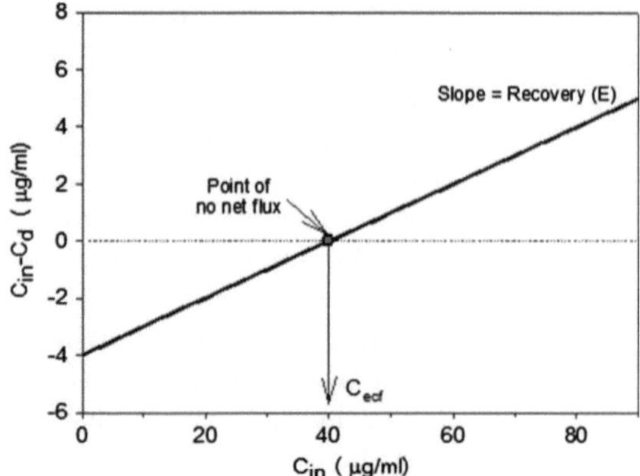

Fig. 4. Typical plot from an NNF experiment for determining E and C_{ecf}.

values are obtained at each time point. To calculate the C_{ecf} and the E values as a function of time, the results must be combined at each time point and subjected to regression analysis as described in the NNF method.

4. Notes

4.1. Uses and Limitations of the Variation of Perfusion Flow Rate Method

To accurately apply this method when used under in vivo conditions, the existence of a constant C_{ecf} is needed in order to measure the parameters of (6.6). This is not a problem when the drug is administered following a zero order kinetics for an enough period of time (for example, via constant intravenous infusions, Alzet® osmotic minipumps …). However, it is not a workable method when C_{ecf} is not at steady state. Moreover, the use of E values obtained under steady-state conditions with this method to correct other C_d values obtained when the same drug is administered by another mode of administration (for example a bolus injection) could lead to errors, according to the above-commented properties of E.

4.2. Uses and Limitations of the Delivery Method

The delivery method is stable over time, allowing flexibility for incorporation to an experimental design. Obviously, the delivery experiment cannot be carried out simultaneously during the MD-PK experiment. The measurement must be conducted before (or after) the start (end) of the pharmacokinetic experiment. It must then be assumed that the E value does not change as a result of the MD-PK experiment. Finally, it must be remembered that the E value calculated in a delivery experiment could be inadequate for correctly converting C_d obtained under transient pharmacokinetic conditions.

4.3. Selection of a Calibrant: The Key Step in the Application of the Retrodialysis Method

Retrodialysis requires the use of a calibrant with physical and biological properties identical to those of the study drug. Specifically, the ideal calibrant should have diffusion and PK properties similar to those of the assayed drug. This requirement can be achieved by using a radiolabeled version of the drug molecule; theoretically, a radiolabeled form of the drug will exhibit the same diffusion coefficient and PK properties as the actual drug substance.

4.4. Uses and Limitations of the NNF Method

The NNF method requires a stable steady-state drug ECF concentration and, unfortunately, a long time to be performed. Another important limitation (in a way similar to the two above-commented methods) is related to the use of the *E* value derived by this method to correctly convert C_d into C_{ecf}. Again, this inconvenience only manifests when the drug administered in the PK experiment is not at steady state.

4.5. DNNF Method for Nonsteady-State Conditions in PK

The DNNF method overcomes the important problem of the time dependency of *E*. It can be used independently of the route and modality for drug administration. However, this is a very complex and time-consuming protocol, with another important disadvantage from an ethical point of view: the requirement of a significantly larger number of experimental animals than other methods. Moreover, authors with a long-standing experience with the use of the DNNF indicate that often this method generates highly variable outcomes (40).

4.6. Do the Different Calibration Methods Offer Comparable Results?

In recent years several authors have examined the suitability and agreement of the different calibration methods both in vitro and in vivo.

The in vitro performance of the different methods has been repeatedly compared. In general, good agreement among the results obtained with the different methods has been reported. For example, the estimates of E, obtained by the variation of flow rate method and the NNF method, have been found to be practically identical (38). In the same manner, the in vitro comparisons of the NNF, variation in flow rate, and slow perfusion rate methods seem to demonstrate a good correlation among the three methods (32).

When comparisons between calibration methods are conducted in the in vivo setting, the results are, in general, also coincident. Some illustrative exceptions have been, however, reported. As examples of good agreement we should mention two classical studies. In the first, Menacherry et al. (32) compared the *E* values for cocaine in brain tissue estimated by the NNF, variation in flow rate, and very slow perfusion rate methods. The authors did not detect statistically significant differences in *E* values with the three methods. Another important comparative study was published by Wang et al. in 1993 (48). These authors compared the retrodialysis and NNF methods for the in vivo calibration of MD probes. Two antiviral nucleosides (zidovudine (AZT) as the drug of interest and 3′-azido-2′,3′-dideoxyuridine (AZdU) as calibrant), which differ structurally by only a methylene group, were used in the experiments. The comparison of the results obtained with the two methods in CSF and in the thalamus ECF of rabbits showed no significant differences. Moreover, concentrations of AZT, measured directly in ventricular CSF, agreed well with those determined by NNF or retrodialysis methods.

However, there have also been studies reporting important disagreements among several calibration methods when applied to particular drugs. As an illustrative example, mention will be made of the work of Song and Lunte (51). These authors compared the delivery and NNF methods in determining the *E* values for caffeine and acetaminophen in muscle and in brain tissue. Both caffeine and acetaminophen cross the blood–brain barrier. However, while caffeine is actively transported by a saturable carrier-mediated process, passive diffusion is the only process used by acetaminophen to cross the blood–brain barrier. No differences were found between the two methods when determinations were made in muscle for either caffeine or acetaminophen. However, the *E* value determined in brain by the delivery method was higher for caffeine and lower for acetaminophen than *E* determined by the NNF method. For caffeine this discrepancy was attributed to the saturable active transport across the blood–brain barrier, which resulted in *E* being dependent upon the concentration of caffeine in brain. These results are illustrative of the difficulties of selecting a calibration method. In our opinion, this study clearly shows that any MD calibration procedure must be validated for each compound in each studied tissue.

Another illustrative example is offered by Sjöberg et al. (52). These authors compared the estimated unbound steady-state concentration of theophylline in blood and brain tissue in rats. They used the tritium method (not commented in this chapter), the low perfusion rate method, and the NNF method. Curiously, all three methods accurately predicted the steady-state theophylline concentrations. However, the *E* value obtained with the NNF method was clearly lower than 100% (the expected in vivo recovery for theophylline at a flow rate of 0.1 μl/min). As the authors pointed out, this observation did not agree with the consistent results in predicting the steady-state concentration of the drug obtained with the three methods. The question of whether active processes, such as uptake and release, influenced the slope of the regression line of the NNF method without affecting the estimation of steady-state drug concentration (as other authors have also suggested (53)) remains to be elucidated.

References

1. Brunner M, Müller M (2002) Microdialysis: an in vivo approach for measuring drug delivery in oncology. Eur J Clin Pharmacol 58(4):227–234
2. Müller M, dela Pena A, Derendorf H (2004) Issues in pharmacokinetics and pharmacodynamics of anti-infective agents: distribution in tissue. Antimicrob Agents Chemother 48: 1441–1453
3. Brunner M, Derendorf H, Müller M (2005) Microdialysis for in vivo pharmacokinetic/pharmacodynamic characterization of anti-infective drugs. Curr Opin Pharmacol 5(5):495–499
4. Langer O, Müller M (2004) Methods to assess tissue-specific distribution and metabolism of drugs. Curr Drug Metab 5(6):463–481
5. Stenken JA, Holunga DM, Decker SA, Sun L (2001) Experimental and theoretical microdialysis studies of in situ metabolism. Anal Biochem 290(2):314–323
6. Hsiao JK, Ball BA, Morrison PF, Mefford IN, Bungay PM (1990) Effects of different semipermeable membranes on in vitro and in vivo performance of microdialysis probes. J Neurochem 54(4):1449–1452
7. Tao R, Hjorth S (1992) Differences in the in vitro and in vivo 5-hydroxytryptamine extraction performance among three common microdialysis membranes. J Neurochem 59(5): 1778–1785
8. Kendrick KM (1989) Use of microdialysis in neuroendocrinology. Methods Enzymol 168:182–205
9. Johnson RD, Justice JB (1983) Model studies for brain dialysis. Brain Res Bull 10(4):567–571
10. Raei N (2000) Influence of the dialysis membrane nature on the pharmacokinetic parameters derived from microdialysis experiments. Tesis de Licenciatura, Universidad de Valencia
11. Jacobson I, Sandberg M, Hamberger A (1985) Mass transfer in brain dialysis devices: a new method for the estimation of extracellular amino acids concentration. J Neurosci Methods 15(3):263–268
12. Rosdahl H, Ungerstedt U, Henriksson J (1997) Microdialysis in human skeletal muscle and adipose tissue at low flow rates is possible if dextran-70 is added to prevent loss of perfusion fluid. Acta Physiol Scand 159(3):261–262
13. Heppert KE, Flora WH, Davies MI (1998) The importance of balancing bile salt concentration to avoid fluid loss when using the microdialysis shunt probe. Curr Separations 17:61–63
14. Moghaddam B, Bunney BS (1989) Ionic composition of microdialysis perfusing solution alters the pharmacological responsiveness and basal outflow of striatal dopamine. J Neurochem 53(2):652–654
15. Cosford RJ, Parsons LH, Justice JB Jr (1994) Effect of tetrodotoxin and potassium infusion on microdialysis extraction fraction and extracellular dopamine in the nucleus accumbens. Neurosci Lett 178:175–178
16. de Lange ECM, Danhof M, de Boer AG, Breimer DD (1994) Critical factors of intracerebral microdialysis as a technique to determine the pharmacokinetics of drugs in rat brain. Brain Res 666(1):1–8
17. Borg N, Ståhle L (1999) Recovery as a function of the osmolality of the perfusion medium in microdialysis experiments. Anal Chim Acta 379:319–325
18. Lindefors N, Amberg G, Ungerstedt U (1989) Intracerebral microdialysis: I. Experimental studies of diffusion kinetics. J Pharmacol Methods 22(3):141–156
19. Benveniste H, Hansen AJ, Ottosen NS (1989) Determination of brain interstitial concentrations by microdialysis. J Neurochem 52(6): 1741–1750
20. Nicholson C, Phillips JM (1981) Ion diffusion modified by tortuosity and volume fraction in the extracellular microenvironment of the rat cerebellum. J Physiol 321:225–257
21. Nicholson C, Syková E (1998) Extracellular space structure revealed by diffusion analysis. Trends Neurosci 21(5):207–215
22. Nicholson C (2001) Diffusion and related transport properties in brain tissue. Rep Prog Phys 64:815–864
23. Bungay PM, Morrison PF, Dedrick RL (1990) Steady-state theory for quantitative microdialysis of solutes and water in vivo and in vitro. Life Sci 46(2):105–119
24. Morrison PF, Bungay PM, Hsiao JK, Ball BA, Mafford IN, Dedrick RL (1991) Quantitative microdialysis. In: Robinson TE, Justice JB Jr (eds) Microdialysis in the neurosciences. Elsevier Science Publishers BV, Amsterdam
25. Bungay PM, Dykstra KH, Morrison PF, Dedrick RL (1991) Comparison of mathematical models of microdialysis. Curr Separations 10:106–117
26. Morrison PF, Bungay PM, Hsiao JK, Ball BA, Mefford IN, Dedrick RL (1991) Quantitative microdialysis: analysis of transients and application to pharmacokinetics in brain. J Neurochem 57(1):103–119
27. Zetterstrom T, Vernet L, Ungerstedt U, Tossman U, Jonzon B, Fredholm BB (1982) Purine levels in the intact rat brain. Studies

with an implanted perfused hollow fibre. Neurosci Lett 29(2):111–115

28. Benveniste H, Hüttemeier PC (1990) Microdialysis: theory and application. Prog Neurobiol 35:195–215
29. Parsons LH, Justice JB Jr (1992) Extracellular concentration and in vivo recovery of dopamine in the nucleus accumbens using microdialysis. J Neurochem 58(1):212–218
30. Höistad M, Chen KC, Nicholson C, Fuxe K, Kehr J (2002) Quantitative dual-probe microdialysis: evaluation of (3H)mannitol diffusion in agar and rat striatum. J Neurochem 81(1):80–93
31. Dykstra KH, Hsiao JK, Morrison PF, Bungay PM, Mefford IN, Scully MM, Dedrich RL (1992) Quantitative examination of tissue concentration profiles associated with microdialysis. J Neurochem 58(3):931–940
32. Menacherry S, Hubert W, Justice JB Jr (1992) In vivo calibration of microdialysis probes for exogenous compounds. Anal Chem 64(6):577–583
33. Parsons LH, Justice JB Jr (1994) Quantitative approaches to in vivo brain microdialysis. Crit Rev Neurobiol 8(3):189–220
34. Chen KC, Höistad M, Kehr J, Fuxe K, Nicholson C (2002) Quantitative dual-probe microdialysis: mathematical model and analysis. J Neurochem 81(1):94–107
35. Tang A, Bungay PM, Gonzales RA (2003) Characterization of probe and tissue factors that influence interpretation of quantitative microdialysis experiments for dopamine. J Neurosci Methods 126(1):1–11
36. Bungay PM, Dedrick RL, Fox E, Balis FM (2001) Probe calibration in transient microdialysis in vivo. Pharm Res 18(3):361–366
37. Stahle L (1991) The use of microdialysis in pharmacokinetics and pharmacodynamics. In: Robinson TE, Justice JB Jr (eds) Microdialysis in the neurosciences. Elsevier Science Publishers BV, Amsterdam
38. Stahle L, Segersvärd S, Ungerstedt U (1991) A comparison between three methods for estimation of extracellular concentrations of exogenous and endogenous compounds by microdialysis. J Pharmacol Methods 25(1):41–52
39. Wages SA, Church WH, Justice JB Jr (1986) Sampling considerations for on-line microbore liquid chromatography of brain dialysate. Anal Chem 58(8):1649–1658
40. Cremers TI, de Vries MG, Huinink KD, van Loon JP, v d Hart M, Ebert B, Westerink BH, De Lange EC (2009) Quantitative microdialysis using modified ultraslow microdialysis: direct rapid and reliable determination of free brain concentrations with the MetaQuant technique. J Neurosci Methods 178(2):249–254
41. Van Belle K, Dzeka T, Sarre S, Ebinger G, Michotte Y (1993) In vitro and in vivo microdialysis calibration for the measurement of carbamazepine and its metabolites in rat brain tissue using the internal reference technique. J Neurosci Methods 49(3):167–173
42. Sauernheimer C, Williams KM, Brune K, Geisslinger G (1994) Application of microdialysis to the pharmacokinetics of analgesics: problems with reduction of dialysis efficiency in vivo. J Pharmacol Toxicol Methods 32(3): 149–154
43. Zhao Y, Liang X, Lunte CE (1995) Comparison of recovery and delivery in vitro for calibration of microdialysis probes. Anal Chim Acta 316:403–410
44. Robinson DL, Lara JA, Brunner LJ, Gonzales RA (2000) Quantification of ethanol concentrations in the extracellular fluid of the rat brain: in vivo calibration of microdialysis probes. J Neurochem 75(4):685–1693
45. Ståhle L (1994) Zidovudine and alovudine as cross-wise recovery internal standards in microdialysis experiments? J Pharmacol Toxicol Methods 31(3):167–169
46. Larsson CI (1991) The use of an "internal standard" for control of the recovery in microdialysis. Life Sci 49(13):73–78
47. Wong SL, Wang Y, Sawchuk RJ (1992) Analysis of zidovudine distribution to specific regions in rabbit brain using microdialysis. Pharm Res 9(3):332–338
48. Wang Y, Wong SL, Sawchuk RJ (1993) Microdialysis calibration using retrodialysis and zero-net flux: application to a study of the distribution of zidovudine to rabbit cerebrospinal fluid and thalamus. Pharm Res 10(10): 1411–1419
49. Bouw MR, Hammarlund-Udenaes M (1998) Methodological aspects of the use of a calibrator in in vivo microdialysis-further development of the retrodialysis method. Pharm Res 15(11):1673–1679
50. Lönnroth P, Jansson PA, Smith U (1987) A microdialysis method allowing characterization of intercellular water space in humans. Am J Physiol 253(2 pt 1):228–231
51. Song Y, Lunte CE (1999) Comparison of calibration by delivery versus no net flux for quantitative in vivo microdialysis sampling. Anal Chim Acta 379(3):251–262
52. Sjöberg P, Olofsson IM, Lundqvist T (1992) Validation of different microdialysis methods for the determination of unbound steady-state concentrations of theophylline in blood and brain tissue. Pharm Res 9(12):1592–1598
53. Rollema H, Westerink B, Drijfhout WJ (1991) Monitoring Molecules in Neuroscience: Proceedings of the 5th International Conference on in Vivo Methods; Krips Repro, Meppel: The Netherlands

Chapter 7

Combination of In Vivo Microdialysis with Selective Electrochemical Detection for Online Continuous Monitoring of Brain Chemistry

Yuqing Lin, Zipin Zhang, and Lanqun Mao

Abstract

This chapter summarizes the recent development of combination of in vivo microdialysis with selective detection, especially electrochemical detection, to form novel online analytical methods for continuously monitoring brain chemistry, without the need for sample collection, pretreatment, or separation. While efficient combination of in vivo microdialysis sampling directly with selective electrochemical detection is envisaged to provide less technically demanding in vivo and online analytical methods for near-real-time monitoring physiologically important neurochemicals in the living animals, which is particularly useful for understanding the molecular basis of brain functions, the complexity in the components of the cerebral systems and the inherent features of the as-developed integrated online analytical systems (i.e., without sampling collection, pretreatment, or separation) virtually puts the selective electrochemical detection into a great challenge with respect to the selectivity, sensitivity, stability, and reproducibility. This chapter focuses on the following three aspects: (1) brief introduction of in vivo and online analytical methods developed by directly combining in vivo microdialysis sampling with selective electrochemical detection; (2) recent developments along with this line both in the method establishment and in their applications in understanding brain chemistry through interfacing electroanalytical chemistry with physiology and pathology; and (3) summary and future developments in this field.

Key words: In vivo microdialysis, In vivo electrochemistry, Online detection, Brain chemistry

1. Introduction

Effective monitoring of the release and uptake of neurochemicals from the extracellular space surrounding neurons has paved a straightforward approach to understanding of the molecular basis of brain functions (1–5). This is because, although the brain is considered to be a single organ, its internal structure and organization is an extraordinarily complicated network of specialized cells called neurons, where each neuron reserves chemical neurotransmitters.

Giuseppe Di Giovanni and Vincenzo Di Matteo (eds.), *Microdialysis Techniques in Neuroscience*, Neuromethods, vol. 75,
DOI 10.1007/978-1-62703-173-8_7,

Brain functions are essentially created, modulated, and carried out by sending messages with neurotransmitters between synapses on each neuron by making direct electrical contacts and through converting action potential (electrical signals) into chemical signals (6–8). During these processes, brain chemistry is a complex system that allows the brain to function efficiently with the use of neurochemicals known as neurotransmitters, neuromodulators, and other kinds of species. It is thus believed that variations in brain chemistry may explain how those chemicals affect the brain functions.

While the invasive methods for neurochemical monitoring with probe implantation into brain tissues have a less spatial resolution, as compared with the noninvasive techniques for imaging such as magnetic resonance imaging (MRI) and positron emission tomography (PET), they exhibit better chemical specificity and higher temporal resolution as well as cheaper instrumentation and are thus more readily suitable for continuously tracing the dynamic changes of the neurochemicals (9–12). Nowadays, it becomes widely accepted that knowledge of the dynamic change of the neurochemicals in extracellular cerebral fluid largely facilitates understanding of brain functions, for example, the mechanism for the neurotransmitters to impact target neurons and ultimately to affect physiological events (13, 14).

In general, there are two strategies for continuously monitoring the dynamic changes of the neurochemicals in vivo (15–20). One is in vivo sensing and biosensing that employs electrochemical/optical probes, electrochemical probes in particular, with a (ultra)micro size directly implantable into brain region to real-time record the change in the levels of the neurochemicals during the physiological processes. The other is in vivo microdialysis sampling coupled to subsequent sample detection. As the main part of in vivo sensing and biosensing techniques, in vivo voltammetry that employs microelectrodes implanted into brain regions to real-time monitor the neurochemicals electrochemically in the cerebral systems has been well documented and reviewed previously (21–31). This kind of methods bear advantages in high temporal resolution and minimal tissue damage due to the small size of the probes employed and have consequently been used to real-time monitor the levels of catecholamines, serotonin, and glucose in the cerebral systems. The second kind of method generally in vitro analyzes the neurochemicals sampled in vivo from brain regions, in which the sampling technique remains significantly critical. Since its establishment in 1974 by Ungersted and Pycock (32–34), in vivo microdialysis has been widely used as in vivo sampling technique with excellent capability to provide information on the neurochemistry of the extracellular microenvironment within a specific region of the brain or other organs (35–40). This technique arose historically in the field of neuroscience and has, in turn, been contributing to this field greatly. So far, microdialysis has been demonstrated to be suitable not only for in vivo sampling of the extracellular space of the brain but also for introducing exogenous biochemicals, drugs, and toxins directly into the regions of interest in pharmacokinetic and pharmacodynamic studies to investigate the effect or fate of these agents on the brain (41–45). In these studies, coupling of in vivo microdialysis with the following sample detection provided the most comprehensive information on the dynamic changes of molecules involved in cellular communication and metabolism at an integrative (whole body) level, preserving the overall physiological and behavioral functions.

The protocols for continuously monitoring the dynamic changes in the neurochemicals in vivo through combination of in vivo microdialysis with the following detection can be fallen into three categories, as typically illustrated in Fig. 1. The first kind of methods (Fig. 1a) essentially include in vivo microdialysis to in vivo sample the microdialysates from cerebral systems, sample collection and storage, sample pretreatment (for example, chemical derivatization employed for amino acids), sample separation (typically, high-performance liquid chromatography (HPLC), and capillary electrophoresis) and sample detection (for example, fluorescence, chemiluminescence, electrochemistry, and UV–vis spectroscopy) (46–49). In this kind of methods, the dialysates sampled in vivo were first subjected to pretreatment such as chemical derivatization, then injected into the separation system, and finally analyzed with the detectors. While this kind of methods include several steps for sample measurements and require sophisticated instrumentation, they have been extensively used to study the neurochemical changes in many physiological and pathological investigations, as described in other chapters. The second kind of methods are mainly based on efficient integrating of in vivo microdialysis sampling, sample collection and separation, as well as detection into one line to form integrated online analytical systems (Fig. 7.1b) (50–66). Unlike the first kind of methods in which the dialysates were manually (in most cases) injected into the separation/detection system, the second kind of methods automatically perform the dialysates flow into the capillary and utilize capillary electrophoresis for sample separation with an improved temporal resolution, as systematically studied by Kennedy et al. (50–59). While the online analysis of the dialysates with this kind of methods was mostly carried out discontinuously, rather than continuously, this elegant strategy yet provided comprehensive information on the near-real-time change in the levels of the neurochemicals, as described in the chapters contributed from Robert Kennedy. Unlike the first two kinds of methods, the third kind of methods efficiently and directly combine in vivo microdialysis with online selective detection (Fig. 7.1c) (67, 68). That is to say, the dialysates in vivo sampled from the cerebral systems are directly perfused into the detector for near-real-time and continuous detection. The procedures for sample

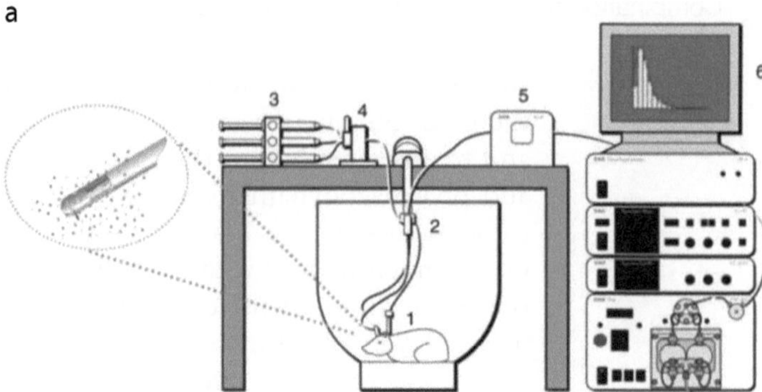

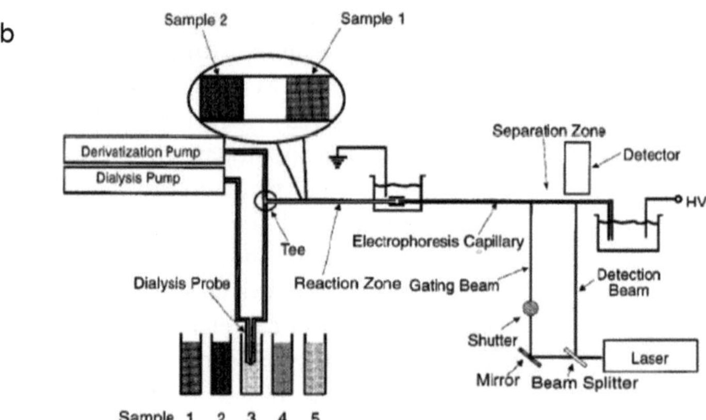

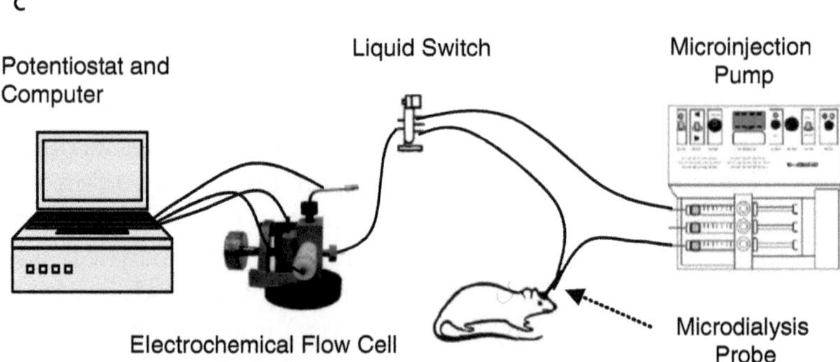

Fig. 1. Typical illustration of three kinds of strategies for monitoring the dynamic changes in the levels of the neurochemicals through combination of in vivo microdialysis with the following detection: (**a**) Arrangement for the neurochemical measurements with combination of in vivo microdialysis sampling with off-line separation and detection. The system was normally formed by in vivo microdialysis (*1*), dual-channel microdialysis swivel, head block tether, and lever arm (*2*), syringe pump (*3*), syringe switch (*4*), fraction collector (*5*), and HPLC separation system with detectors (*6*). The figure was adapted from (48), and reproduced by permission of the Royal Society of Chemistry. (**b**) Microdialysis coupled online with capillary electrophoresis and laser-induced fluorescence detection (CE-LIF). Figure shows the fluidic connections for sampling, online derivatization, and transfer to separation capillary. Direction of flow was from the pumps, through the dialysis probe, into the tee, and onto the reaction and separation capillaries. The figure was reprinted with permission from (59). Copyright 1999 American Chemical Society. (**c**) Online analytical methods for continuously monitoring neurochemicals in the brain of freely moving rats through directly integrating in vivo microdialysis with online selective electrochemical detection. The figure was reprinted from (122). Copyright (2009), with permission from Elsevier.

collection, pretreatment, and separation involved in the first two kinds of methods are avoided in this kind of methods. This kind of methods are complementary with the first two protocols in terms of the kinds of neurochemicals studied. For example, the involvement of separation procedure in the first two protocols generally makes them very useful for selective detection of multiple neurochemicals with similar electrochemical and optical properties, such as amino acids and catecholamines (49, 69–74). The third one that can continuously and selectively track one kind of neurochemicals, such as glutamate (75) and ascorbate (76), and may be extended for simultaneously monitoring several kinds of neurochemicals, e.g., glucose and lactate (77) provided the uses of highly selective and cross talk-free sensing and biosensing array. As mentioned above, the avoidance of sample collection, pretreatment, and separation in the third kind of methods greatly simplifies the instrumentation and determination procedures and largely shortens the analytical time and, as a consequence, makes the methods more efficient for continuously tracing the change in the levels of the neurochemicals involved in the physiological and pathological processes. This chapter mainly aims at the third kind of protocols for continuously monitoring neurochemicals through direct combination of in vivo microdialysis with online selective electrochemical detection, with regard to the establishment of this kind of online analytical methods and their implications in physiological and pathological investigations. Emphasis of this chapter is given both on the challenges and their solutions in the method development and on the recent advances in this field.

2. Recent Developments of Online Analytical Methods

As mentioned above, the online analytical methods developed by combining in vivo microdialysis directly with selective electrochemical detection essentially abandon the procedures for sample collection, pretreatment, and separation. Although this abandonment greatly simplifies the instrumentation and procedures and shortens the analysis time, while still providing near-real-time information on the dynamic change in the neurochemicals, it substantially puts the selective electrochemical detection into a great challenge, which can be roughly summarized as follows. (1) Since the methods do not include a sample separation procedure, they have to exhibit a good selectivity against other kinds of electroactive neurochemicals coexisting in extracellular cerebral fluid, e.g., ascorbic acid (AA), uric acid (UA), catecholamine, and their metabolites. The selectivity for the neurochemicals has generally been achieved by the uses of biorecognition elements such as enzymes and of functional nanostructures with unique electronic and structural properties,

(2) since most kinds of neurochemicals are present in the cerebral systems with relatively low levels and the levels are further diluted with in vivo microdialysis sampling. These facts actually require the electrochemical methods employed for the continuous monitoring of the neurochemicals to have a high sensitivity with a very low limit of detection; and (3) due to the chemical complexity of the dialysates, the electrode surface is normally rationally functionalized with, for example, biorecognition elements or functional nanostructures, to achieve the selectivity for the online continuous monitoring of the neurochemicals. This, along with the long-term operation of the electrochemical detector in a continuous-flow system during animal experimentation, strongly necessitates the electrochemical methods to have a good stability and reproducibility. This requirement is quite different from that for the electrodes used in the former two kinds of methods (Fig. 1a, b) since, in most cases, no critical surface functionalization is applied for the electrodes used in those methods. The challenges and their solutions inherent in the electrochemical methods for the selective detection directly coupled with in vivo microdialysis essentially vary with the kinds of the neurochemicals of interest and will be demonstrated in more details in the following parts.

2.1. Neurotransmitters

2.1.1. Glutamate

Glutamate acts as one of the main excitatory neurotransmitters and plays an important role in the functioning of mammalian nervous system (78, 79). Highly sensitive and selective measurement of this kind of neurotransmitter is strongly required for the investigations on various physiological and pathological phenomena, such as the signal transmission mechanism involving learning and memory at the synaptic or cell level, and neurological diseases such as ischemia. The concentration of glutamate in the extracellular space of the brain varies between 4 and 350 μM with the dialysate level varying between 1 and 10 μM (80, 81). Similar to that for other kinds of amino acids, the measurement of glutamate has been carried out by pre-column derivatization, for example, with the fluorophore *o*-phthaldialdehyde, and HPLC separation followed by fluorescence detection (46, 49). While the involvement of HPLC separation endowed this method with a high selectivity, the measurement of glutamate was yet complicated in chemical derivatization and required column switching with multiple HPLC pumps. Recently, glutamate was also measured by capillary electrophoresis with a high sensitivity, but the method still required sample derivatization (51–53, 82).

As mentioned above, the continuous monitoring of glutamate in the cerebral systems through directly integrating in vivo microdialysis sampling with selective electrochemical detection virtually necessitates high selectivity for the detector. Since glutamate in itself is electrochemically inactive, its selective electrochemical detection has been mainly based on the uses of enzymes (i.e., glutamate oxidase

(GluOx) and glutamate dehydrogenase (GluDH)) both as the biorecognition elements and as the elements to transit glutamate into electrochemically detectable species to produce current signal readout through the mechanisms shown below:

$$\text{Glutamate} + O_2 + H_2O \xrightarrow{\text{GluOx}} \alpha\text{-Ketoglutarate} + NH_4^+ + H_2O_2 \quad (1)$$

$$\text{Glutamate} + NAD^+ \xrightarrow{\text{GluDH}} \alpha\text{-Ketoglutarate} + NH_4^+ + \text{NADH}. \quad (2)$$

Although both GluOx (1) and GluDH (2) can generally be used to construct electrochemical biosensors as the detector for continuously monitoring glutamate, GluOx has been employed more frequently than GluDH since GluOx is more robust and more readily available from commercial sources. Moreover, the uses of GluOx to catalyze the oxidation of glutamate with oxygen as the electron acceptor produces hydrogen peroxide (H_2O_2) and H_2O_2 can be electrochemically oxidized or reduced to generate electronic signal readout for the measurements of glutamate. The uses of GluDH to catalyze the oxidation of glutamate necessitate the presence of NAD^+ cofactor that is difficult to confine onto electrode surface to finally form an integrative electrochemical biosensor as the online selective detector for continuously monitoring glutamate in a continuous-flow system.

Zilkha et al. developed an online analytical method for continuously monitoring glutamate in the dorsolateral striatum and the parietal cortex in the rat brain by combining in vivo microdialysis with selective electrochemical detection based on GluOx-based electrochemical biosensor for glutamate. GluOx was co-immobilized onto electrode surface with the electropolymerized film of 1,2-diaminobenzene and the detection was carried out through the direct oxidation of the produced H_2O_2 (83). The interference from electroactive AA was eliminated with the polymerized film of 1,2-diaminobenzene used. The resulted high sensitivity and fast response time of the detector substantially made the analytical method very suitable for investigating conditions that produce rapid changes in brain extracellular glutamate, which was illustrated by monitoring the changes in extracellular glutamate subsequent to cardiac arrest, and K^+-induced local depolarization.

Niwa et al. established an effective online analytical method for continuously monitoring glutamate by integrating in vivo microdialysis with a small-volume L-glutamate oxidase enzymatic reactor (0.75 mm i.d. and 2.5 cm long) that was coupled with an electrochemical detector in a thin-layer radial flow cell with an active volume of 70–340 nL (75). They used glassy carbon bulk or carbon film ring-disk electrodes modified with Os poly(vinylpyridine) mediator containing horseradish peroxidase (Os-gel-HRP) as the detector for the determination of H_2O_2 produced from the enzymatic reactor. The overall efficiency of glutamate detection with the sensor reached 94% under optimum conditions due to an efficient enzymatic reaction in the reactor and high conversion efficiency in the radial flow cell, as reported by the authors. The sensitivity for the detection was 24.3 nA/μM and the detection limit was 7.2 nM (S/N = 3). The effect of other kinds of the neurochemicals, such as AA, was minimized effectively by applying a low potential to the electrode for H_2O_2 detection and via the ring-disk electrode geometry by using the disk for pre-oxidation. The validity of this method for online continuous monitoring of glutamate was demonstrated by monitoring glutamate concentration changes at the submicromolar level caused by KCl stimulation of a single nerve cell and micromolar glutamate concentration increases caused by electrical stimulation of a brain slice.

To eliminate the interference from AA and thus develop an online analytical method for continuous monitoring of glutamate, Boutelle et al. proposed an effective approach to removal of AA by introducing a pre-oxidation system in the upstream of amperometric biosensor for glutamate to consume AA in the sample (84). The biosensor was prepared by co-immobilizing GluOx with the electropolymerized film of *o*-phenylenediamine. The ability of the pre-oxidation system, combined with property of the electropolymerized film to block out electroactive interferents like AA and UA, eventually resulted in a virtually interference-free glutamate assay with a low detection limit of 0.3 μM in the presence of physiological levels of interferents, a sensitivity of 4 nA/μM, and a linear region up to 30 μM. By doing so, the authors were able to establish a new online analytical system that was proved to be very effective for continuously monitoring glutamate in the brain with a temporal resolution of <1 min.

As an advance for eliminating AA interference, Hayashi et al. developed a microfluidic device consisting of two glass plates integrated with a nanoliter-volume enzyme pre-reactor and an enzyme-modified electrode for the highly selective continuous measurement of glutamate (85). The working electrode was modified with a bilayer film of Os-gel-HRP/GluOx on one plate. The other plate had a thin-layer flow channel integrated with a pre-reactor with a number of micropillars (20 mm in diameter, 20 mm high, and separated from each other by a 20 mm gap) modified with ascorbate oxidase (AAOx) to biochemically oxidize AA. The use of AAOx-based biochemical peroxidation of AA to replace the previous electrochemical pre-oxidation step, on one hand, simplified the experimental procedure and, on the other hand, effectively eliminated the great interference from AA since the high operation potential employed for electrochemical pre-oxidation step might produce unknown electroactive species, which also caused interference at the detection electrode. As a result, the authors successfully developed an online analytical method that could detect 1 μM glutamate continuously at a low flow rate, even when AA concentration was 100 mM.

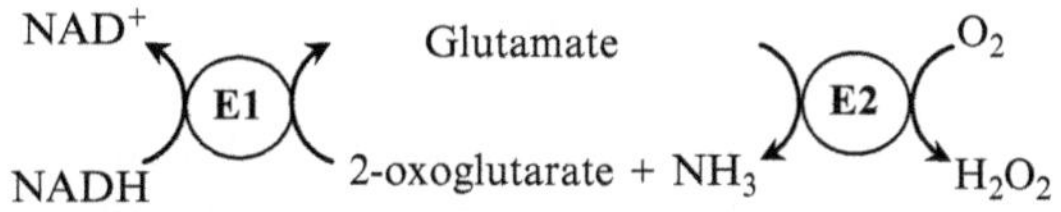

Fig. 2. Enzymatic recycling model of glutamate with GluOx and GluDH co-immobilized into a reactor to enhance the sensitivity for the online continuous measurement of glutamate. E1 = GluDH, E2 = GluOx.

Different from the determination scheme demonstrated above, Yao et al. reported an online method for continuously monitoring glutamate with an enhanced sensitivity through substrate recycling (86) (Fig. 2). The system consisted of a microdialysis probe, immobilized enzyme reactor, and poly(1,2-diaminobenzene)-coated platinum electrode. The enzyme reactor was prepared by the co-immobilization of GluOx and GluDH in a PTFE reactor to recycle glutamate. The produced H_2O_2 was detected with a poly (1,2-diaminobenzene)-coated platinum electrode positioned in a downstream, without interference from oxidizable species, such as AA, in the sample and NADH added to the carrier buffer. Since the cycle was also initiated with 2-oxoglutarate, saccharopine dehydrogenase reactor was positioned in series before the amplifier reactor to remove 2-oxoglutarate in the dialysate. With such a protocol, the authors were able to obtain 160-fold increase in sensitivity for glutamate measurement, as compared with the unamplified responses.

2.1.2. Dopamine

Dopamine (DA) belongs to one kind of the most important neurotransmitters called catecholamines and constitutes about 80% of the catecholamine content in the brain. It is well established that DA plays a central role in aspects of incentive motivation, such as learning and memory (87–89). Reliable measurement of extracellular DA concentration is thus important in understanding the physiological processes in the central nervous systems. Among the methods reported so far for DA monitoring, electrochemical methods including HPLC (47, 90) or CE (50, 54) separation coupled with electrochemical detection and in vivo voltammetry that employs microelectrode techniques represent the most attractive due to the good electrochemical property of this kind of neurotransmitter.

The main challenge underlying the online continuous monitoring of DA by directly combining in vivo microdialysis with selective electrochemical detection is the method selectivity since other kinds of electroactive species, such as AA, UA, and 3,4-dihydroxyphenylacetic acid (DOPAC, DA metabolite), are oxidized at potentials very close to those for DA oxidation. Moreover, unlike other kinds of the neurochemicals, such as glutamate and glucose, there are no enzymes that can specifically catalyze the DA oxidation or reduction. While monoamine oxidase (MAO) catalyzes the oxidation of DA, the activity of this kind of enzyme is not specific: other kinds of amine-containing neurotransmitters including serotonin, norepinephrine, and epinephrine can also be oxidized under the catalysis of MAO. On the other hand, although the good electrochemical properties of some kinds of recently developed nanostructures, such as carbon nanotubes (CNTs), substantially enabled the selective detection of some kinds of physiologically important species, such as AA (31), the structural similarity between DA and other kinds of catecholamines unfortunately invalidates such a protocol for selective DA detection through an oxidative electrochemical approach.

By taking advantage of the sequential chemical reaction properties of DA including oxidation (3), deprotonation (4), intramolecular cyclization (5), and disproportionation reactions (6), to finally give 5,6-dihydroxyindoline quinone, and of the two-electron and two-proton reduction of the formed 5,6-dihydroxyindoline quinone (reaction 7), Mao et al. recently developed a new electrochemical method for selective determination of DA based on the reduction of 5,6-dihydroxyindoline quinone at a negative potential that was totally separated from that for the redox processes of AA and DOPAC (91).

(-2H$^+$-2e, or Laccase/O_2) (3)

NH_3^+ OH OH ⇌ NH_3^+ O O

-H$^+$ (4)

NH_3^+ O O ⇌ NH_2 O O

k_{cyc} (5)

NH_2 O O → N H OH OH

NH_3^+ O O + N H OH OH → NH_3^+ OH OH + N H O O (6)

(-2H$^+$-2e, or Laccase/O_2) (7)

N H OH OH ⇌ N H O O

To construct an electrochemical biosensor for selective determination of DA, laccase, one kind of multi-copper blue oxidases, was confined onto CNTs and used as the biocatalyst to drive the first oxidation of DA (3) into its quinone form and thus initialize the sequential reactions of DA (4)–(5) to finally give 5,6-dihydroxyindoline quinone. In addition, laccase also catalyzes the oxidation of AA and DOPAC into electroinactive species with the concomitant reduction of O_2. As a consequence, the combined uses of the chemical properties inherent in DA and the multifunctional catalytic properties of laccase as well as the excellent electrochemical properties of CNTs substantially enabled the prepared biosensors to be well competent for the selective determination of DA with the coexistence of physiological levels of AA and DOPAC.

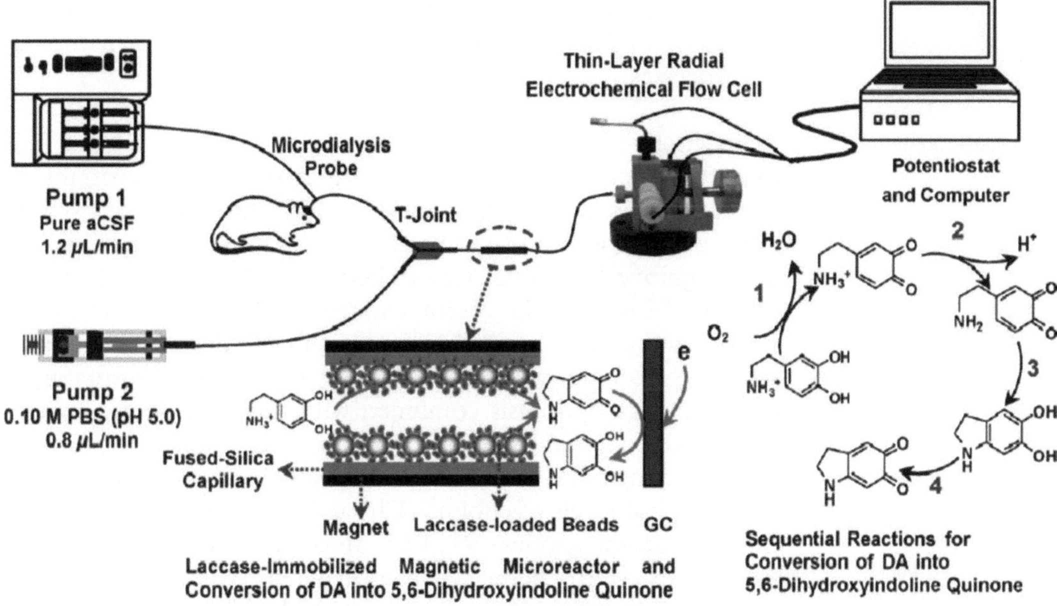

Fig. 3. Illustration of online analytical methods for continuous measurement of DA based on a non-oxidative electrochemical approach with integration of in vivo microdialysis, laccase-immobilized magnetic microreactor, and electrochemical detection. The figure was reprinted from (92). Copyright (2010), with permission from Elsevier.

To successfully combine the selective electrochemical biosensor with in vivo microdialysis to form a new online method for continuously monitoring DA, Mao et al. immobilized laccase enzyme onto magnetite nanoparticles and the nanoparticles were confined into a fused-silica capillary through an external magnetic field to fabricate a magnetic microreactor (92). The microreactor was placed in the upstream of the thin-layer electrochemical flow cell to efficiently catalyze the oxidation of DA into its quinonoid form and thereby initialize the sequential reactions including deprotonation, intramolecular cyclization, disproportionation, and/or oxidation to finally give 5,6-dihydroxyindoline quinone. The electrochemical reduction of the produced 5,6-dihydroxyindoline quinone at bare glassy carbon electrode was used as the signal readout for the DA measurement (Fig. 3). The authors found that the uses of the laccase-immobilized microreactor well eliminated the interference from AA and DOPAC. Moreover, the successful transition of the mechanism for DA detection from the conventional oxidative electrochemical approach to the non-oxidative one substantially enabled the online measurements to be interference-free from physiological levels of UA, 5-hydroxytryptamine, norepinephrine, and epinephrine. The current response was linear with DA concentration within a concentration range from 1 to 20 μM with a sensitivity of 3.97 nA/μM. The detection limit, based on a signal-to-noise ratio of three, was calculated to be 0.3 μM. The high selectivity and the good linearity as well as the high stability of the online analytical method made it very potential for continuous monitoring of cerebral DA release in physiological and pathological processes.

2.1.3. Acetylcholine

Acetylcholine (ACh) has been known as one kind of the most important neurotransmitters that are involved in neurotransmission processes in both the peripheral and central nervous systems (93). Besides, it also plays important roles in neurological diseases and cognitive functions. Researchers have demonstrated that the decrease of ACh in the cerebral system made individuals prone to various nerve disorders including Alzheimer's disease and multiple sclerosis. To date, the methods using liquid chromatography or capillary electrophoresis combined with different detectors, typically electrochemical detector, have been developed for selective measurement of ACh in the cerebral systems with coexistence of physiological levels of choline (Ch) and other kinds of physiologically important species (94, 95). Since ACh is electrochemically inactive, its electrochemical detection generally necessitated a post-column enzyme reactor to convert ACh into electrochemically detectable species. This, along with the involvement of separation procedure for differentiating ACh in the extracellular fluid, on one hand, complicated the analytical systems and, on the other hand, lowered the temporal resolution of the analysis. In this regard, online continuous monitoring of ACh in brain tissue, without sample separation, is highly desired.

Niwa et al. reported a small-volume online sensor that could continuously monitor extracellular ACh selectively in a brain tissue culture by using a small enzymatic pre-reactor (22 μL inner volume) immobilized with choline oxidase (ChOx) and catalase in series to remove choline (Ch) from the sample and thereby eliminate its interference (8)–(9) (96). Carbon electrodes were modified with an acetylcholine esterase (AChE), ChOx, and Os-gel-HRP to selectively detect ACh, through the reactions (10)–(11). The sensor sensitivity was 43.7 nA/μM for ACh under optimized conditions. Almost no response was seen when 100 μM Ch was continuously injected. The detection limit for ACh with the sensor was comparable to that obtained using liquid chromatography with electrochemical detection combined with an enzymatic reactor. The application of such an analytical system was demonstrated by continuously monitoring ACh release from cultured brain tissue. The extracellular ACh increase of 20 nM was observed continuously with the online sensor combined with a microcapillary sampling probe located very close to the tissue after electrical stimulation to cultured rat hippocampal tissue:

$$\mathrm{Ch} + \mathrm{O_2} + 2\mathrm{H_2O} \xrightarrow{\mathrm{ChOx}} \text{Betaine aldehyde} + 2\mathrm{H_2O_2} \quad (8)$$

$$2\mathrm{H_2O_2} \xrightarrow{\mathrm{Catalase}} 2\mathrm{H_2O} + \mathrm{O_2} \quad (9)$$

$$\mathrm{ACh + H_2O \xrightarrow{AChE} Ch + CH_3COOH} \quad (10)$$

$$\mathrm{Ch + O_2 + 2H_2O \xrightarrow{ChOx} Betaine\ aldehyde + 2H_2O_2}. \quad (11)$$

2.1.4. γ-Aminobutyric Acid

γ-Aminobutyric acid (GABA) is the main inhibitory neurotransmitter in the mammalian central nervous system and plays a critical role in regulating neuronal excitability throughout the nervous system (97). Similarly to the glutamate assays mentioned above, GABA measurement has been performed by pre-column derivatization with the fluorophore *o*-phthaldialdehyde, HPLC separation, followed by fluorescence detection because it is insensitive either to electrochemical or spectroscopic methods (46, 49). On the other hand, different from other kinds of the neurochemicals like glutamate, ACh, or glucose that could be easily converted into electrochemically detectable species through an enzymatic pathway, it is difficult to construct GABA biosensors in combination with an enzymatic reaction because neither oxidase nor dehydrogenase can be available for this kind of neurotransmitter.

For the first time, Niwa et al. developed an online analytical method that could continuously monitor GABA in the cerebral system with the uses of Gabase and GluOx. Gabase mainly contains two kinds of enzymes, γ-aminobutyrate ketoglutarate aminotransferase (GABA-T) and succinic semialdehyde dehydrogenase (SSDH) (98). This kind of enzymes convert GABA into glutamate (12). Glutamate could then be specifically oxidized into α-ketoglutarate and ammonia in the presence of GluOx (13). The enzymatic reaction produces H_2O_2 that was electrochemically detectable either through direct oxidation or HRP-catalyzed reduction reactions. To construct an online biosensor that was used in a continuous-flow system for continuously monitoring GABA, the authors immobilized GluOx and catalase into a small-volume enzymatic reactor to completely deplete glutamate endogenously existing in the cerebral systems (Fig. 4). A glassy carbon (GC) electrode modified with a top-layer film consisting of gabase and GluOx and a bottom-layer film containing Os-gel-HRP was used as the biosensor for GABA. The as-prepared tri-enzyme-based biosensor showed a low detection limit (0.1 μM) and good linearity (0.1–10 μM) toward GABA in a continuous-flow system. Moreover, the system exhibited a sensitivity of 1.56 nA/μM for GABA, and almost no response toward 10 μM glutamate under a continuous-flow condition. AA interference was also well suppressed either by further spin-coating of Nafion film on the surface of the biosensor or shifting the operating potential toward a negative direction, e.g., −0.10 V (vs. Ag/AgCl). GABA was measured without adding α-ketoglutarate when the glutamate concentration in the sample was high because the enzymatic reaction of glutamate in the pre-reactor generates α-ketoglutarate that was needed for the enzymatic transformation of GABA to glutamate. The excellent

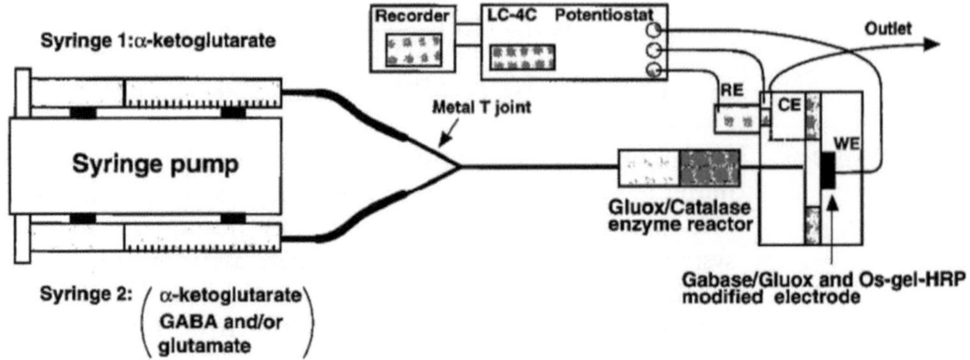

Fig. 4. Schematic illustration of the online analytical system for continuously monitoring GABA with the uses of Gabase and GluOx for the selective detection of GABA. The figure was reprinted with permission from (98). Copyright 1998 American Chemical Society.

properties of this method strongly suggest its potential application in continuous measurement of GABA in cerebral microdialysate:

$$\text{GABA} + \alpha\text{-Ketoglutarate} \xrightarrow[\text{SSDH}]{\text{GABA - T}} \text{SuccinicSemiadehyde (SSA)} + \text{Glutamate} \quad (12)$$

$$\text{Glutamate} + O_2 + H_2O \xrightarrow{\text{GluOx}} \alpha\text{-Ketoglutarate} + NH_4^+ + H_2O_2. \quad (13)$$

2.2. Glucose and Lactate

Glucose is the primary fuel used by most cells in the body to generate the energy that is needed to carry out cellular functions (99, 100). Lactate is a three-carbon compound that is produced when insufficient oxygen is present in cells to break down pyruvate to acetyl CoA. Glucose extracellular levels reflect a sensitive balance between glucose delivery from blood and glucose consumption by the brain (99, 100). This balance is maintained by a sophisticated chemical control mechanism that can itself go wrong. In the injured brain, the efficiency of this coupling between likely abnormal glucose use and possibly inadequate glucose supply is vital. Consequently, extracellular glucose levels give a sensitive indication at any moment of whether energy demand is being met. Lactate derives from pyruvate, a metabolite in the main glucose metabolic pathway. Lactate production occurs pathologically during the failure of oxygen supply, but it also occurs in the presence of oxygen as a quick method of producing energy (101). Simultaneous measurements of glucose and lactate in the central nervous system are of great physiological and pathological importance because the concentration of both species in extracellular fluid has been widely used as an index of brain activity closely linked to brain energy metabolism and further as a diagnostic tool in acute human brain insults, such as stroke and head trauma (99–101).

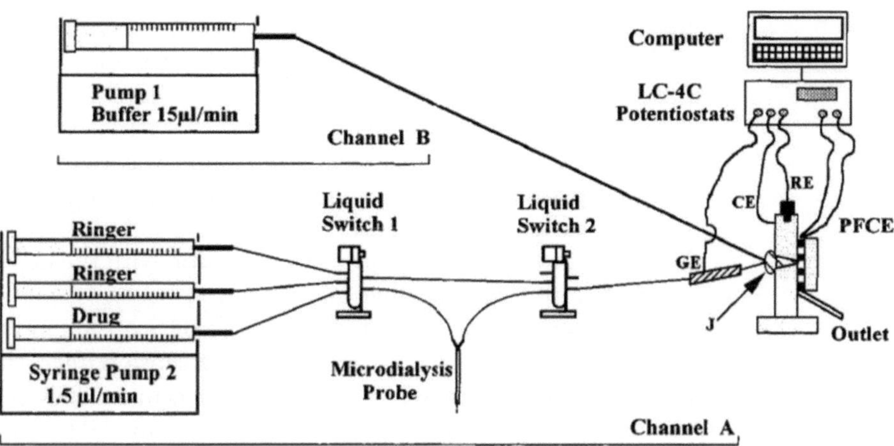

Fig. 5. Representation of online analytical system for simultaneous and continuous measurements of lactate and glucose in the rat brain with oxidase-based electrochemical biosensors as the selective detectors. The figure was reprinted with permission from (77). Copyright 1998 American Chemical Society.

The basal concentrations of glucose and lactate in cerebral systems are at millimolar level. Electrochemical biosensors for the determination of both species are generally based on the uses of enzymes, i.e., glucose oxidase (GOx) and lactate oxidase (LOx), or glucose dehydrogenase (GDH) and lactate dehydrogenase (LDH), which specifically catalyze the oxidation of glucose and lactate, in the presence of dissolved oxygen or NAD^+ as electron donor. Although both kinds of enzymes, i.e., oxidases and dehydrogenases, can be generally used for biosensor construction, the biosensors based on oxidases are more frequently employed for the practical online electrochemical measurements, as compared with those based on dehydrogenases. This is because the latter kind of biosensors necessitate the presence of NAD^+ cofactor that is difficult to confine onto electrode surface, as mentioned above.

By using oxidases and ring and split-disk plastic film carbon electrodes (PFCEs), Osborone et al. developed an online analytical method for simultaneously and continuously monitoring glucose and lactate in the rat brain through integration of in vivo microdialysis and online selective electrochemical detection (77, 102) (Fig. 5). To construct the online biosensors for glucose and lactate, PFCEs were coated initially with Os-gel-HRP and then with a second coating of oxidase enzyme to produce enzyme bilayer (oxidase/Os-gel-HRP). The electrodes were subsequently over-coated with cellulose acetate for use in the determination of glucose and lactate at 0 V (vs. Ag/AgCl). A low-volume platinum generator electrode maintained at +0.30 V (vs. Ag/AgCl) was positioned upstream of the detector to pre-oxidize AA and provide diagnostic information about the flow of the dialysate. The uses of split-disk electrode eventually enabled different kinds of oxidase enzymes to be immobilized on each half of a split-disk, Os-gel-HRP-coated PFCE to facilitate the electrochemically independent yet continuous online measurements of these two analytes from a single dialysate, with minimal cross talk between partner split-disk electrodes. The utility of this analytical system was demonstrated by the quantitative online continuous assay of the changes in dialysate striatum extracellular glucose and lactate from a conscious rat during local stimulation of neurons by perfusion with the depolarizing agent, Veratridine, and physical restraint.

Niwa et al. fabricated a microfluidic device integrated with a micro-pre-reactor and glucose- and lactate-sensing microelectrodes in a dual channel that was coupled to in vivo microdialysis for continuously monitoring glucose and lactate (103, 104). The device consists of two glass plates bonded together with UV-curable resin. The microelectrodes were made of carbon film, and a thin-layer flow channel and a flow separator between the glucose and lactate electrodes. One carbon electrode was modified with a bilayer of GOx/Os-gel-HRP for glucose biosensing and the other was modified with a bilayer of LOx/Os-gel-HRP for lactate biosensing. The interference from AA was totally eliminated by incorporating AAOx into the device upstream of the thin-layer flow channel. Besides, no cross talk between the glucose and lactate biosensors was observed under the conditions employed by the authors. By combining the device with a microdialysis probe, a linear relationship between the currents and the concentrations of glucose and lactate was achieved within a concentration range from 5 μM to 5 mM. The validation of such a method for continuously monitoring glucose and lactate was illustrated by monitoring the decrease in the glucose concentration and the increase in the lactate concentration in the rat brain following perfusion of veratridine solution into the rat cortex.

As demonstrated above, the selective electrochemical detection with the oxidase-based electrochemical biosensors has been generally accomplished with the detection of H_2O_2 produced from the enzyme reactions as the signal readout. To maintain the selectivity and efficient electrocatalysis for the H_2O_2, HRP has been widely employed and such an HRP-based bioelectrocatalytic scheme has been well coupled to the oxidase-based biocatalytic system toward the target to generate a large variety of electrochemical biosensors that could be used as online selective detection to form new analytical systems for continuously monitoring neurochemicals. However, due to the difficulties in conducting HRP onto electrode and the similar bioelectrochemical catalytic activity of HRP toward O_2 and H_2O_2 through a direct electron transfer of this kind of peroxidase, an electron transfer mediator has been always involved in the bioelectrocatalysis of H_2O_2 in the oxidase-based biosensing systems. While the strategies for HRP-based bioelectrocatalysis toward H_2O_2 have been employed for continuously monitoring some kinds of neurochemicals, such as ACh (96), glutamate (75), and glucose/lactate (77), with oxidase-based electrochemical biosensors as the online selective detectors, the involvement of such strategies in the establishment of online analytical methods essentially made these methods technically complicated, requiring co-immobilization of redox mediators and HRP onto the electrode surface, and suffer from the risks with interference from AA since HRP has been reported to be able to catalyze the reaction between AA and H_2O_2.

To circumvent such problem, Mao et al. recently demonstrated an effective method to catalyze the reduction of H_2O_2 through a non-HRP pathway at a potential more positive than that for O_2 reduction (67). They used Prussian blue (PB) to accomplish such a pursuit due to the mostly highest catalytic activity of PB toward H_2O_2 reduction, among almost all kinds of electrocatalysts reported so far. As a result, the use of such kind of the so-called artificial peroxidase (i.e., PB) to replace the "natural peroxidase" (i.e., HRP) essentially offered a facile but reliable and durable electrochemical method for simultaneous and online measurements of glucose and lactate in rat brain, combining with oxidases (i.e., GOx and LOx). The simultaneous and online measurements were virtually interference-free from AA and other electroactive species coexisting in the brain. Moreover, the dual-oxidase/PB-based biosensors suffered from little cross talk and exhibited a good stability and reproducibility. With such an online analytical method, the basal levels of glucose and lactate in the microdialysate from the striatum of the freely moving rats were determined to be 200 ± 30 and 400 ± 50 μM ($n = 3$), respectively.

Boutelle et al. successfully extended the online analytical method for continuously monitoring glucose and lactate for clinical brain-injury patients, not only for laboratory investigations (105–111). The assay was based on mediated bi-enzyme beds with electrochemical detection and was specifically designed for use with head trauma patients in neurointensive care (Fig. 6). The simultaneous measurements of the levels of glucose and lactate in the dialysate were demonstrated to have a high time resolution (30-s or even 15-s sampling intervals). The rapid nature of this dual-online analytical system made it valuable in studying the roles of energy metabolites in the brain extracellular fluid and as an important new method for routine clinical monitoring of patients in the neurointensive care unit. The advantage of online monitoring was that the results were continuously displayed for use by the clinical staff in guiding patient care. As an example for clinical application, the authors tried the clinical measurements with such a system during neurosurgery to clip an aneurysm.

While the oxidase-based electrochemical biosensors have been proved to be well competent as the selective detector in the online analytical systems, such a detecting system is essentially O_2 dependant.

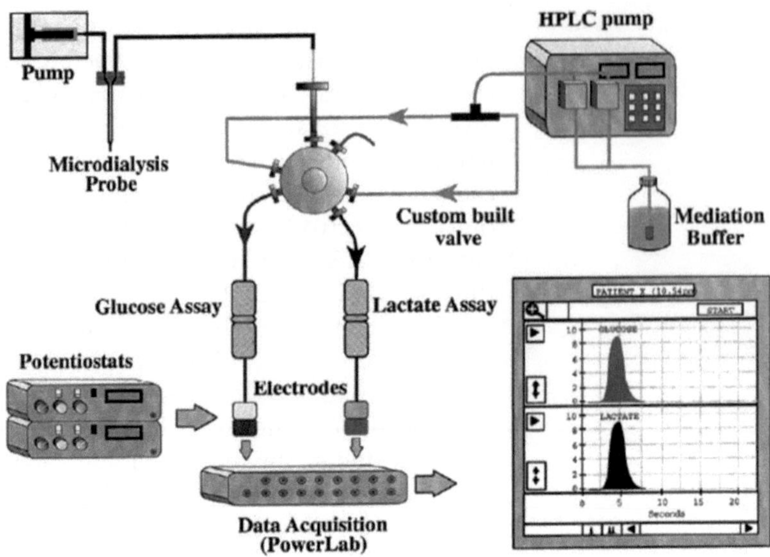

Fig. 6. Diagram of the online analytical system for simultaneous detection of glucose and lactate in the dialysate from patients. The figure was reprinted from (105). Copyright (2002), with permission from Elsevier.

This property of the analytical systems unfortunately makes them not well competent for continuously monitoring neurochemicals under some kinds of pathological conditions, for typical example, brain ischemia/reperfusion, since brain ischemia/reperfusion normally occurs with a large fluctuation in the brain anoxia, resulting in the change in the O_2 level in the brain (112, 113). Such a change substantially renders difficulties in applying oxidase-based electrochemical biosensors for monitoring glucose and lactate following the cerebral ischemia/reperfusion because the current responses of this kind of biosensors based on the detection of O_2 consumption, H_2O_2 production, or mediator regeneration are all O_2 dependent. To address this problem, Mao et al. recently developed the dehydrogenase-based electrochemical biosensors for glucose and lactate online measurements (68). The two biosensors were developed onto the dual split-disk plastic carbon film electrodes with methylene green adsorbed onto single-walled carbon nanotubes (SWNTs) as the electrocatalyst for the oxidation of β-nicotinamide adenine dinucleotide (NADH) at a low potential of 0.0 V (vs. Ag/AgCl). Artificial cerebrospinal fluid (aCSF) containing NAD^+ was externally perfused from a second pump and online mixed with the brain microdialysates to minimize the variation of pH that occurred following the cerebral ischemia/reperfusion and to supply NAD^+ cofactor and O_2 for the enzymatic reactions of dehydrogenases and ascorbate oxidase, respectively, as shown in Fig. 7. The online analytical method was demonstrated to exhibit a high selectivity against the electrochemically active species endogenously existing in the

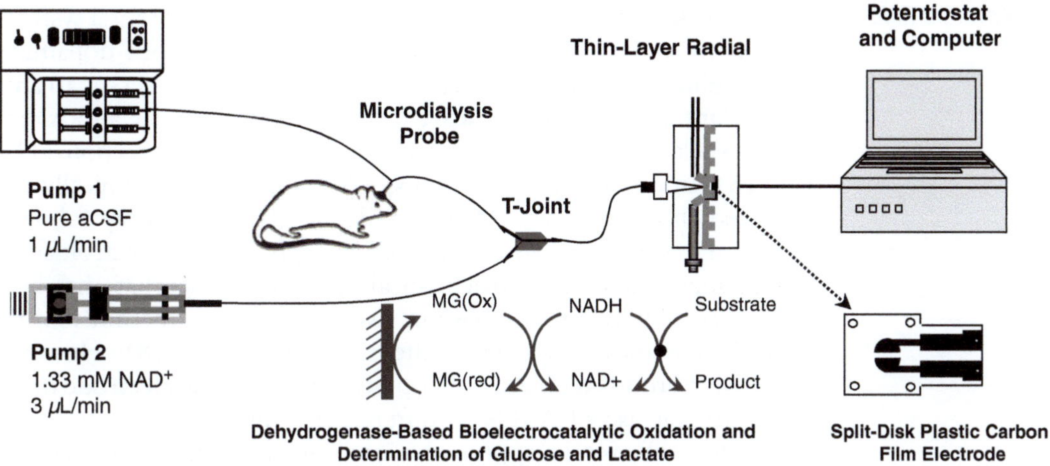

Fig. 7. Diagram of a physiologically relevant online analytical system for continuous and simultaneous monitoring of striatum glucose and lactate following global cerebral ischemia/reperfusion with dehydrogenase-based electrochemical biosensors as the selective detectors.

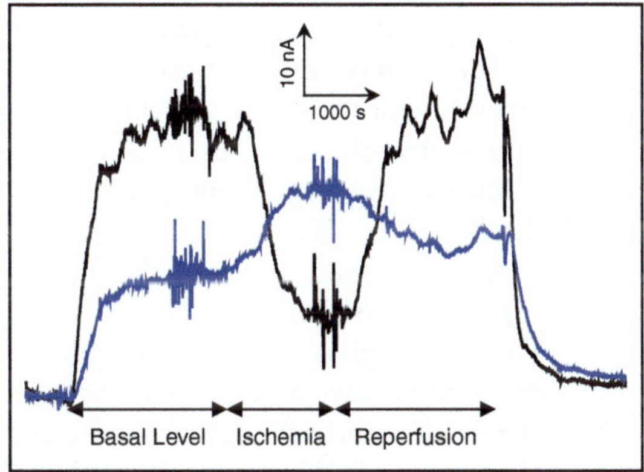

Fig. 8. Typical amperometric responses recorded with the online analytical system with dehydrogenase-based glucose (*black curve*) and lactate (*blue curve*) biosensors as the selective detectors for the microdialysates continuously sampled from the rat striatum following the global cerebral ischemia/reperfusion. The figure was reprinted with permission from (68). Copyright 2009 American Chemical Society.

cerebral systems and a high tolerance against the variation of pH and O_2 following cerebral ischemia/reperfusion. The online analytical system was further used for continuously monitoring glucose and lactate in the rat striatum following the global ischemia/reperfusion (Fig. 8). When the animals were administrated into 20 min of global ischemia, the extracellular glucose was decreased to be $23.0 \pm 4.0\%$ ($n = 3$) of the basal level. This level was gradually restored to be $119.3 \pm 55.0\%$ ($n = 3$) of the basal level after 40 min of reperfusion. Contrarily, the level of lactate was gradually increased to be $213.3 \pm 46.6\%$ ($n = 3$) of the basal level following 20 min of the global ischemia and then restored to be $106.0 \pm 23.0\%$ ($n = 3$) of the basal level during 40 min of reperfusion.

2.3. Ascorbic Acid

As one kind of most important neurochemicals in cerebral systems, AA not only functions as a neuromodulator of both dopamine- and glutamate-mediated neurotransmission but also acts as an antioxidant and free radical scavenger in the intracellular antioxidant network and plays an important role in neuro-protection (114, 115). In physiological solutions, AA can be electrochemically oxidized through a two-electron and one-proton pathway followed by an irreversible hydrolysis process to produce an electroinactive product. Such an electrochemical property of ascorbate has been successfully used to establish in vivo voltammetric methods for the real-time measurements in the rat brain (116–118). However, those methods required either pretreatment to the electrode or the additional use of AAOx to differentiate the current response for AA from the total response obtained. In addition, despite a combination of sample separation, for example, HPLC with direct electrochemical detection has also been used for the determination of AA in the brain microdialysates (119, 120); such an off-line analytical method still suffers from limitations of low time resolution and much instrumentation as well as the degradation of AA during the sampling and separation procedure.

Mao et al. reported that the use of SWNTs, especially those after vacuum heat treatment at 500 °C, was able to greatly enhance the electron transfer kinetics of AA oxidation at a low potential (ca. −0.05 V vs. Ag/AgCl) and possess a strong ability against electrode fouling (76). Based on this finding, they developed a novel online analytical method by directly integrating in vivo microdialysis sampling with a selective electrochemical detector based on an SWNT-modified glassy carbon electrode (Fig. 1c). The method was demonstrated to have a high selectivity against catecholamines and their metabolites and other electroactive species of physiological levels, which was studied in vitro. The authors have also in vivo studied the selectivity, stability, and repeatability of this method and found that the online analytical method with heat-treated SWNT-modified GC electrode as the electrochemical detector was excellent, as displayed in Fig. 9. The validity of this method for continuously monitoring cerebral AA was demonstrated by measuring the change of AA level in the rat brain during global brain ischemia.

With such an online analytical method, Mao et al. recently studied the changes in extracellular AA in different brain ischemia/reperfusion models. As shown in Fig. 10 (left panel) (121), in the global two-vessel occlusion (2-VO) ischemia model, the striatum AA did not change with statistic significance until 60 min after occlusion and was decreased to be 91 ± 3% ($n = 5$, $P < 0.05$) of the basal level (8.05 ± 023 μM) at the time point of 60 min after occlusion (Left panel, A). In the 2-VO ischemia/reperfusion model, AA remained unchanged during the 10 min of ischemia, and was sharply increased to be 267 ± 74% ($n = 5$, $P < 0.05$) of the basal level after the initial 15 min of reperfusion, and then decreased to be

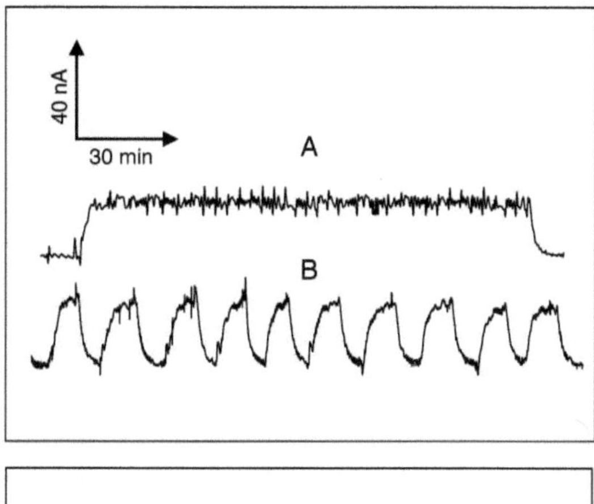

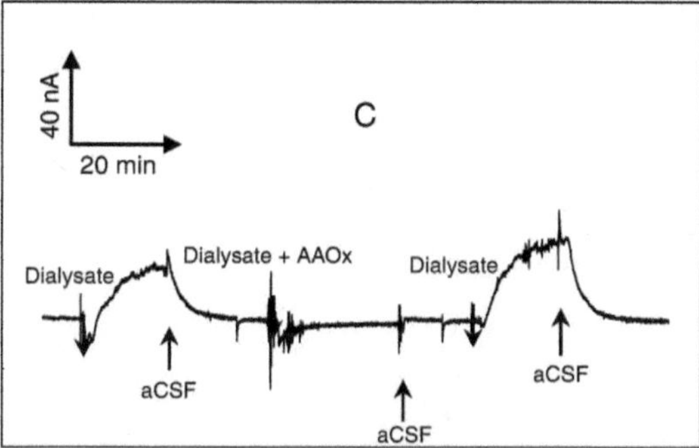

Fig. 9. Typical current–time responses for the brain dialysates recorded continuously (**a**) and repeatedly (**b**) of free moving rats and (**c**) for the brain dialysate with the brain perfused with or without AAOx in aCSF. A 47.2 U mL^{-1} of AAOx was added into the aCSF, and the prepared mixture was used as the perfusion solution. When the brain microdialysate perfused with aCSF containing no AAOx was switched to flow into the electrochemical cell (as indicated in the figure), the increase in the current response recorded was due to the oxidation of AA in the brain microdialysate. Such a response was decreased to the baseline level when the aCSF was switched to flow into the electrochemical cell (also as indicated in the figure). Three figures were reprinted with permission from (76). Copyright 2005 American Chemical Society.

122 ± 33% ($n = 5$, $P < 0.05$) of the basal level after 50 min of reperfusion (Left panel, B). Extracellular AA was largely increased after 5 min of left middle cerebral artery occlusion (LMCAO) and was then gradually increased to be 257 ± 49% ($n = 5$, $P < 0.05$) of the basal level after 60 min of LMCAO ischemia (Left panel, C). In the LMCAO ischemia/reperfusion model, AA was greatly increased during 10 min of ischemia and then gradually increased to be 309 ± 69% ($n = 5$, $P < 0.05$) of the basal level after the consecutive 50 min of reperfusion (Left panel, D). The results demonstrated here may be useful for understanding the neurochemical processes in the acute period of cerebral ischemia and could thus be important for neuroprotective therapeutics for cerebral ischemic injury.

In addition, the authors also studied the dynamic regional changes of extracellular AA level in four different brain regions in

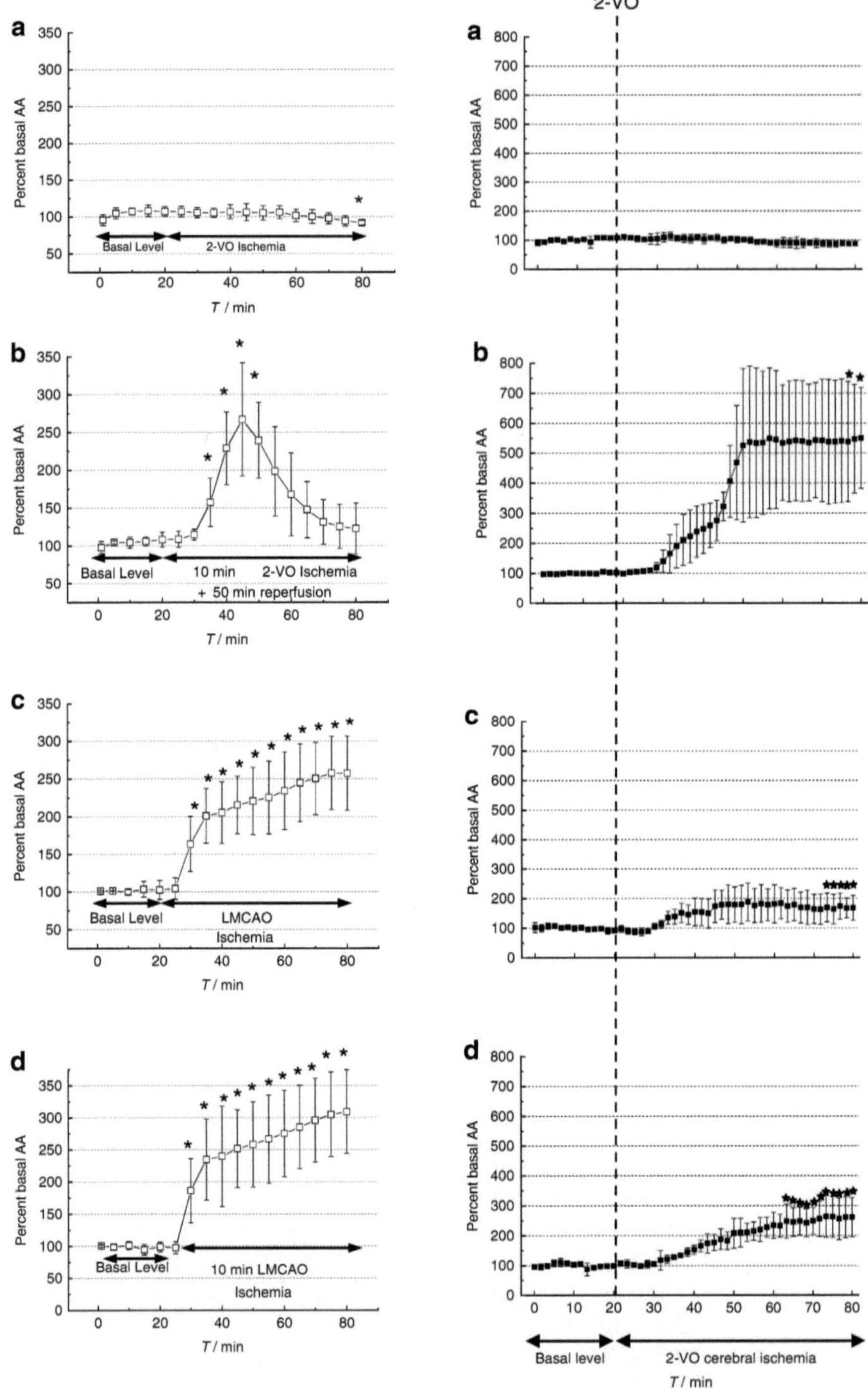

Fig. 10. *Left panel*: Statistical results of the level of striatum AA in different global and focal ischemia/reperfusion models of 2-VO ischemia (**a**), 2-VO ischemia/reperfusion (**b**), LMCAO ischemia (**c**), and LMCAO ischemia/reperfusion (**d**). The conditions for different models were indicated in the figure. The figure was reprinted from (121). Copyright (2008), with permission from Elsevier. *Right panel*: Statistical results of AA levels in striatum (**a**), cortex (**b**), dorsal hippocampus (**c**), and ventral hippocampus (**d**) during the 2-VO global cerebral ischemia. The conditions administrated for different regions were indicated in the figure. The time interval for data acquisition was 100 s. *Asterisks* indicate significant difference compared with pre-ischemia physiological period with the two-tailed Student's *t* test ($P < 0.05$). The figure was reprinted from (122). Copyright (2009), with permission from Elsevier.

1 h after global cerebral ischemia induced by two-vessel occlusion (2-VO), with the same online analytical method (122). The regional distributions of physiological AA levels in the microdialysates from striatum, cortex, dorsal hippocampus, and ventral hippocampus were 2.97 ± 0.06, 3.98 ± 0.09, 3.02 ± 0.47, and 3.80 ± 0.29 μM, respectively. As shown in Fig. 10 (right panel), in 1 h after 2-VO cerebral ischemia, the microdialysate AA levels in the above four regions varied in a different manner; the striatum AA was slowly decreased to $86.49 \pm 5.53\%$ of the basal level ($n = 3$, $P < 0.05$) (right panel, A) while the AA levels in the cortex, dorsal hippocampus, and ventral hippocampus were increased to $549.80 \pm 167.86\%$ ($n = 3$, $P < 0.05$) (right panel, B), $167.81 \pm 41.85\%$ ($n = 4$, $P < 0.05$) (right panel, C), and $261.24 \pm 65.00\%$ ($n = 3$, $P < 0.05$) (right panel, D), in relative to their respective basal levels, respectively. The recorded spatiotemporal regional changes in the extracellular AA levels essentially reflect the intricate neurochemical changes during the acute period of the global cerebral ischemia and may thus be useful for understanding the neurochemical processes of the global cerebral ischemia.

2.4. H_2O_2

H_2O_2 represents one of the most important reactive oxygen species (ROS) and is implicated in a number of neurological disease states as well as serves a critical role in normal cell function (123). H_2O_2 is also gaining increasing recognition as a rapid neuromodulatory signaling molecule. Despite a growing interest in H_2O_2, there are few experimental tools available to directly monitor dynamic H_2O_2 fluctuations in intact brain for understanding the chemical essence of per-oxidative damage and its implication in brain function and human disease. Molecule imaging with H_2O_2-responsive fluorophores was a powerful, and commonly used, approach for examining the diverse roles that H_2O_2 plays in complex biological environments (124, 125). However, conventional fluorescent probes lack selectivity for H_2O_2 over other kinds of ROS and were thus difficult to use for dynamic in vivo measurements. On the other hand, unlike many other neurochemicals mentioned above, H_2O_2 is electroactive and can be theoretically detected directly. However, the serious specificity problems associated with the direct electrochemical measurements unfortunately made in vivo voltammetric or online amperometric measurements of H_2O_2 in the cerebral systems almost impossible. In addition, the changes in the endogenous levels of O_2 and pH that are essentially associated with the formation of H_2O_2 under pathophysiological conditions also render difficulty in the electrochemical determination of H_2O_2. Similarly to the online biosensors mentioned above, HRP, upon wired with electrode through electron transfer mediators or promoters, can be relatively useful for selective determination of H_2O_2 in most cases; yet such HRP-based amperometric biosensors are not well competent for continuous and online measurement of

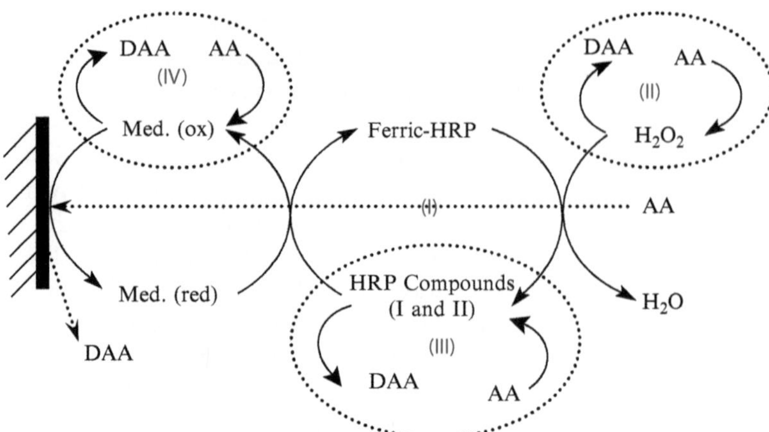

Fig. 11. Schematic depiction of electron-transfer process and chemical reactions of AA in electrocatalytic reduction of H_2O_2 at a mediated HRP-based electrode. The figure was reprinted with permission from (126). Copyright 2002 American Chemical Society.

cerebral H_2O_2 mainly because of the great interference from AA because AA interferes with the measurement of H_2O_2 through the reactions given in Fig. 11 (126). As shown, AA can be directly oxidized at the substrate electrode to produce dehydroascorbic acid (DAA) (reaction I). Besides, AA can chemically react with H_2O_2 under the catalysis of metal ions, e.g., Fe^{3+} or Cu^{2+} (reaction II) or by HRP (reaction III). Moreover, AA can possibly react with some mediators used for shuttle electron transfer of HRP, e.g., Ospoly (vinylpyridine) (reaction IV). All these by-reactions substantially made it difficult to simply use the HRP-based amperometric biosensor as the online detector for continuous measurement of cerebral H_2O_2.

Mao et al. have developed an online analytical system for continuous biosensing of H_2O_2 through integrating in vivo microdialysis with selective electrochemical detector based on enzyme-modified ring-disk carbon film electrode (Fig. 12) (126). The central disk electrode was coated with AAOx to pre-oxidize AA. The ring electrode was co-immobilized HRP with electropolymerized film of pyrrole since the polypyrrole (PPy) could be used to promote or mediate the electronic communication between HRP and electrode at a low potential (0.0 V). The HRP/PPy-modified ring electrode was over-coated with a thin film of polyphenol (PPh) to further improve the selectivity against AA. On the other hand, the homogeneous reaction between AA and H_2O_2 was suppressed by removing heavy metal ion catalysts with the chelating agent, ethylenediamine-tetraacetic acid (EDTA), introduced in external phosphate-buffered solution. The external buffer also delivered an excess of dissolved O_2 needed for the AAOx-catalytic reaction and stabilized the pH level of the electrode environment. Further, unlike other mediates used for shuttle electron transfer

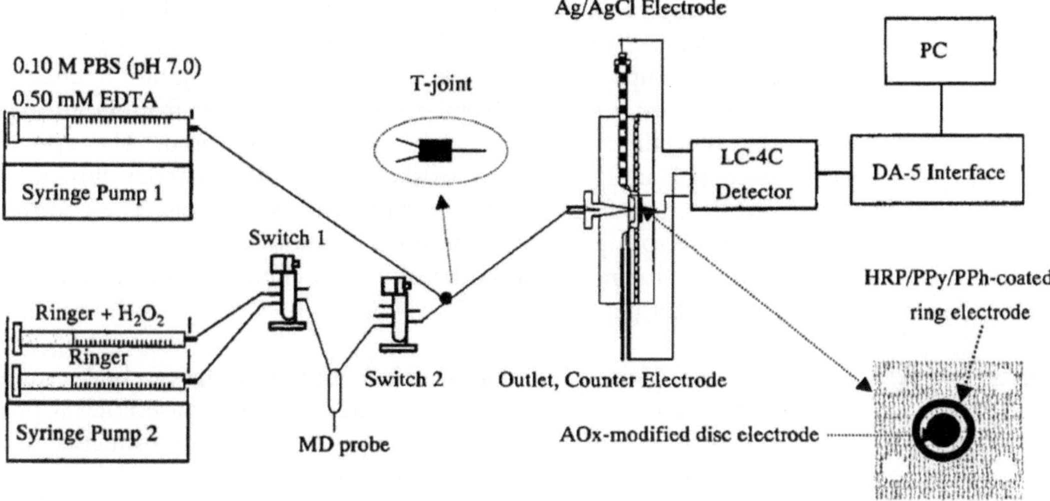

Fig. 12. Illustration of online analytical system for continuously monitoring H_2O_2 with HRP/PPy/PPh-modified ring and AAOx-modified disk PCFE positioned in a thin-layer flow cell. The figure was reprinted with permission from (126). Copyright 2002 American Chemical Society.

between HRP and electrode, PPy used did not react with AA chemically, overcoming the limitations inherent in the HRP-based amperometric H_2O_2 biosensors. This procedure eventually enabled trace levels of H_2O_2 to be readily monitored without interference from physiological levels of AA, UA, electroactive neurotransmitters, and their principal metabolites, in a continuous-flow system.

2.5. Quinones

It has been reported that brain ischemia generally occurs with response to complicated redox chemical processes. For example, brain ischemia occurs with the release of high amounts of neurotransmitters from nerve terminals into the extracellular space (127, 128). During reperfusion, the released catecholic species such as DA, norepinephrine, epinephrine, and their metabolites, e.g., DOPAC, may be oxidized enzymatically or non-enzymatically into endogenous *o*-quinones by oxygen present in the reperfusing fluid. Since the resulting quinones are as electroactive as the primary dopaquinone (DOQ), they can be monitored electrochemically. Unlike other kinds of neurochemicals such as glucose, lactate, and glutamate, quinones are unstable species because of risks of polymerization and reduction and present at micromolar levels. As a consequence, the measurement of such kind of neurochemicals requires the implementation of an in vivo microdialysis sampling closely connected to an online sensitive detection. Michotte et al. developed an online analytical method for continuously monitoring endogenous quinones through combining in vivo microdialysis sampling with an amperometric detector with a glassy carbon working electrode operating at −0.20 V (vs. Ag/AgCl) (129). The instrumental setup comprised a syringe pump pulse-damper consisting of an air bubble and a silica capillary, which permitted considerable reduction of background current fluctuations and allowed improved detection limits. This configuration enabled micromolar amounts of total quinones, generated from DA during the reperfusion period, to be readily monitored. The applied potential and the use of working electrode material eliminated interferences from AA, H_2O_2, riboflavin, and thiols. The continuous monitoring of endogenous quinones was demonstrated for the first time in an animal model of brain ischemia induced by a vasoconstrictor peptide.

2.6. Ca^{2+} and Mg^{2+}

Ca^{2+} and Mg^{2+} are two kinds of divalent ions in the cerebral systems and play critical roles in the physiological and pathological processes. For example, Ca^{2+} is an important signal transduction element and is required for many functions in the central nervous systems including gene expression, neurotransmitter release, neurite outgrowth regulation, synaptogenesis, and synaptic transmission (130, 131). Mg^{2+} is an important mediator and regulator of Ca^{2+} signaling, and plays a classical role in defining the properties of adenosine triphosphate (132, 133). While many methods have previously been reported for in vitro measurements of Ca^{2+} and Mg^{2+}, simple but effective measurements of both species in the cerebral systems remain a long-standing challenge. Although atomic absorption spectroscopy (AAS) and inductively coupled plasma-mass spectrometry (ICPMS) have previously been employed for both discontinuous and continuous measurements of Ca^{2+} and Mg^{2+} in the rat brain (134, 135), these methods were still limited by an expensive instrumentation, or a poor time resolution caused by the sample collection for AAS measurements, or complex desalting procedures for ICPMS measurements. While the excellent properties of electrochemical methods have substantially made them particularly attractive for effectively monitoring brain chemistry, the poor electroactivity of Ca^{2+} and Mg^{2+} unfortunately renders great difficulties in applying such kind of methods to monitor both species in the cerebral systems.

Very recently, Mao et al. developed a simple but very effective electrochemical method for continuous monitoring of electrochemically inactive Ca^{2+} and Mg^{2+} in the rat brain based on the enhancement of both kinds of divalent cations toward the electrocatalytic oxidation of β-nicotinamide adenine dinucleotide (NADH) with organic redox dyes, such as toluidine blue O (TBO), as the electrocatalyst (136). The current enhancement from the oxidation of NADH was thus used to constitute an electrochemical method for the measurements of Ca^{2+} and Mg^{2+} in a continuous-flow system with the polyTBO-modified electrode as the detector. As shown in Fig. 13, upon being integrated with in vivo microdialysis, the electrochemical method was successfully applied in investigating on cerebral Ca^{2+} and Mg^{2+} of living

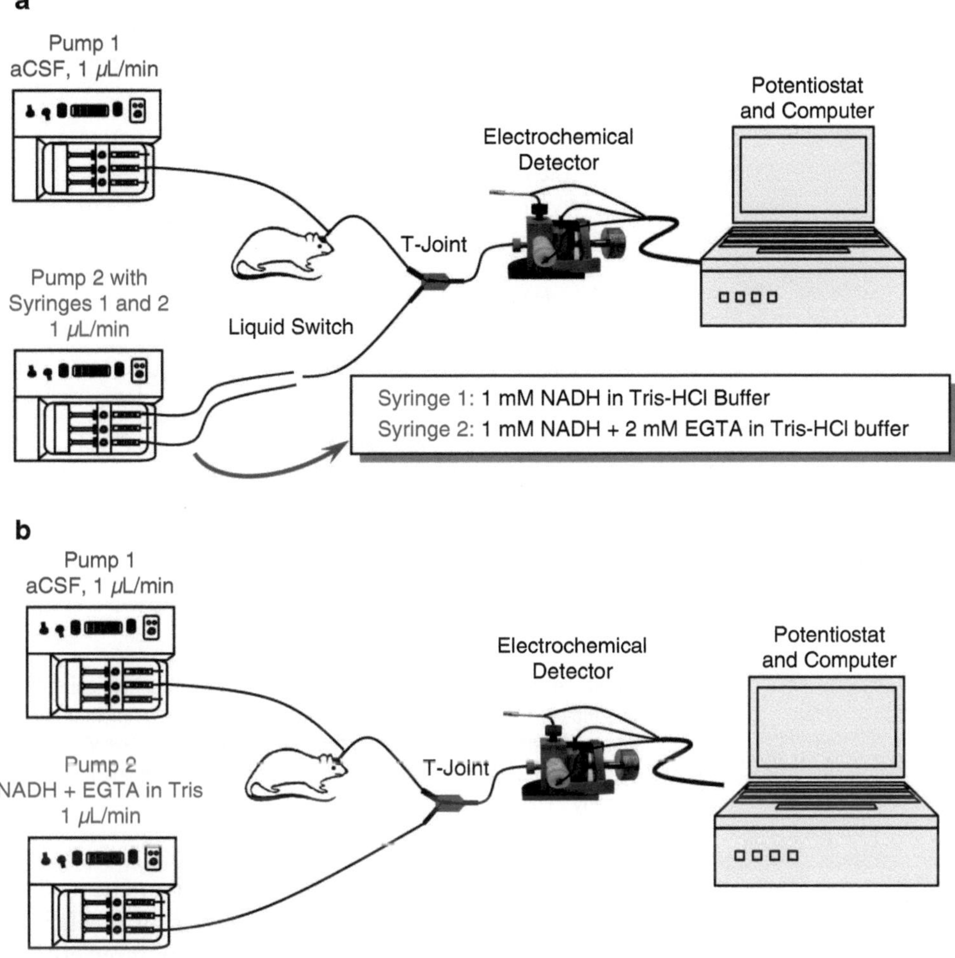

Fig. 13. Illustration of the online electrochemical method for (**a**) simultaneous measurements of Ca^{2+} and Mg^{2+} and (**b**) continuous monitoring of Mg^{2+} in the rat brain by efficiently integrating electrochemical detection with in vivo microdialysis. The figure was reprinted with permission from (136). Copyright 2010 American Chemical Society.

animals in two aspects: (1) online simultaneous measurements of the basal levels of Ca^{2+} and Mg^{2+} in the brain of the freely moving rats by using ethylene glycol-*bis* (2-aminoethylether) tetraacetic acid (EGTA) as the selective masking agent for Ca^{2+} to differentiate the net current responses selectively for Ca^{2+} and Mg^{2+} and (2) online continuous monitoring of the cerebral Mg^{2+} following the global ischemia by using Ca^{2+}-masking agent (i.e., EGTA) to completely eliminate the interference from Ca^{2+}.

As demonstrated by the authors, the integration of polyTBO-based electrochemical detector with in vivo microdialysis offered a new approach to simultaneous measurements of Ca^{2+} and Mg^{2+} in rat brain through differentiation of the net current response for each species from the total response for a mixture containing both

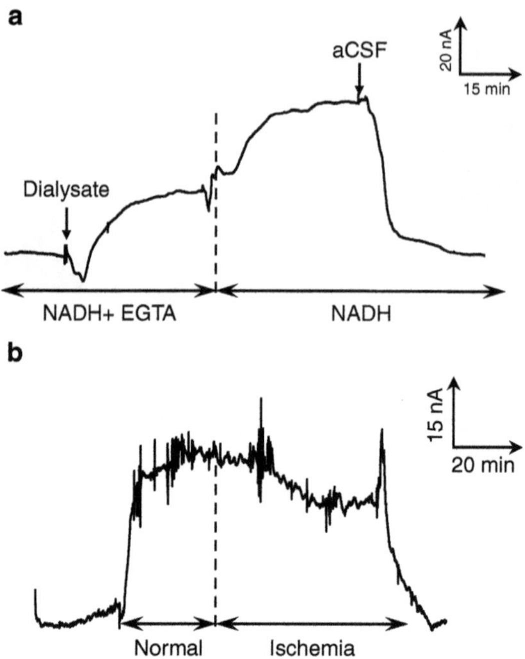

Fig. 14. Current–time responses obtained with the online electrochemical method for (**a**) simultaneous measurements of Ca^{2+} and Mg^{2+} and (**b**) continuous monitoring of Mg^{2+} in the rat brain by efficiently integrating electrochemical detection with in vivo microdialysis. The two figures were reprinted with permission from (136). Copyright 2010 American Chemical Society.

Ca^{2+} and Mg^{2+} by using EGTA as the masking reagent to selectively chelate Ca^{2+}. Meanwhile, the selective monitoring of Mg^{2+} became possible when EGTA was used to eliminate the interference from Ca^{2+}. Both methods were subsequently used for simultaneously monitoring the basal levels of Ca^{2+} and Mg^{2+} in the rat brain and for selectively monitoring the change in the Mg^{2+} level following brain ischemia, as shown in Fig. 14.

3. Summary and Future Developments

The strategies for online continuous monitoring of neurochemicals through combining in vivo microdialysis sampling with selective electrochemical detection have been widely discussed in this chapter from basic buildup, challenges, and latest advancements in both method establishment and practical applications in understanding of the molecular basis of physiological and pathological events. Compared with the in vivo sensing and biosensing methods, in vivo voltammetry in particular, based on implantation of in vivo electrochemical/optical probes into brain tissues to real-time record the changes in the neurochemicals, the online analytical methods discussed here may be more easily adopted by neuroscientists because they do not require a stronger background in chemistry, for example, electrochemistry, and can monitor a larger variety of neurochemicals. Compared with the methods based on in vivo microdialysis and off-line detection, the online analytical methods essentially avoid the procedures employed for sample collection, pretreatment, and separation and are thus less technically demanding, but with an enhanced temporal resolution. While the rational design of electrode/electrolyte interface to achieve selectivity for the electrochemical detection involved in the online analytical methods yet remains challenging to neuroscientists, the online analytical methods with electrochemical sensors and biosensors as the selective detectors remain very potentially attractive for continuously monitoring neurochemicals.

The future developments in the online analytical methods for continuously monitoring brain chemistry with electrochemical sensors and biosensors as the selective detectors may lie in the following aspects. (1) Since the electrochemical methods demonstrated so far for online selective detection are yet limited only for several kinds of neurochemicals, the methodology for selective detection should be further developed in the future studies not only continuously based on electrochemical mechanism but also through other mechanisms including fluorescence, chemiluminescence, electrogenerated chemiluminescence, UV–vis spectroscopy, and so forth. For example, new electrochemical detectors can be developed through the uses of recently developed micro/nanostructures, for example, graphene nanosheets and metal nanoparticles that have been demonstrated to possess rich surface chemistry and excellent electrochemical properties. In addition, the selective electrochemical detector can also be established through the exploitation of new kinds of biorecognition elements such as aptamers since, as one kind of artificial oligonucleotides with specific binding affinity toward a variety of targets ranging from small molecules, proteins, and even to cells, aptamers have been used as the recognition elements for biosensing applications. The selective detection can also be extended to non-electrochemical mechanisms. This is the case since not all neurochemicals are electrochemically active (for example, most kinds of amino acids and peptides) or can be changed to be electrochemically detectable either through chemical (e.g., chemical reactions) or biochemical (e.g., biochemical transition with biorecognition elements) approaches. For a typical example, we have recently demonstrated that the exploitation of both the optical properties and the designable surface chemistry of gold nanoparticles could form a new optical mechanism to visualize the change in cerebral glucose

(137). Compared with the electrochemical detection of glucose in the cerebral systems, this colorimetric method bears simplicity in instrumentation and is more readily to achieve the selectivity. This earlier attempt strongly suggests that the exploitation of non-electrochemical mechanisms could also pave effective approaches to selective detection in the online analytical methods for continuous measurements of neurochemicals in the cerebral systems; (2) while early attempts have demonstrated the capability of the online analytical methods for the simultaneous and continuous measurements of two kinds of neurochemicals, this potentiality remains to be largely explored in the future studies. This trend is essentially activated by the facts that the physiological event basically occurs with a complicated chemical process, in which several kinds of neurochemicals are certainly involved. In these studies, simultaneous and continuous measurements of multiple neurochemicals remain very essential. Such pursuit could be presumably accomplished through developing electrochemical/optical biosensors in a micrototal analysis system (μTAS) by implementing microelectromechanical systems (MEMS) and nanotechnology. There is a great potential to apply MEMS and biosensor technologies to automatize and further improve the speed and feasibility of online electrochemical detection for continuously monitoring neurochemicals. This will include micro-fabrication of biosensor arrays in a microfluidic system and mission-orientated synthesis of multifunctional nanostructures and biorecognition elements that are highly selective and relatively stable for biosensing applications, and (3) unlike the methods widely investigated and finally established by analytical chemists, the online analytical methods discussed here are intentionally developed for neurochemical investigations, which is very similar to other kinds of neurochemical methods, such as in vivo voltammetry and in vivo microdialysis coupled with off-line detection. For this purpose, close collaboration will be definitely needed in the future studies to further bridge the gap between neuroscience and analytical chemistry.

Acknowledgement

This work is financially supported by NSF of China (Grant Nos., 90813032, 20935005, 20975104 for L. Mao, and 21045001 for Y. Lin), National Basic Research Program of China (973 Program, 2007CB935603 and 2010CB933502), and Chinese Academy of Sciences (KJCX2-YW-W25, Y2010015).

References

1. Venton BJ, Wightman RM (2003) Psychoanalytical electrochemistry: dopamine and behavior. Anal Chem 75:414A–421A
2. Stuart JN, Hummon AB, Sweedler JV (2004) The chemistry of thought: neurotransmitters in the brain. Anal Chem 76:121A–128A
3. Dale N, Hatz S, Tian F, Llaudet E (2005) Listening to the brain: microelectrode biosensors for neurochemicals. Trends Biotechnol 23:420–428
4. Marco G, Devauchelle B, Berquin P (2009) Brain functional modeling, what do we measure with fMRI data? Neurosci Res 64:12–19
5. Miller G (2009) On the origin of the nervous system. Science 325:24–26
6. Coutinho V, Knöpfel T (2002) Metabotropic glutamate receptors: electrical and chemical signaling properties. Neuroscientist 8:551–561
7. Zhu C, Wu LQ, Wang X, Lee JH, English DS, Ghodssi R, Raghavan SR, Payne GF (2007) Reversible vesicle restraint in response to spatiotemporally controlled electrical signals: a bridge between electrical and chemical signaling modes. Langmuir 23:286–291
8. Verkhratsky A (2009) Astrocytes in (patho) physiology of the nervous system. In: Haydon PG, Parpura V (eds) Neurotransmitter receptors in astrocytes. Springer, New York
9. Stefan H, Hummel C, Scheler G, Genow A, Druschky K, Tilz C, Kaltenhäuser M, Hopfengärtner R, Buchfelder M, Romstöck J (2003) Magnetic brain source imaging of focal epileptic activity: a synopsis of 455 cases. Brain 126:2396–2405
10. Gross BA, Hanna DM (2010) Artificial neural networks capable of learning spatiotemporal chemical diffusion in the cortical brain. Pattern Recogn 43:3910–3921
11. Wilson JRF, Green A (2009) Acute traumatic brain injury: a review of recent advances in imaging and management. Eur J Trauma Emerg Surg 35:176–185
12. Kemp GJ (2000) Non-invasive methods for studying brain energy metabolism: what they show and what it means. Dev Neurosci 22:418–428
13. Garris PA (2010) Advancing neurochemical monitoring. Nat Methods 7:106–108
14. Ostrovskii AMA (2010) Current trends in modern brain science. Her Russ Acad Sci 80: 187–198
15. Georganopoulou DG, Carley R, Jones DA, Boutelle MG (2000) Development and comparison of biosensors for in-vivo applications. Faraday Discuss 116:291–303
16. Troyer KP, Heien ML, Venton BJ, Wightman RM (2002) Neurochemistry and electroanalytical probes. Curr Opin Chem Biol 6: 696–703
17. Khan AS, Michael AC (2003) Invasive consequences of using micro-electrodes and microdialysis probes in the brain. Trends Anal Chem 22:503–508
18. Zhang MN, Mao LQ (2005) Enzyme-based amperometric biosensors for continuous and on-line monitoring of cerebral extracellular microdialysate. Front Biosci 10:345–352
19. Watson CJ, Venton BJ, Kennedy RT (2006) In vivo measurements of neurotransmitters by microdialysis sampling. Anal Chem 78: 1391–1399
20. Adams RN (1976) Probing brain chemistry with electroanalytical technique. Anal Chem 48:1128A–1137A
21. Wang J (1999) Electroanalysis and biosensors. Anal Chem 71:328R–332R
22. Wilson GS, Gifford R (2005) Biosensors for real-time in vivo measurements. Biosens Bioelectron 20:2388–2403
23. Huffman ML, Venton BJ (2009) Carbon-fiber microelectrodes for in vivo applications. Analyst 13:18–24
24. Crespi F (2010) Wireless in vivo voltammetric measurements of neurotransmitters in freely behaving rats. Biosens Bioelectron 25: 2425–2430
25. Tse RS, Wong SC, Yuen CP (1980) Determination of deuterium/hydrogen ratios in natural waters by fourier transform nuclear magnetic resonance spectrometry. Anal Chem 52:2445–2448
26. Feng JX, Brazell M, Renner K, Kasser R, Adams RN (1987) Electrochemical pretreatment of carbon fibers for in vivo electrochemistry: effects on sensitivity and response time. Anal Chem 59:1863–1867
27. Zhang XJ, Zhang WM, Zhou XY, Ogorevc B (1996) Fabrication, characterization, and potential application of carbon fiber cone nanometer-size electrodes. Anal Chem 68:3338–3343
28. Adams RN (1976) Probing brain chemistry with electroanalytical techniques. Anal Chem 48:1126A–1138A
29. Stamford JA, Kruk ZL, Millar J (1984) Regional differences in extracellular ascorbic

acid levels in the rat brain determined by high speed cyclic voltammetry. Brain Res 299: 289–295

30. Phillips PEM, Stuber GD, Heien MLAV, Wightman RM, Carelli RM (2003) Subsecond dopamine release promotes cocaine seeking. Nature 422:614–618
31. Zhang MN, Liu K, Xiang L, Lin YQ, Su L, Mao LQ (2007) Carbon nanotube-modified carbon fiber microelectrodes for in vivo voltammetric measurement of ascorbic acid in rat brain. Anal Chem 79:6559–6565
32. Parry TJ, Carter TL, McElligott JG (1990) Physical and chemical considerations in the in vitro calibration of microdialysis probes for biogenic amine neurotransmitters and metabolites. J Neurosci Methods 32:175–183
33. Tisdall MM, Smith M (2006) Cerebral microdialysis: research technique or clinical tool. Br J Anaesth 97:18–25
34. Ungerstedt U, Pycock C (1974) Functional correlates of dopamine neurotransmission. Bull Schweiz Akad Med Wiss 30:44–55
35. Bourne JA (2003) Intracerebral microdialysis: 30 years as a tool for the neuroscientist. Clin Exp pharmacol Physiol 30:16–24
36. Guihen E, O'Connor WT (2010) Capillary and microchip electrophoresis in microdialysis: recent applications. Electrophoresis 31:55–64
37. Boret H, Fesselet J, Meaudre E, Gaillard PE, Cantais E (2006) Cerebral microdialysis and $P_{ti}O_2$ for neuro-monitoring before decompressive craniectomy. Acta Anaesthesiol Scand 50:252–254
38. Peña A, Liu P, Derendorf H (2000) Microdialysis in peripheral tissues. Adv Drug Deliv Rev 45:189–216
39. Davani S, Chocron S, Muret P, Mersin N, Etievent JP, Kantelip JP (2003) Myocardial microdialysis. Importance and potential in cardiovascular research. Pathol Biol 51:39–43
40. Goodman JC, Robertson CS (2009) Microdialysis: is it ready for prime time? Curr Opin Crit Care 15:110–117
41. Obrenovitch TP, Zilkha E (2001) Microdialysis coupled to online enzymatic assays. Methods 23:63–71
42. Korf J, Huininka KD, Posthuma-Trumpie GA (2010) Ultraslow microdialysis and microfiltration for in-line, on-line and off-line monitoring. Trends Biotechnol 28:150–158
43. Hutchinson PJ, O'Connell MT, Nortje J, Smith P, Al-Rawi PG, Gupta AK, Menon DK, Pickard JD (2005) Cerebral microdialysis methodology-evaluation of 20 kDa and 100 kDa catheters. Physiol Meas 26:423–428
44. Alavijeh MS, Palmer AM (2010) Measurement of the pharmacokinetics and pharmacodynamics of neuroactive compounds. Neurobiol Dis 37:38–47
45. Li YJ, Peris J, Zhong L, Derendorf H (2006) Microdialysis as a tool in local pharmacodynamics. AAPS J 8:E222–E235
46. Kehr J (1998) Determination of glutamate and aspartate in microdialysis samples by reversed-phase column liquid chromatography with fluorescence and electrochemical detection. J Chromatogr B 708:27–38
47. Li N, Guo JZ, Liu B, Yu YQ, Cui H, Mao LQ, Lin YQ (2009) Determination of monoamine neurotransmitters and their metabolites in a mouse brain microdialysate by coupling high-performance liquid chromatography with gold nanoparticle-initiated chemiluminescence. Anal Chim Acta 645: 48–55
48. Davies MI, Lunte CE (1997) Microdialysis sampling coupled on-line to microseparation techniques. Chem Soc Rev 26:215–222
49. Nandi P, Lunte SM (2009) Recent trends in microdialysis sampling integrated with conventional and microanalytical systems for monitoring biological events: a review. Anal Chim Acta 651:1–14
50. Lada MW, Kennedy RT (1996) Quantitative in vivo monitoring of primary amines in rat caudate nucleus using microdialysis coupled by a flow-gated interface to capillary electrophoresis with laser-induced fluorescence detection. Anal Chem 68:2790–2797
51. Lada MW, Vickroy TW, Kennedy RT (1997) High temporal resolution monitoring of glutamate and aspartate in vivo using microdialysis on-line with capillary electrophoresis with laser-induced fluorescence detection. Anal Chem 69:4560–4565
52. Cellar NA, Burns ST, Meiners JC, Chen H, Kennedy RT (2005) Microfluidic chip for low-flow push-pull perfusion sampling in vivo with on-line analysis of amino acids. Anal Chem 77:7067–7073
53. Sandlin ZD, Shou MS, Shackman JG, Kennedy RT (2005) Microfluidic electrophoresis chip coupled to microdialysis for in vivo monitoring of amino acid neurotransmitters. Anal Chem 77:7702–7708
54. Shou MS, Ferrario CR, Schultz KN, Robinson TE, Kennedy RT (2006) Monitoring dopamine in vivo by microdialysis sampling and on-line CE-laser-induced fluorescence. Anal Chem 78:6717–6725
55. Cellar NA, Kennedy RT (2006) A capillary–PDMS hybrid chip for separations-based sensing

of neurotransmitters in vivo. Lab Chip 6: 1205–1212

56. Wang M, Roman GT, Schultz K, Jennings C, Kennedy RT (2008) Improved temporal resolution for in vivo microdialysis by using segmented flow. Anal Chem 80:5607–5615
57. Ferrario CR, Shou M, Samaha AN, Watson CJ, Kennedy RT, Robinson TE (2008) The rate of intravenous cocaine administration alters c-fos mRNA expression and the temporal dynamics of dopamine, but not glutamate, overflow in the striatum. Brain Res 1209:151–156
58. Wang M, Roman GT, Perry ML, Kennedy RT (2009) Microfluidic chip for high efficiency electrophoretic analysis of segmented flow from a microdialysis probe and in vivo chemical monitoring. Anal Chem 81: 9072–9078
59. Thompson JE, Vickroy TW, Kennedy RT (1999) Rapid determination of aspartate enantiomers in tissue samples by microdialysis coupled on-line with capillary electrophoresis. Anal Chem 71:2379–2384
60. Hogan BL, Lunte SM, Stobaugh JF, Lunte CE (1994) On-line coupling of in vivo microdialysis sampling with capillary electrophoresis. Anal Chem 66:596–602
61. Zhou SY, Zuo H, Stobaugh JF, Lunte CE, Lunte SM (1995) Continuous in vivo monitoring of amino acid neurotransmitters by microdialysis sampling with on-line derivatization and capillary electrophoresis separation. Anal Chem 67:594–599
62. Zhou JX, Heckert DM, Zuo H, Lunte CE, Lunte SM (1999) On-line coupling of in vivo microdialysis with capillary electrophoresis/electrochemistry. Anal Chim Acta 379:307–317
63. O'Brien KB, Esguerra M, Miller RF, Bowser MT (2004) Monitoring neurotransmitter release from isolated retinas using online microdialysis-capillary electrophoresis. Anal Chem 76:5069–5074
64. José RJ, María DLC (2006) Coupling microdialysis to capillary electrophoresis. Trends Anal Chem 25:563–571
65. Klinker CC, Bowse MT (2007) 4-Fluoro-7-nitro-2,1,3-benzoxadiazole as a fluorogenic labeling reagent for the in vivo analysis of amino acid neurotransmitters using online microdialysis-capillary electrophoresis. Anal Chem 79:8747–8754
66. Li Z, Zharikova A, Bastian J, Esperon L, Hebert N, Mathes C, Rowland NE, Peris J (2008) High temporal resolution of amino acid levels in rat nucleus accumbens during operant ethanol self-administration: involvement of elevated glycine in anticipation. J Neurochem 106:170–181
67. Lin YQ, Liu K, Yu P, Xiang L, Li XC, Mao LQ (2007) A facile electrochemical method for simultaneous and on-line measurements of glucose and lactate in brain microdialysate with prussian blue as the electrocatalyst for reduction of hydrogen peroxide. Anal Chem 79:9577–9583
68. Lin YQ, Zhu NN, Yu P, Su L, Mao LQ (2009) Physiologically relevant online electrochemical method for continuous and simultaneous monitoring of striatum glucose and lactate following global cerebral ischemia/reperfusion. Anal Chem 81:2067–2074
69. Kennedy RT, Watson CJ, Haskins WE, Powell DH, Strecker RE (2002) In vivo neurochemical monitoring by microdialysis and capillary separations. Curr Opin Chem Biol 6:659–665
70. Parrot S, Sauvinet V, Riban V, Depaulis A, Renaud B, Denoroy L (2004) High temporal resolution for in vivo monitoring of neurotransmitters in awake epileptic rats using brain microdialysis and capillary electrophoresis with laser-induced fluorescence detection. J Neurosci Methods 140:29–38
71. Devall AJ, Blake R, Langman N, Smith CGS, Richards DA, Whitehead KJ (2007) Monolithic column-based reversed-phase liquid chromatography separation for amino acid assay in microdialysates and cerebral spinal fluid. J Chromatogr B 848:323–328
72. Anouti S, Vandenabeele-Trambouze O, Koval D, Cottet H (2008) Heart-cutting two-dimensional capillary electrophoresis for the on-line purification and separation of derivatized amino acids. Anal Chem 80:1730–1736
73. Buck K, Ferger B (2008) Intrastriatal inhibition of aromatic amino acid decarboxylase prevents L-DOPA-induced dyskinesia: a bilateral reverse in vivo microdialysis study in 6-hydroxydopamine lesioned rats. Neurobiol Dis 29:210–220
74. Poinsot V, Gavard P, Feurer B, Couderc F (2010) Recent advances in amino acid analysis by CE. Electrophoresis 31:105–121
75. Niwa O, Torimitsu K, Morita M, Osborne P, Yamamoto K (1996) Concentration of extracellular l-glutamate released from cultured nerve cells measured with a small-volume online sensor. Anal Chem 68:1865–1870
76. Zhang MN, Liu K, Gong KP, Su L, Chen Y, Mao LQ (2005) Continuous on-line monitoring of extracellular ascorbate depletion in the rat striatum induced by global ischemia with carbon nanotube-modified glassy carbon

electrode integrated into a thin-layer radial flow cell. Anal Chem 77:6234–6242

77. Osborne PG, Niwa O, Kato T, Yamamoto K (1998) Plastic film carbon electrodes: enzymatic modification for on-line, continuous, and simultaneous measurement of lactate and glucose using microdialysis sampling. Anal Chem 70:1701–1706
78. Collingridge GL, Lester RA (1989) Excitatory amino acids receptors in the vertebrate central nervous system. Pharmacol Rev 40:143–210
79. Wahl F, Obrenovitch TP, Hardy AM, Plotkine M, Boulu R, Symon L (1994) Extracellular glutamate during focal cerebral ischemia in rats: time course and calcium dependency. J Neurochem 63:1003–1011
80. Benveniste H, Huttemeier PC (1990) Microdialysis-theory and application. Prog Neurobiol 35:195–215
81. Obrenovitch TP, Urenjak J, Richards DA, Ueda Y, Curzon G, Symon L (1993) Extracellular neuroactive amino acids in the rat brain striatum during moderate and severe transient ischemia. J Neurochem 61:178–186
82. Poinsot V, Bayle C, Couderc F (2003) Recent advances in amino acid analysis by capillary electrophoresis. Electrophoresis 24:4047–4062
83. Zilkha E, Obrenovitch TP, Koshy A, Kusakabe H, Bennetto HP (1995) Extracellular glutamate: on-line monitoring using microdialysis coupled to enzyme-amperometric analysis. J Neurosci Methods 60:1–9
84. Berners MOM, Boutelle MG, Fillenz M (1994) On-line measurement of brain glutamate with an enzyme/polymer-coated tubular electrode. Anal Chem 66:2017–2021
85. Hayashi K, Kurita R, Horiuchi T, Niwa O (2003) Selective detection of l-glutamate using a microfluidic device integrated with an enzyme-modified pre-reactor and an electrochemical detector. Biosens Bioelectron 18:1249–1255
86. Yao T, Suzuki S, Nakahara T, Nishino H (1998) Highly sensitive detection of l-glutamate by on-line amperometric micro-flow analysis based on enzymatic substrate recycling. Talanta 45:917–923
87. Vallone D, Picetti R, Borrelli E (2000) Structure and function of dopamine receptors. Neurosci Biobehav Rev 24:125–132
88. Suri RE, Bargas J, Arbib MA (2001) Modeling functions of striatal dopamine modulation in learning and planning. Neuroscience 103: 65–85
89. Denenberg VH, Kima DS, Palmiter RD (2004) The role of dopamine in learning, memory, and performance of a water escape task. Behav Brain Res 148:73–78
90. Jung MC, Shi GY, Borland L, Michael AC, Weber SG (2006) Simultaneous determination of biogenic monoamines in rat brain dialysates using capillary high-performance liquid chromatography with photoluminescence following electron transfer. Anal Chem 78:1755–1760
91. Xiang L, Lin YQ, Yu P, Su L, Mao LQ (2007) Laccase-catalyzed oxidation and intramolecular cyclization of dopamine: a new method for selective determination of dopamine with laccase/carbon nanotube-based electrochemical biosensors. Electrochim Acta 52:4144–4152
92. Lin YQ, Zhang ZP, Zhao LZ, Wang X, Yu P, Su L, Mao LQ (2010) A non-oxidative electrochemical approach to online measurements of dopamine release through laccase-catalyzed oxidation and intramolecular cyclization of dopamine. Biosens Bioelectron 25:1350–1355
93. Hasselmo ME, Bower JM (1993) Acetylcholine and memory. Trends Neurosci 16:218–222
94. Huang T, Yang L, Gitzen J, Kissinger PT, Vreeke M, Heller AJ (1995) Detection of basal acetylcholine in rat-brain microdialysate. J Chromatogr B Biomed Appl 670:323–327
95. Nirogi R, Mudigonda K, Kandikere V, Ponnamaneni R (2010) Quantification of acetylcholine, an essential neurotransmitter, in brain microdialysis samples by liquid chromatography mass spectrometry. Biomed Chromatogr 24:39–48
96. Niwa O, Horiuchi T, Kurita R, Torimitsu K (1998) On-line electrochemical sensor for selective continuous measurement of acetylcholine in cultured brain tissue. Anal Chem 70:1126–1132
97. Walker JE (1983) Glutamate, GABA, and CNS disease: a review. Neurochem Res 8:521–550
98. Niwa O, Kurita R, Horiuchi T, Torimitsu K (1998) Small-volume on-line sensor for continuous measurement of γ-aminobutyric acid. Anal Chem 70:89–93
99. Helbok R, Schmidt JM, Kurtz P, Hanafy KA, Fernandez L, Stuart RM, Presciutti M, Ostapkovich ND, Connolly ES, Lee K, Badjatia N, Mayer SA, Claassen J (2010) Systemic glucose and brain energy metabolism after subarachnoid hemorrhage. Neurocrit Care 12:317–323
100. Clausen F, Hillered L, Gustafsson J (2011) Cerebral glucose metabolism after traumatic brain injury in the rat studied by 13C-glucose and microdialysis. Acta Neurochir 153: 653–658

101. Marklund N, Salci K, Ronquist G, Hillered L (2006) Energy metabolic changes in the early post-injury period following traumatic brain injury in rats. Neurochem Res 31:1085–1093
102. Osborne PG, Niwa O, Kato T, Yamamoto K (1997) On-line, continuous measurement of extracellular striatal glucose using microdialysis sampling and electrochemical detection. J Neurosci Methods 77:143–150
103. Osborne PG, Niwa O, Kato T, Yamamoto K (1996) On-line, real time measurement of extracellular brain glucose using microdialysis and electrochemical detection. Curr Sep 15:19–23
104. Kurita R, Hayashi K, Xu F, Yamamoto K, Kato T, Niwa O (2002) Microfluidic device integrated with pre-reactor and dual enzyme-modified microelectrodes for monitoring in vivo glucose and lactate. Sens Actuators B 87:296–303
105. Jones DA, Parkin MC, Langemann H, Landolt H, Hopwood SE, Strong AJ, Boutelle MG (2002) On-line monitoring in neurointensive care enzyme-based electrochemical assay for simultaneous, continuous monitoring of glucose and lactate from critical care patients. J Electroanal Chem 538–539: 243–252
106. Parkin MC, Hopwood SE, Strong AJ, Boutelle MG (2003) Resolving dynamic changes in brain metabolism using biosensors and on-line microdialysis. TrAC Trends Anal Chem 22:487–497
107. Parkin MC, Hopwood SE, Jones DA, Hashemi P, Landolt H, Fabricius M, Lauritzen M, Boutelle MG, Strong AJ (2005) Dynamic changes in brain glucose and lactate in pericontusional areas of the human cerebral cortex, monitored with rapid sampling on-linemic rodialysis: relationship with depolarisation-like events. J Cereb Blood Flow Metab 25:402–413
108. Hopwood SE, Parkin MC, Bezzina EL, Boutelle MG, Strong AJ (2005) Transient changes in cortical glucose and lactate levels associated with peri-infarct depolarisations, studied with rapid-sampling microdialysis. J Cereb Blood Flow Metab 25:391–401
109. Deeba S, Corcoles EP, Hanna BG, Pareskevas P, Aziz O, Boutelle MG, Darzi A (2008) Use of rapid sampling microdialysis for intraoperative monitoring of bowel ischemia. Dis Colon Rectum 51:1408–1413
110. Hashemi P, Bhatia R, Nakamura H, Dreier JP, Graf R, Strong AJ, Boutelle MG (2009) Persisting depletion of brain glucose following cortical spreading depression, despite apparent hyperaemia: evidence for risk of an adverse effect of Leão's spreading depression. J Cereb Blood Flow Metab 29:166–175
111. Boutelle MG, Fellows LK, Cook C (1992) Enzyme packed bed system for the on-line measurement of glucose, glutamate, and lactate in brain microdialysate. Anal Chem 64:1790–1794
112. Liu SM, Shi HL, Liu WL, Furuichi T, Timmins GS, Liu KJ (2004) Interstitial pO_2 in ischemic penumbra and core are differentially affected following transient focal cerebral ischemia in rats. J Cereb Blood Flow Metab 24:343–349
113. Rossi DJ, Brady JD, Mohr C (2007) Astrocyte metabolism and signaling during brain ischemia. Nat Neurosci 10:1377–1386
114. Rice ME (2000) Ascorbate regulation and its neuroprotective role in the brain. Trends Neurosci 23:209–216
115. Grünewald RA (1993) Ascorbic acid in the brain. Brain Res Rev 18:123–133
116. Deakin MR, Kovach PM, Stutts KJ, Wightman RM (1986) Heterogeneous mechanisms of the oxidation of catechols and ascorbic-acid at carbon electrodes. Anal Chem 58: 1474–1480
117. Derrington AM, Lennie P, Wright MJ (1979) The mechanism of peripherally evoked responses in retinal ganglion cells. J Physiol 289:299–310
118. Lovick TA, Hilton SM (1985) Vasodilator and vasoconstrictor neurones of the ventrolateral medulla in the cat. Brain Res 331: 353–357
119. Hou Y, Wu CF, Yang JY, Tu L, Gu PF, Bi XL (2005) Differential effects of clozapine on ethanol-induced ascorbic acid release in mouse and rat striatum. Neurosci Lett 380:83–87
120. Cheng FC, Yang LL, Yang DY, Tsai TH, Lee CW, Chen SH (2000) Monitoring of extracellular pyruvate, lactic acid, and ascorbic acid during cerebral ischemia: a microdialysis study in awake gerbils. J Chromatogr A 870:389–394
121. Liu K, Lin Y, Xiang L, Yu P, Su L, Mao L (2008) Comparative study of change in extracellular ascorbic acid in different brain ischemia/reperfusion models with in vivo microdialysis combined with on-line electrochemical detection. Neurochem Int 52:1247–1255
122. Liu K, Lin YQ, Yu P, Mao LQ (2009) Dynamic regional changes of extracellular ascorbic acid during global cerebral ischemia: Studied with in vivo microdialysis coupled

with on-line electrochemical detection. Brain Res 1253:161–168

123. Lei B, Adachi N, Arai T (1998) Measurement of the extracellular H_2O_2 in the brain by microdialysis. Brain Res Protoc 3:33–36
124. Van de Bittner GC, Dubikovskaya EA, Bertozzi CR, Chang CJ (2010) In vivo imaging of hydrogen peroxide production in a murine tumor model with a chemoselective bioluminescent reporter. Proc Natl Acad Sci 107:21316–21321
125. Miller EW, Chang CJ (2007) Fluorescent probes for nitric oxide and hydrogen peroxide in cell signaling. Curr Opin Chem Biol 11:620–625
126. Mao LQ, Osborne PG, Yamamoto K, Kato T (2002) Continuous on-line measurement of cerebral hydrogen peroxide using enzyme-modified ring-disk plastic carbon film electrode. Anal Chem 74:3684–3689
127. Obrenovitch TP, Richards DA (1995) Extracellular neurotransmitter changes in cerebral ischemia. Cerebrovasc Brain Metab Rev 7:1–54
128. Schapira AHV (1995) Oxidative stress in Parkinson's disease. Neuropathol Appl Neurobiol 21:3–9
129. Pravda M, Bogaert L, Sarre S, Ebinger G, Kauffmann JM, Michotte Y (1997) On-line in vivo monitoring of endogenous quinones using microdialysis coupled with electrochemical detection. Anal Chem 69:2354–2361
130. Clapham DE (1995) Calcium signaling. Cell 80:259–268
131. Mattson MP (2007) Calcium and neurodegeneration. Aging Cell 6:337–350
132. Killilea DW, Ames BN (2008) Magnesium deficiency accelerates cellular senescence in cultured human fibroblasts. Proc Natl Acad Sci U S A 105:5768–5773
133. Slutsky I, Abumaria N, Wu LJ, Huang C, Zhang L, Li B, Zhao X, Govindarajan A, Zhao MG, Zhuo M, Tonegawa S, Liu G (2010) Enhancement of learning and memory by elevating brain magnesium. Neuron 65:165–177
134. Chung YT, Ling YC, Yang CS, Sun YC, Lee PL, Lin CY, Hong CC, Yang MH (2007) In vivo monitoring of multiple trace metals in the brain extracellular fluid of anesthetized rats by microdialysis-membrane desalter-ICPMS. Anal Chem 79:8900–8910
135. Lin MC, Huang YL, Liu HW, Yang DY, Lee CP, Yang LL, Cheng FC (2004) On-line microdialysis-graphite furnace atomic absorption spectrometry in the determination of brain magnesium levels in gerbils subjected to cerebral ischemia/reperfusion. J Am Coll Nutr 23:561S–565S
136. Zhang ZP, Zhao LZ, Lin YQ, Yu P, Mao LQ (2010) Online electrochemical measurements of Ca^{2+} and Mg^{2+} in rat brain based on divalent cation enhancement toward electrocatalytic NADH oxidation. Anal Chem 82:9885–9891
137. Jiang Y, Zhao H, Lin YQ, Zhu NN, Ma YR, Mao LQ (2010) Colorimetric detection of glucose in rat brain using gold nanoparticles. Angew Chem Int Ed 49:4800–4804

Chapter 8

Application of Spinal Microdialysis in Freely Moving Rats

Vincent Umbrain, Lin Shi, Jan Poelaert, and Ilse Smolders

Abstract

Two different spinal microdialysis approaches in freely moving rats are demonstrated. The assessment of the stability and the influences of related factors on spinal microdialysis are discussed. Using spinal microdialysis we demonstrate its contribution on spinal cord neurotransmission knowledge with practical examples studying pain. In particular, (1) we assessed basal (glu) and prostaglandin E_2 (PGE_2) levels in dorsal horn and in cerebrospinal fluid, (2) we evaluated the influences of related factors during the experimental setting, and (3) we investigated formalin-induced releases of Glu and PGE_2 following peripheral inflammation.

Key words: Spinal microdialysis, Pain, In vivo microdialysis, Cerebrospinal fluid glutamate, Prostaglandin E_2

1. Introduction

Microdialysis in pharmacokinetics research offered many benefits such as more frequent data points, clean samples, no loss of body fluid, and consumption of fewer experimental animals per study (1). This valuable tool has been increasingly introduced to preclinical pain research, which allows us to correlate nociceptive behavior with neurotransmitter release in pain-related region.

1.1. Principle of Microdialysis

Microdialysis is an in vivo sampling technique based on a size-selective diffusion of the analyte through a semipermeable membrane. The sampling is accomplished by implanting a probe consisting of a short length of hollow-fiber dialysis membrane affixed to narrow bore inlet and outlet conduits. The driving force for the movement of molecules across the membrane is the concentration gradient established between the extracellular fluid and the fluid within the probe lumen. Small molecules diffuse into (recovery) or out of (delivery) the probe while large molecules such as proteins and molecules bound to proteins are excluded (Fig. 1).

Giuseppe Di Giovanni and Vincenzo Di Matteo (eds.), *Microdialysis Techniques in Neuroscience*, Neuromethods, vol. 75,
DOI 10.1007/978-1-62703-173-8_8,

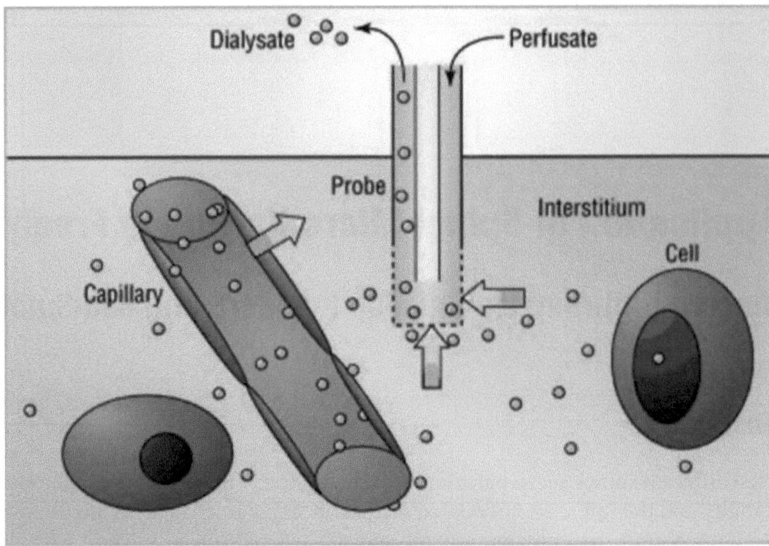

Fig. 1. Diagram of a microdialysis probe (adapted from Muller 2002). The semipermeable membrane at the probe tip allows exchange of soluble molecules between the probe and the surrounding tissue. When the probe is implanted into tissue interstitium, molecules continuously diffuse out of the interstitial space fluid into the perfusion medium. Samples are continuously collected and analyzed by standard chemical analytical techniques.

Because proteins do not pass through the probe membrane into the dialysate, microdialysis samples do not need further sample cleanup and no enzymatic degradation takes place. The quality of data, quantity of useful information, and reduction in the number of animals and labor requirements all point towards the advantages of using this method in an experimental setting (2, 3).

1.2. Methodology of Spinal Microdialysis

Application of microdialysis to the spinal cord has been employed in three approaches: (1) a concentric dialysis probe inserts into the spinal gray matter after a partial laminectomy (4, 5), (2) a linear probe passes transversally through the dorsal spinal cord through two opposing holes overlying vertebrae (6, 7), and (3) a triple loop probe inserts into the intrathecal space via the atlanto-occipital membrane where cerebrospinal fluid (CSF) is sampled (8). The latter two approaches are more reliable when used in awake animals.

1.2.1. Surgical Preparation of Spinal Microdialysis

(a) Microdialysis in spinal dorsal horn in situ
Microdialysis with a conventional concentric microdialysis probe introduced through the dorsal surface of the spinal cord was mainly applied in anesthetized animals (5, 9, 10). A small laminectomy was performed at spinal level T_{13}. The microdialysis membrane, usually with a length of 2.0 mm and a diameter of 250 μm, was inserted aiming caudally with an oblique angle of approximately 45° into the dorsal spinal cord immediately lateral to the dorsal vein. Sampling was usually started 2 h

after surgery. As the microdialysis sampling is performed after a short recovery of surgery the potential for physiological consequences of the extensive surgery and trauma associated with the implant may intervene with the neurotransmitter releases following nociceptive input.

(b) Microdialysis in spinal dorsal horn transversely
A microdialysis fiber introduced transversely through the rat spinal cord dorsal horn enabled experiments on anesthetized as well as awake animals. In briefly, two small holes were drilled through the lateral surface of vertebra T_{13} at the level of the dorsal horn. A dialysis probe with a 2–3 mm active length membrane was placed transversely through the spinal dorsal horn. This transverse microdialysis system was firstly described by Skilling and Smullin (6) and provided the ability to locate precisely the spinal terminals from which the neurotransmitter release occurs (11–13). However, its utility is limited by the difficult surgical preparation, the construction of the dialysis probe, and the movement of the awake rats.

(c) Microdialysis in lumbar CSF
The surgical procedures were essentially identical to the insertion of an intrathecal catheter as originally described by Yaksh (14) and the microdialysis probe was subarachnoidally introduced via the atlanto-occipital membrane (8). The loop of the catheter was placed at the rostral margin of the lumbar enlargement (Fig. 2) and the triple dialysis catheter permits simultaneous intrathecal drug delivery and dialysis by single catheterization. It also allows performing the perfusion concurrently with behavioral assessment in the anesthetized rat. Because of the

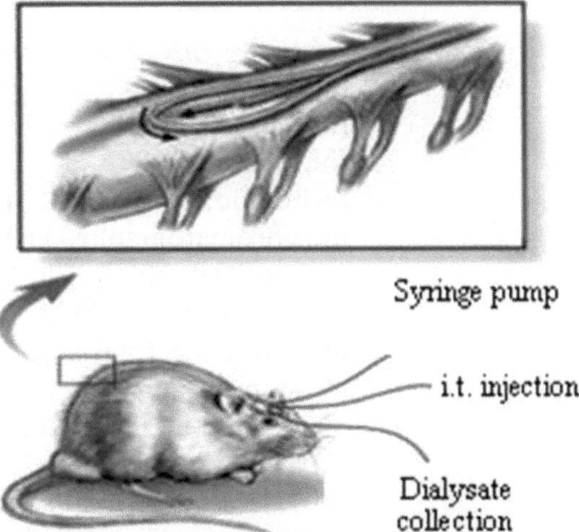

Fig. 2. Loop probe in the intrathecal space. (Modified with permission from the drawing published by Marsil Scientific, http://www.marsilsci.com/index.html).

minimal surgical preparation and dissection, this technique is less traumatic and the recovery of the animal is accelerated. It has been shown that intrathecal microdialysis provides an effective system to assess the effects of spinally delivered agents on the release of Glu and other neurotransmitters (15). However, a specific limitation of the loop dialysis system is the inability to locate precisely the spinal terminals from which the neurotransmitter release occurs.

1.2.2. Microdialysis Setting and the Influencing Factors

When microdialysis sampling is conducted in an awake animal, the animals are placed in a freely moving chamber, and the specialized containment system may prevent the tangling of the fluid tubing. The probe is connected to a microdialysis pump and an isotonic perfusion fluid flows through the interior of the probe at a constant flow rate. The dialysate is then collected in aliquots and analyzed. Microdialysis is not typically carried out at equilibrium conditions implying that the concentration of an analyte in the sample will be some fraction of the actual concentration in the surrounding extracellular fluid. The relationship between the analytic concentration in the dialysate and that in the extracellular fluid may be thought of as extraction efficiency and it is often called "relative recovery." The relative recovery is mostly affected by the components of the perfusion fluid, the perfusion speed, the membrane itself, and the temperature.

(a) *Perfusion fluid.* Although a variety of fluids have been used for perfusion in microdialysis (16), the probe is usually perfused with a physiological solution such as Ringer's or artificial CSF. The constituents of perfusion media such as Ca^{2+}, K^{+}, Mg^{2+}, and pH affect the recovery of a certain substance. The ideal perfusion solution resembles as close as possible to the extracellular fluid.

(b) *Perfusion rate.* The perfusion flow rate is inversely related to relative recovery and is directly proportional to the absolute recovery. A low flow rate results in a higher concentration in the microdialysate (17). This is particularly important if the technique used for analysis, for example high-performance liquid chromatography (HPLC), is optimized for a small sample amount (10–20 μl). For Radioimmunoassay (RIA) or Enzyme immunoassay (ELISA), large sample volumes are normally used and the total amount of substance per sample may be more important than a highly concentrated low-volume sample. Thus, the flow rate and the sampling interval can preferably be adjusted to yield a higher amount of the substance (18). In studies on neurotransmission or neuropeptide release in spinal cord people usually use a flow rate between 5 and 10 μl/min, while 2 μl/min is mostly recommended in brain microdialysis (1).

(c) *Membrane property.* Membrane materials may interact with transported substances and affect the transport. The concentrations of the substances in dialysates increase in proportion to length and diameter of the dialysis membrane.

(d) *Temperature.* An increase in temperature enhances the recovery rate by 1–2%/°C. To prevent fluctuation in diffusion, uniform temperature maintenance at 37°C (±0.5°C) is recommended.

1.3. Detection of Neurotransmitters in Microdialysis Samples

In studies of nociceptive mechanisms in spinal cord, glutamate (Glu)/aspartate (Asp) and PGE_2 have been detected in microdialysates obtained from the intrathecal space (19) and from dorsal horn (7, 12). HPLC or microbore Liquid Chromatography coupled to an electrochemical or fluorimetric detector are the most commonly used methods (20–23) to detect amino acids and monoamines. ELISAs have been used successfully to detect prostanoids in dialysates (24).

1.4. Application of Spinal Microdialysis in Pain Research

Numerous microdialysis studies have been conducted in spinal cord tissue or fluid for estimating neurotransmitter release in a number of animal pain models. Skilling and Smullin (6) successfully demonstrated for the first time the release of Glu and Asp in response to formalin-induced acute nociceptive stimulation. By means of spinal cord microdialysis it has been demonstrated that Glu and NO are sequentially released in the dorsal horns of the spinal cord in response to peripheral nociceptive stimulation (25, 26). Following carrageenan-induced paw inflammation (27), PGE_2 levels increased 20-fold in CSF in association with the peripheral inflammatory response. The COX-1-specific inhibitor did not inhibit this response. By contrast, a COX-2-specific inhibitor significantly attenuated the elevated PGE_2 concentrations in CSF. The microdialysis technique was recently used to sample amino acids from the extracellular fluid of the spinal dorsal horn of patients in whom surgery in the dorsal root entry zone was performed (28). This technique offers new possibilities for clinical research on neurotransmitters involved in some relevant pathological states, especially in chronic pain.

1.5. Application of Spinal Microdialysis in Pharmacokinetics

While most researchers use microdialysis to sample the neurotransmitters or neuromediators following nociceptive input, Hoizey et al. (29) implanted microdialysis probes into the spinal cord of rats to monitor the pharmacokinetics of gacyclidine enantiomers. A recent study (30) determined the intrathecal bioavailability of a mixture of bupivacaine and lidocaine using microdialysis in either the epidural or intrathecal space in rabbits. Ummenhofer and his team simultaneously implanted multiple probes in the CSF at various locations along the spine of pigs enabling the researchers to compare the spinal distribution and clearance of four intrathecally administered opioids (31).

1.6. Aims of the Study

1. To validate and compare two spinal microdialysis approaches in awake, freely moving rats.
2. To assess basal Glu and PGE_2 levels in dorsal horn and in CSF.
3. To evaluate the influences of related factors during the experiment.
4. To investigate formalin-induced releases of Glu and PGE_2 following peripheral inflammation.

2. Materials and Methods

The experimental protocol was approved by the Bioethical Committee for Animal Experimentation of the Vrije Universiteit Brussel and was in accordance with the guidelines for animal experimentation of IASP. Adult male Wistar rats (300–350 g; B&K Universal Limited; England) were housed in groups of four, with free access to food and water. The experiments were carried out between 8:00 and 18:00 h to minimize diurnal rhythmical variability.

2.1. Materials

2.1.1. Spinal Microdialysis Probes

A linear microdialysis (LM) probe consists of a short length (3 mm) of hollow dialysis fiber extended by flexible, plastic, narrow-bore inlet and outlet tubes (Fig. 3, BAS, West Lafayette, Indiana). The diameter (OD) is 320 μm and the molecular weight cutoff is 32 kDa.

A triple loop microdialysis probe consists of an intrathecal tubing attached with inlet and outlet tubes including an active loop dialysis fiber of 4 cm with 200 μm inner diameter and 300 μm outer diameter (Fig. 4, Marsil scientific, San Diego, USA). The molecular weight cutoff is 11 kDa. This triple catheter permits simultaneous acute intrathecal drug delivery and chronic intrathecal dialysis simultaneously.

2.1.2. "Raturn" Awake Animal System

The animal bowl is placed on a turntable, which is in contact with a drive mechanism (Bioanalytical System Inc, USA, Fig. 5). The drive mechanism is activated by a pair of optical sensors mounted on a counterbalanced arm above the animal. The rat moves up and down freely, and can circle in either direction. The drive mechanism is activated only by rotational movement which would otherwise twist and disconnect the tubings.

2.2. Surgical Preparation

2.2.1. Spinal Dorsal Horn Microdialysis

Male Wistar rats were anesthetized with sodium pentobarbital (60 mg/kg, i.p.) for surgical preparation. The skin was incised above the vertebral column from T_9-L_5 and the muscle tissue was cleared away from vertebrae T_{12}-L_1 (Fig. 6). Two small holes were drilled through the lateral surface of vertebra T_{13} at the level of the

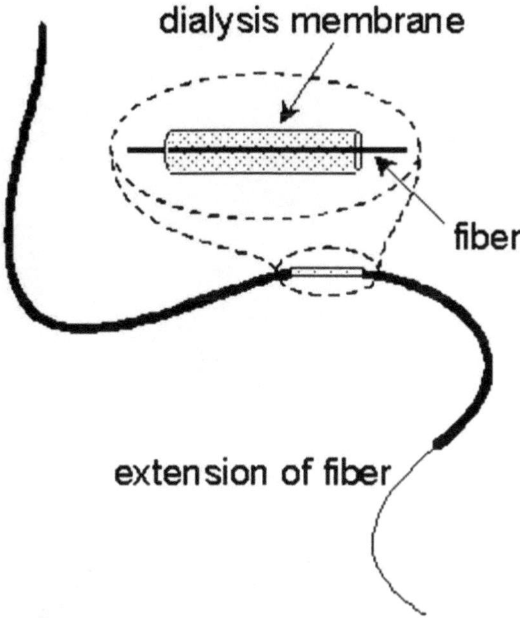

Fig. 3. Linear tissue probe. (Adapted from Heppert 1999).

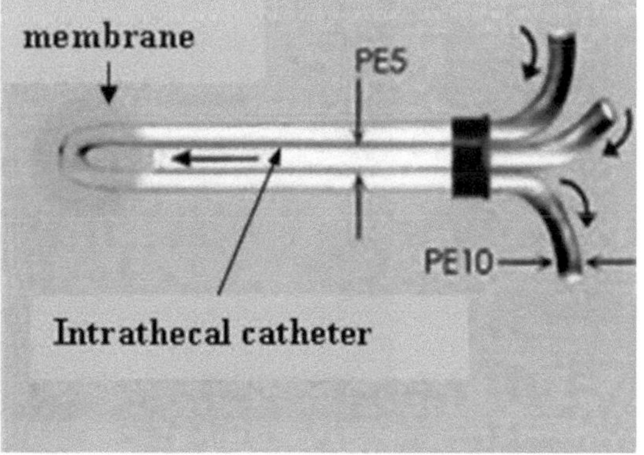

Fig. 4. Triple loop microdialysis probe. (Modified from Marsala 1995).

dorsal horn. A dialysis linear probe with a 3 mm active length membrane (LM-3) was placed transversely through the spinal dorsal cord as described previously (6, 13). The dialysis membrane was covered with epoxy glue, except for the part located in the spinal cord. The ends of the dialysis tube were connected to polyethylene tubing that was exteriorized and fastened in the neck. All rats received 50 μl of buprenorphine S.C. (0.3 mg/ml) for postoperative analgesia.

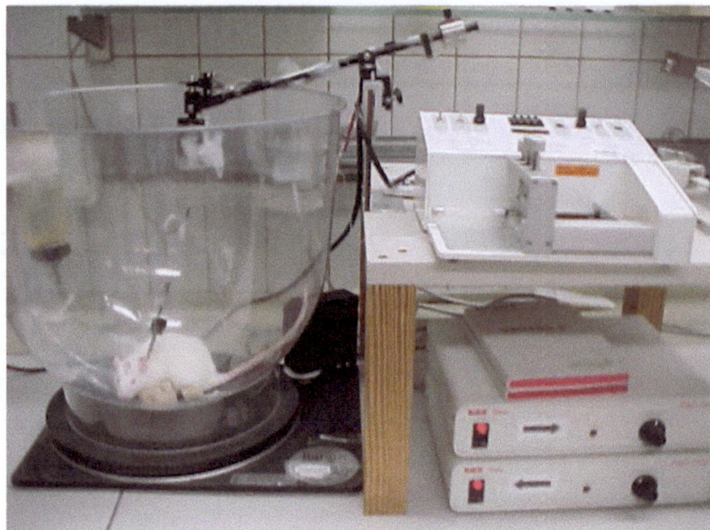

Fig. 5. "Raturn" freely moving animal system.

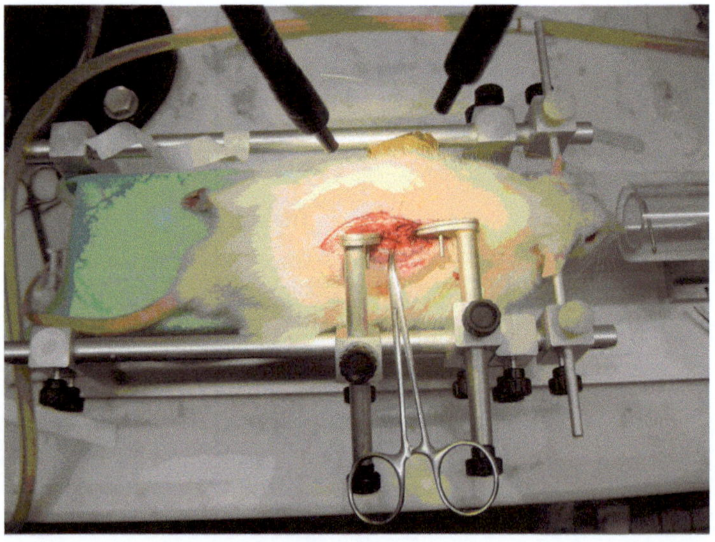

Fig. 6. Implantation of a linear probe in spinal dorsal horn.

2.2.2. Lumbar CSF Microdialysis

A spinal triple loop catheter was introduced subarachnoidally via the atlanto-occipital membrane as described by Marsala et al. (8) under similar pentobarbital anesthesia condition (Fig. 7). In briefly, the loop of the catheter was placed at the rostral margin of the lumbar enlargement. The free ends of the catheter were externalized through the skin at the top of the skull. All rats received 50 μl of buprenorphine S.C. (0.3 mg/ml) for postoperative analgesia.

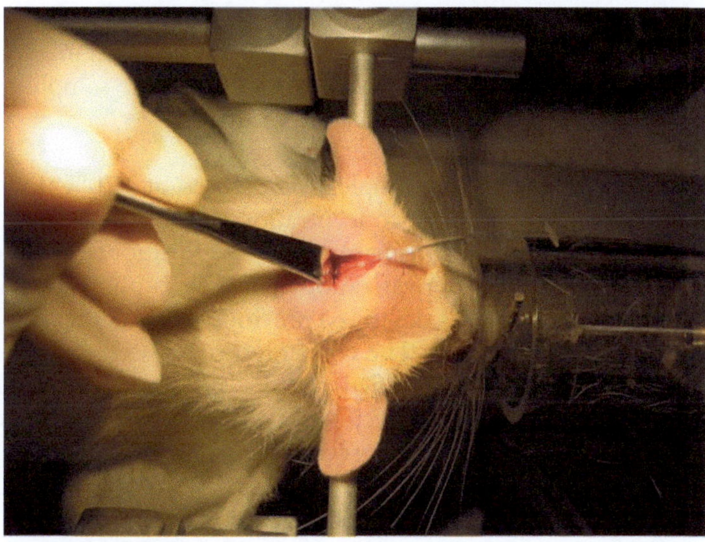

Fig. 7. Implantation of a triple loop probe.

2.3. Validation of Experimental Setting

2.3.1. Baseline Sampling

The rats were placed in a large chamber (Freely Moving System, BAS/Microdialysis, West Lafayette, Indiana) and were allowed to recover from surgery overnight. The dialysis probes were connected to a microdialysis pump (CMA 100, CMA/Microdialysis, Stockholm, Sweden) and were perfused with modified Ringer's solution at a flow rate of 5 μl/min. After an equilibration phase of 1 h, six baseline dialysates were collected every 10 min for 1 h.

2.3.2. Influence of Flow Rate/Perfusing Solution on Basal Glu Levels

Two subgroups were set to analyze the influence of altering the perfusing solution and the flow rate on basal Glu levels.

1. Artificial CSF (NaCl 124 mM, KCl 3.0 mM, $CaCl_2$ 1.35 mM, Na_2HPO_4 0.242 mM, $NaHCO_3$ 20 mM, $MgCl_2$ 1.1 mM) at a flow rate 5 μl/min.
2. Modified Ringer's solution (NaCl 147 mM, KCl 4 mM, $CaCl_2$ 2.3 mM) at flow rate of 2 μl/min.

2.3.3. Influence of Rotation on Basal Glu Levels

Rat activities, particularly the excitatory behavior induced by formalin injection, activated the rotational movement of the microdialysis Raturn bowl. The rotation itself might possibly result in neurotransmitter change. Therefore, we tested the influence of the rotation on Glu release. Following the baseline sampling, the animal bowl was manually rotated 20 times, which is much more than the rotational movement induced by the rat activities during the experiment. The samples were further collected at 10-min interval for 1 h.

2.3.4. Formalin-Induced Glu and PGE_2 Changes

After the six baseline microdialysate collections with modified Ringer's solution at the flow rate of 5 μl/min, 50 μl of 5% formalin was subsequently injected into the plantar surface of the hind paw of the rat, and 10-min samples were further collected for an additional 90 min. In the sham group, 50 μl of saline was injected following 60 min of baseline sampling and dialysates were further collected for 90 min.

2.4. Detection of Glu and PGE_2

All samples were stored at −70°C for subsequent analysis of Glu. The concentrations of Glu were analyzed by microbore Liquid Chromatography with fluorescence detection after precolumn derivatization with ortho-phtalaldehyde and β-mercaptoethanol, as described previously (20).

The concentration of PGE_2 in the microdialysate samples was quantified with a commercially available Correlate-EIA PGE_2 (competitive immunoassay) kit in accordance with the manufacturer's protocol (Assay Design, Inc., USA). The concentration of PGE_2 was calculated from the measured optical density by means of four-parameter logistic regression. A standard curve was constructed between 39.4 and 5,000 pg/ml.

2.5. Verification of Probe Positioning

After each experiment, the animals were killed with an overdose of pentobarbital. The part of the spinal cord containing the microdialysis membrane was dissected. The position of the microdialysis membrane in the dorsal horn or in CSF was confirmed by injecting methylene blue.

2.6. Data Analysis

Data are reported as mean value ± SE and were analyzed with SPSS 11.5 for Windows. Statistical differences in Glu or PGE_2 level in time within one group were assessed by general linear model for repeated measures. Statistical significance of differences was considered if $P < 0.05$.

3. Results

3.1. Baseline Concentrations of Glu and PGE_2

Basal concentrations of Glu (0.82 ± 0.09 μM) and PGE_2 (66.27 ± 17.11 pg/ml) were found in the spinal dorsal horn with the linear probe ($n = 7$) perfused with modified Ringer's solution at a flow rate of 5 μl/min (Fig. 8). Larger but also stable baseline concentrations of Glu (5.96 ± 0.22 μM) and PGE_2 (318.28 ± 37.16 pg/ml) were observed with the loop probes ($n = 7$) in the CSF of awake rats.

3.2. Influence of the Flow Rate on Basal Glu Level

As shown in Fig. 9, decreasing the flow rate from 5 to 2 μl/min increased extracellular Glu concentrations by 222.7 ± 27.3% with a linear probe and by 215.9 ± 15.2% with a loop probe. Since microdialysis is a kinetic dialysis process, lower flow rates will result in higher recovery of Glu across the dialysis membrane.

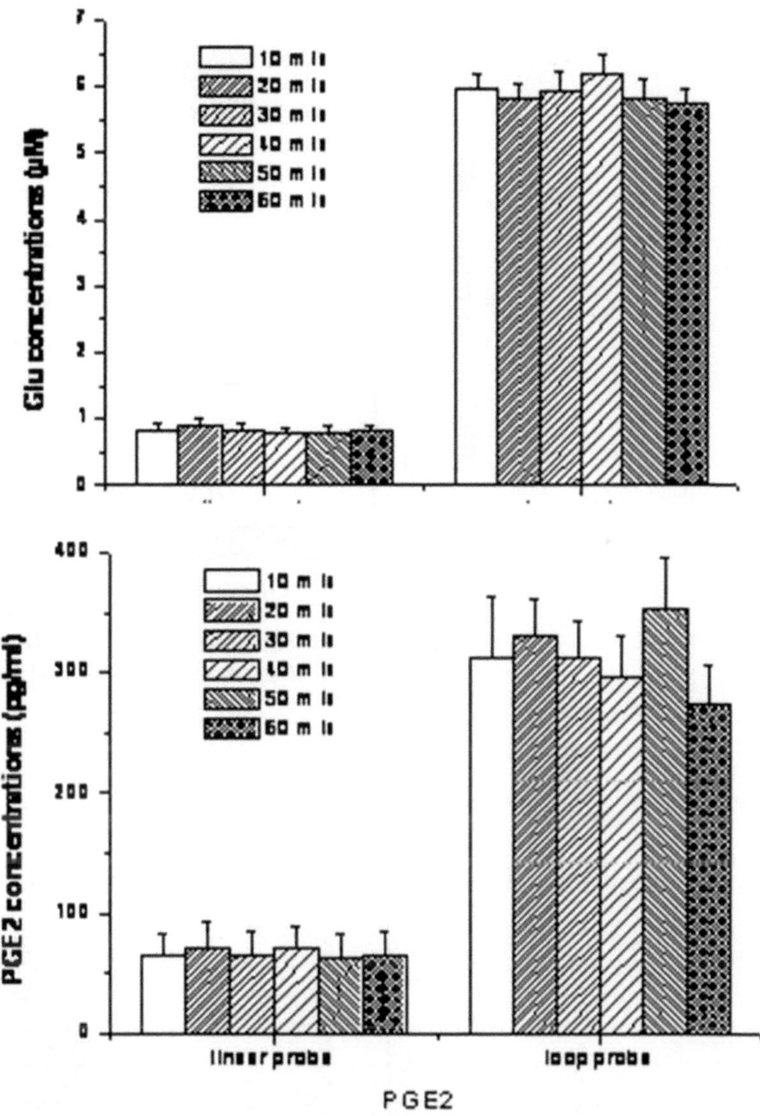

Fig. 8. Basal Glu and PGE_2 levels in the extracellular fluid of the spinal dorsal horn (*left panel*, $n=7$) and in the CSF (*right panel*, $n=7$) during 1-h sampling at 10-min interval. Probes were perfused with modified Ringer's solution at a flow rate of 5 μl/min. The basal concentrations of Glu and PGE_2 were significantly different ($P<0.01$) with two different probes.

3.3. Influence of the Perfusion Fluid on Basal Glu Level

As we applied the LM-3 probe for the first time to the spinal dorsal horn of conscious rats, we further investigated the influence of altering the perfusing fluid on basal Glu concentration. Perfusion with artificial CSF reduced baseline Glu levels to 61.5 ± 9.5% in dorsal horn ($n=4$, Fig. 10).

3.4. Influence of the Bowl Rotation on Basal Glu Level

When the microdialysis "Raturn" bowl was manually rotated 20 times, we noticed a slight but not significant increase in the basal Glu concentration sampled with either the linear or the loop probe (Fig. 11).

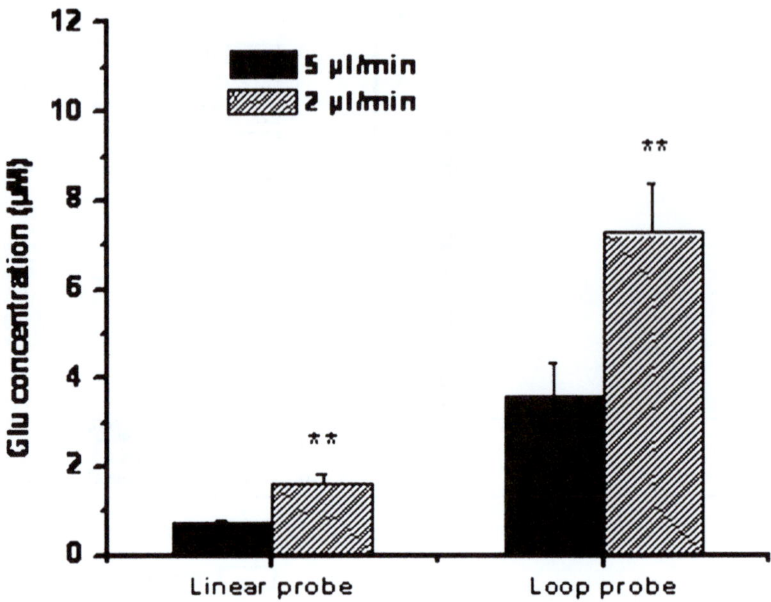

Fig. 9. Glu concentrations change with altering the flow rate. *Dark box*: 5 µl/min, *Grey box*: 2 µl/min; $n=5$; $^{**}P<0.01$ versus the baseline level obtained at 5 µl/min.

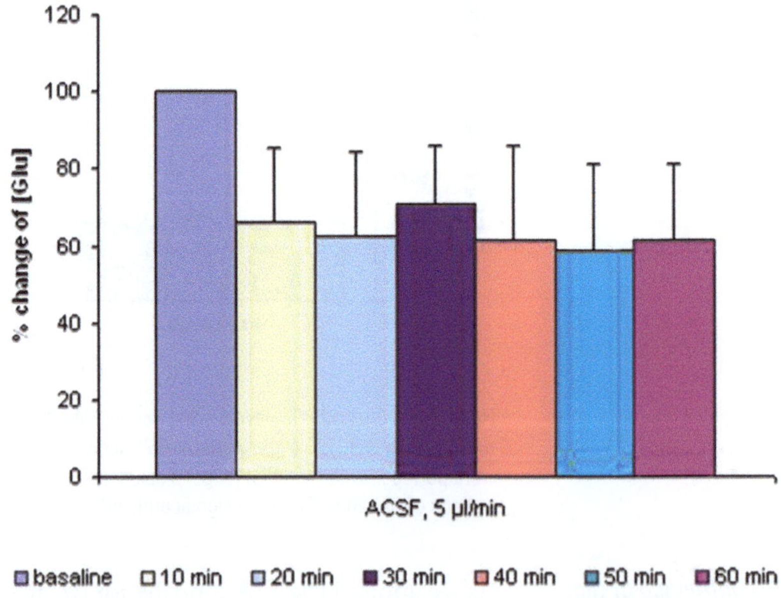

Fig. 10. Change of Glu level with the alteration in perfusing solution. Baseline Glu level was obtained by perfusing the probe with modified Ringer's solution at a flow rate of 5 µl/min and was set as 100% ($n=4$). Perfusion solution was changed from Ringer's to artificial CSF (ACSF) 1 h after baseline sampling.

3.5. Formalin-Induced Glu and PGE_2 Releases

Injection of 50 µl of 5% formalin at the plantar surface of one hind paw of the rats resulted in an immediate increase in Glu to 154.1 ± 10.7% ($P<0.05$, Fig. 12-A_1) and in PGE_2 to 300.3 ± 29.1% ($P<0.01$, Fig. 12-B_1) of their baseline in the spinal dorsal horn by

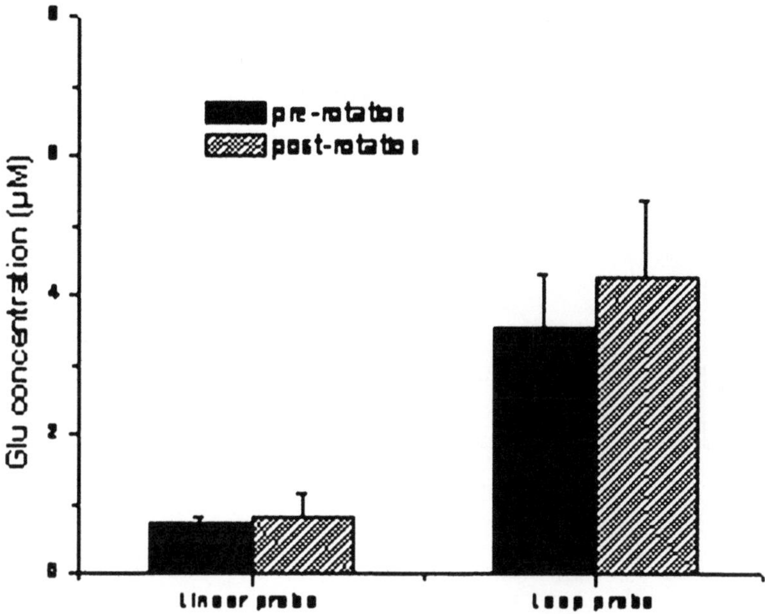

Fig. 11. Influence of rotation on Glu release. Rotation: Rotating manually the bowl for 20 times. No statistical differences were found either with the probe or with the loop probe ($n=5$).

using a linear probe. Most rats showed the typical nociceptive behavior following formalin injection, as flinching, biting, and moving. Those activities activated the rotational movement of the bowl. A similar profile and time course in Glu and PGE_2 responses following formalin inflammation were observed in the CSF with the loop microdialysis probe ($174.7 \pm 10.2\%$ in Glu and $326.2 \pm 50.8\%$ in PGE_2, respectively, Fig. 12-A_2, B_2).

3.6. Verification of the Probe Positioning

Correct probe positioning was obtained in 12 out of 16 operated rats with LM-3 probe in the dorsal horn, and in 19 out of 20 rats with the loop microdialysis probe in lumbar CSF. No significant inflammatory morphologic changes were found around the dialyzed area of the spinal cord after 2–3 days following implantation of each probe (Fig. 13).

4. Discussion

Using two different methods of spinal microdialysis in awake freely moving animals, stable baseline levels of Glu and PGE_2 were obtained in the spinal dorsal horn and in the lumbar CSF during the 1-h sampling period. Although the basal concentrations of Glu and PGE_2 in the dialysates were significantly different for the two approaches, the profile, time course of the response, and the relative change in Glu and PGE_2 induced by formalin remained similar.

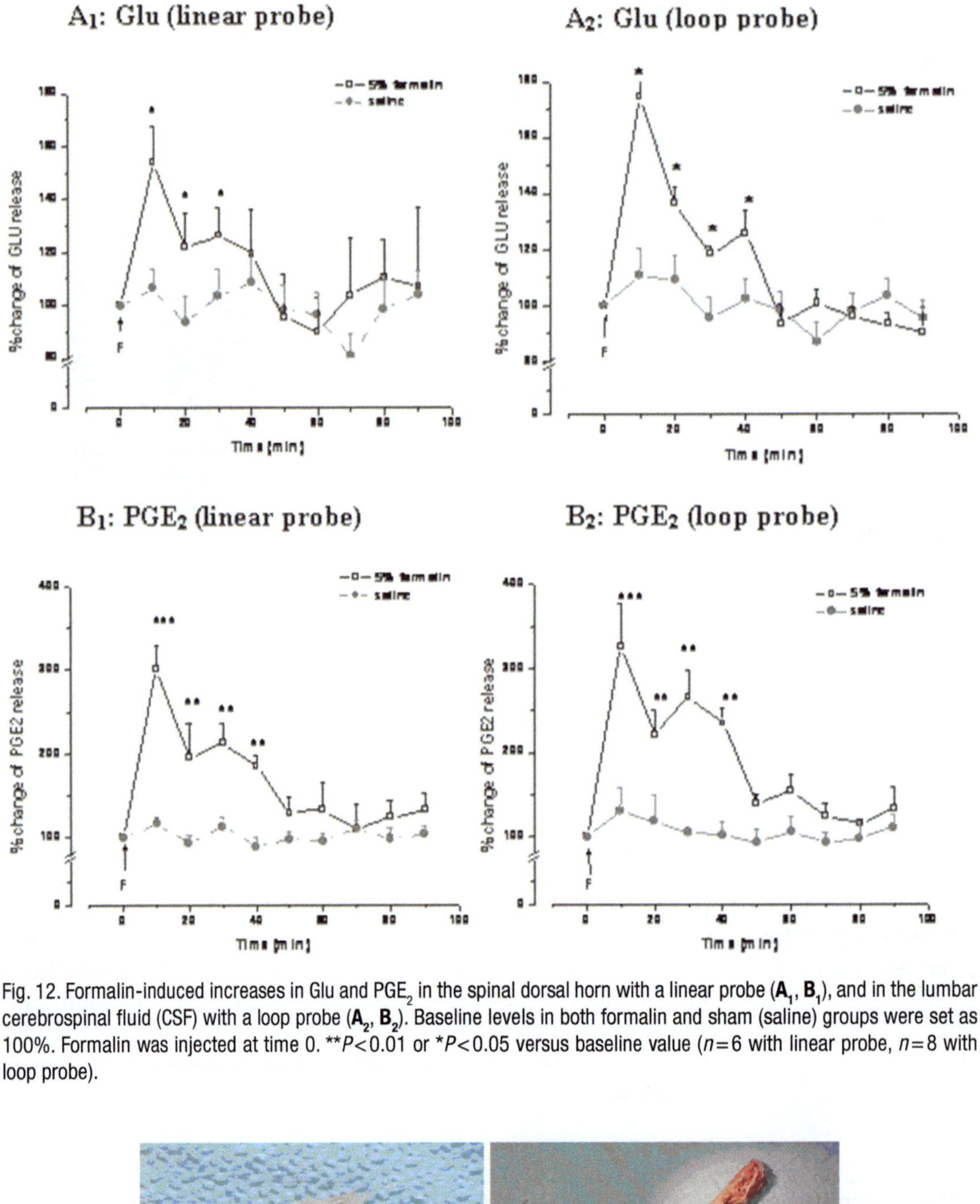

A1: Glu (linear probe)

A2: Glu (loop probe)

B1: PGE2 (linear probe)

B2: PGE2 (loop probe)

Fig. 12. Formalin-induced increases in Glu and PGE_2 in the spinal dorsal horn with a linear probe (**A_1**, **B_1**), and in the lumbar cerebrospinal fluid (CSF) with a loop probe (**A_2**, **B_2**). Baseline levels in both formalin and sham (saline) groups were set as 100%. Formalin was injected at time 0. $^{**}P<0.01$ or $^{*}P<0.05$ versus baseline value ($n=6$ with linear probe, $n=8$ with loop probe).

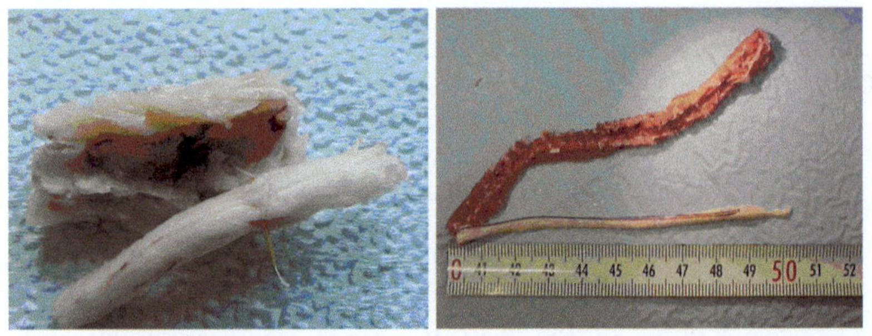

Fig. 13. Verification of the probe positioning (*arrow*). (**a**) Linear probe in the spinal dorsal horn; (**b**) loop probe in the intrathecal space.

The spinal cord microdialysis in the dorsal horn is a useful tool to investigate changes of neurotransmitter release at their precise location (11, 32). However, its complicated surgical preparation and especially the construction of the probe limit its routine use. Each laboratory manufactures its own probes which may contribute to large inter-laboratory variations (6, 26, 33, 34). The linear tissue probe used in the present study is designed for in vivo sampling from peripheral tissues such as the dermis and subcutaneous tissue, and from the liver and other organs (35). The probe consists of a short length of hollow dialysis fiber attached to narrow-bore inlet and outlet tubes. The 3 mm active length of the dialysis membrane allowed it to be located within the spinal dorsal cord. To the best of our knowledge, we are the first to validate this LM-3 linear probe to sample the extracellular fluid of the spinal dorsal horn in awake and freely moving rats. The basal Glu and PGE_2 concentrations in most of the tested rats were lower (<1 μM) when compared to results obtained with home-constructed probes and protocols within the different laboratories (6, 12, 36). Differences in membrane recovery, flow rate, and perfusing solution may account for this. Because dialysis is a kinetic process, decreasing the flow rate should result in increased extracellular Glu concentrations, which was indeed found in our experiments. Taken together, reliable measurements of Glu and PGE_2 release in the dorsal horn can be obtained with the LM-3 linear tissue microdialysis probe using the modified Ringer's solution at a flow rate of 5 μl/min.

The microdialysis of the CSF with a loop probe using the modified method of Marsala (8) was also set up in our laboratory for comparing the profile of the release of Glu in the CSF (37). The 4 cm active membrane of the loop probe offered a much larger recovery than the 3 mm one of the linear probe. The triple dialysis catheter permits simultaneous intrathecal drug delivery and dialysis by single catheterization. Because of the minimal surgical preparation and dissection, the recovery of the animal is accelerated, and this technique represents a less traumatic intervention. The basal as well as the evoked releases of Glu and PGE_2 assessed with the loop probe are more reproducible, and consistent with the similar profile and time course of Glu release when compared to the LM-3 probe following formalin injection. In agreement with previous studies, loop probe microdialysis of the CSF provides an effective system to assess the effects of spinally delivered agents on the release of Glu and other neurotransmitters (15, 38–40). However, a specific limitation of the loop dialysis system is the inability to locate precisely the spinal terminals from which the Glu release occurs. Major advantages of this technique include preservation of spinal tissue integrity and the ability to perform the spinal perfusion concurrently with behavioral assessments in the awake rat.

As mentioned above, marked increases in Glu and PGE_2 concentrations were observed with both approaches at the early phase of the formalin injection. Although the activity-induced rotational activity of the bowl did increase Glu release by approximately 10%, the increase did not reach statistical difference. The significant increases in Glu as well as in PGE_2 were thus specifically elicited by the formalin inflammation (41–43). It is known that paw injection of formalin causes a prolonged activation of primary afferent C-nociceptors, resulting in enhanced activity of nociceptive dorsal horn neurons (44, 45). Following peripheral tissue injury Glu as well as other excitatory neurotransmitters are released from primary afferents and dorsal horn neurons (11, 26). Our results provide kinetic evidence of increased Glu and PGE_2 release from the dorsal horn and in the CSF after paw injection of formalin in freely moving animals.

In conclusion, the microdialysis of the dorsal horn or of the CSF are both effective systems to assess the alteration in Glu as well as PGE_2 release following peripheral nociceptive input. The transverse microdialysis system through the spinal cord dorsal horn provided the ability to locate precisely the spinal terminals from which the neurotransmitter release occurs, whereas spinal loop dialysis in CSF is a robust approach performing the perfusion concurrently with behavioral assessment in the conscious rat. In daily laboratory work the loop probe system in the CSF is more reproducible when the effect of drug on neurotransmitter change is being investigated.

References

1. Khan SH, Shuaib A (2001) The technique of intracerebral microdialysis. Methods 23:3–9
2. Janle EM, Kissinger PT (1996) Microdialysis and ultrafiltration. Adv Food Nutr Res 40: 183–196
3. Muller M (2002) Science, medicine, and the future: microdialysis. BMJ 324:588–591
4. Lindefors N, Brodin E, Ungerstedt U (1987) Microdialysis combined with a sensitive radioimmunoassay. A technique for studying in vivo release of neuropeptides. J Pharmacol Methods 17:305–312
5. Brodin E, Linderoth B, Gazelius B, Ungerstedt U (1987) In vivo release of substance P in cat dorsal horn studied with microdialysis. Neurosci Lett 76:357–362
6. Skilling SR, Smullin DH, Beitz AJ, Larson AA (1988) Extracellular amino acid concentrations in the dorsal spinal cord of freely moving rats following veratridine and nociceptive stimulation. J Neurochem 51:127–132
7. Sluka KA, Westlund KN (1993) Spinal cord amino acid release and content in an arthritis model: the 6 effects of pretreatment with non-NMDA, NMDA, and NK1 receptor antagonists. Brain Res 627:89–103
8. Marsala M, Malmberg AB, Yaksh TL (1995) The spinal loop dialysis catheter: characterization of use in the unanesthetized rat. J Neurosci Methods 62:43–53
9. Gustafsson H, de Araujo LG, Schott E, Stiller CO, Alster P, Wiesenfeld-Hallin Z, Brodin E (1999) Measurement of cholecystokinin release in vivo in the rat spinal dorsal horn. Brain Res Brain Res Protoc 4:192–200
10. Afrah AW, Fiska A, Gjerstad J, Gustafsson H, Tjolsen A, Olgart L, Stiller CO, Hole K, Brodin E (2002) Spinal substance P release in vivo during the induction of long-term potentiation in dorsal horn neurons. Pain 96:49–55
11. Sluka KA, Westlund KN (1992) An experimental arthritis in rats: dorsal horn aspartate and glutamate increases. Neurosci Lett 145:141–144
12. Sorkin LS, McAdoo DJ (1993) Amino acids and serotonin are released into the lumbar spinal cord of the anesthetized cat following intradermal capsaicin injections. Brain Res 607:89–98
13. Scheuren N, Neupert W, Ionac M, Neuhuber W, Brune K, Geisslinger G (1997) Peripheral noxious stimulation releases spinal PGE2 during the first phase in the formalin assay of the rat. Life Sci 60:L-300

14. Yaksh TL, Rudy TA (1976) chronic catheterization of spinal subarachnoid space. Physiol Behav 17: 1031–1036
15. Hua XY, Calcutt NA, Malmberg AB (1997) Neonatal capsaicin treatment abolishes formalin-induced spinal PGE2 release. Neuroreport 8:2325–2329
16. Benveniste H, Huttemeier PC (1990) Microdialysis–theory and application. Prog Neurobiol 35:195–215
17. Benveniste H (1989) Brain microdialysis. J Neurochem 52:1667–1679
18. Malmberg AB, Hamberger A, Hedner T (1995) Effects of prostaglandin E2 and capsaicin on behavior and cerebrospinal fluid amino acid concentrations of unanesthetized rats: a microdialysis study. J Neurochem 65:2185–2193
19. Malmberg AB, Yaksh TL (1995) Cyclooxygenase inhibition and the spinal release of prostaglandin E2 and amino acids evoked by paw formalin injection: a microdialysis study in unanesthetized rats. J Neurosci 15:2768–2776
20. Smolders I, Sarre S, Michotte Y, Ebinger G (1995) The analysis of excitatory, inhibitory and other amino acids in rat brain microdialysates using microbore liquid chromatography. J Neurosci Methods 57:47–53
21. Lunte SM, Lunte CE (1996) Microdialysis sampling for pharmacological studies: HPLC and CE analysis. Adv Chromatogr 36:383–432
22. Davies MI, Cooper JD, Desmond SS, Lunte CE, Lunte SM (2000) Analytical considerations for microdialysis sampling. Adv Drug Deliv Rev 45:169–188
23. Parent M, Bush D, Rauw G, Master S, Vaccarino F, Baker G (2001) Analysis of amino acids and catecholamines, 5-hydroxytryptamine and their metabolites in brain areas in the rat using in vivo microdialysis. Methods 23:11–20
24. Muth-Selbach US, Tegeder I, Brune K, Geisslinger G (1999) Acetaminophen inhibits spinal prostaglandin E2 release after peripheral noxious stimulation. Anesthesiology 91:231–239
25. Okuda K, Sakurada C, Takahashi M, Yamada T, Sakurada T (2001) Characterization of nociceptive responses and spinal releases of nitric oxide metabolites and glutamate evoked by different concentrations of formalin in rats. Pain 92:107–115
26. Vetter G, Geisslinger G, Tegeder I (2001) Release of glutamate, nitric oxide and prostaglandin E2 and metabolic activity in the spinal cord of rats following peripheral nociceptive stimulation. Pain 92:213–218
27. Smith CJ, Zhang Y, Koboldt CM, Muhammad J, Zweifel BS, Shaffer A, Talley JJ, Masferrer JL, Seibert K, Isakson PC (1998) Pharmacological analysis of cyclooxygenase-1 in inflammation. Proc Natl Acad Sci U S A 95:13313–13318
28. Mertens P, Ghaemmaghami C, Bert L, Perret-Liaudet A, Guenot M, Naous H, Laganier L, Later R, Sindou M, Renaud B (2001) Microdialysis study of amino acid neurotransmitters in the spinal dorsal horn of patients undergoing microsurgical dorsal root entry zone lesioning. Technical note. J Neurosurg 94:165–173
29. Hoizey G, Kaltenbach ML, Dukic S, Lamiable D, Lallemand A, D'Arbigny P, Millart H, Vistelle R (2000) Distribution of gacyclidine enantiomers in spinal cord extracellular fluid. Pharm Res 17:148–153
30. Clement R, Malinovsky J, Le Corre P, Dollo G, Chevanne F, Le Verge R (2000) Spinal biopharmaceutics of bupivacaine and lidocaine by microdialysis after their simultaneous administration in rabbits. Int J Pharm 203:227–234
31. Ummenhofer WC, Arends RH, Shen DD, Bernards CM (2000) Comparative spinal distribution and clearance kinetics of intrathecally administered morphine, fentanyl, alfentanil, and sufentanil. Anesthesiology 92:739–753
32. Sorkin LS, Moore JH (1996) Evoked release of amino acids and prostanoids in spinal cords of anesthetized rats: changes during peripheral inflammation and hyperalgesia. Am J Ther 3:268–275
33. Zahn PK, Sluka KA, Brennan TJ (2002) Excitatory amino acid release in the spinal cord caused by plantar incision in the rat. Pain 100:65–76
34. Zhang XY, Zhang JW, Chen B, Bai ZT, Shen J, Ji YH (2002) Dynamic determination and possible mechanism of amino acid transmitter release from rat spinal dorsal horn induced by the venom and a neurotoxin (BmK I) of scorpion Buthus martensi Karsch. Brain Res Bull 58:27–31
35. Heppert KE, Davies MI (1999) Simultaneous determination of caffeine from blood, brain and muscle using microdialysis in an awake rat and the effect of caffeine on rat activity. Curr Sep 18:3–7
36. Sluka KA, Willis WD (1998) Increased spinal release of excitatory amino acids following intradermal injection of capsaicin is reduced by a protein kinase G inhibitor. Brain Res 798:281–286
37. Shi L, Smolders I, Sarre S, Michotte Y, Zizi M, Camu F (2004) Formalin-induced spinal glutamate release in freely moving rats: comparison of two spinal microdialysis approaches. Acta Anaesthesiol Belg 55:43–48
38. Hua XY, Chen P, Marsala M, Yaksh TL (1999) Intrathecal substance P-induced thermal hyperalgesia and spinal release of prostaglandin E2 and amino acids. Neuroscience 89:525–534

39. Hua XY, Moore A, Malkmus S, Murray SF, Dean N, Yaksh TL, Butler M (2002) Inhibition of spinal protein kinase Calpha expression by an antisense oligonucleotide attenuates morphine infusion-induced tolerance. Neuroscience 113:99–107
40. Nishiyama T (2000) Interaction between intrathecal morphine and glutamate receptor antagonists in formalin test. Eur J Pharmacol 395:203–210
41. Coderre TJ, Melzack R (1992) The contribution of excitatory amino acids to central sensitization and persistent nociception after formalin-induced tissue injury. J Neurosci 12: 3665–3670
42. Malmberg AB, Yaksh TL (1995) The effect of morphine on formalin-evoked behaviour and spinal release of excitatory amino acids and prostaglandin E2 using microdialysis in conscious rats. Br J Pharmacol 114:1069–1075
43. Dirig DM, Hua XY, Yaksh TL (1997) Temperature dependency of basal and evoked release of amino acids and calcitonin gene-related peptide from rat dorsal spinal cord. J Neurosci 17:4406–4414
44. Dickenson AH, Sullivan AF (1987) Subcutaneous formalin-induced activity of dorsal horn neurones in the rat: differential response to an intrathecal opiate administered pre or post formalin. Pain 30:349–360
45. Coderre TJ, Fundytus ME, McKenna JE, Dalal S, Melzack R (1993) The formalin test: a validation of the weighted-scores method of behavioural pain rating. Pain 54:43–50

Chapter 9

Monitoring Extracellular Monoamines with In Vivo Microdialysis in Awake Rats: A Practical Approach

Roberto W. Invernizzi

Abstract

The microdialysis technique for the measurement of brain extracellular levels of monoamines has become very popular over the last 2–3 decades, particularly in laboratories involved in neuropharmacology studies. Nevertheless, microdialysis of monoamines is a challenging technique for several reasons and to get reliable results it is necessary to develop sufficient skill and be ready to solve technical and analytical problems that may occur. This chapter is intended to help obtain reliable and reproducible results by describing in detail practical aspects of probe construction and implantation into the rat brain and measurement of noradrenaline, dopamine, and serotonin with HPLC coupled to electrochemical detection, set up, optimized, and validated in this author's laboratory.

Key words: Microdialysis, High-performance liquid chromatography, Electrochemical detection, Monoamines, Noradrenaline, Dopamine, Serotonin, In vivo release

1. Introduction

Brain monoamines, namely, noradrenaline (norepinephrine; NA), dopamine (DA), and serotonin (5-hydroxytryptamine; 5-HT), are neurotransmitters controlling basic brain functions such as sensory-motor integration, neuroendocrine regulation, reward response, arousal, attention, cognitive performance, mood, sleep, and appetite, just to cite a few. In addition, monoamines play a major role in the mechanism of action of drugs used for the treatment of the main psychiatric (depression, schizophrenia) and neurological (Parkinson) diseases. Soon after their presence in brain neurons was identified, in the 1960s (1), several biochemical ex vivo methods were developed to allow the measurement of tissue levels, metabolism, and turnover of brain monoamines. Although these techniques importantly contributed to clarify the mechanism underlying, for instance, Parkinson's disease neuropathology, their

Giuseppe Di Giovanni and Vincenzo Di Matteo (eds.), *Microdialysis Techniques in Neuroscience*, Neuromethods, vol. 75, DOI 10.1007/978-1-62703-173-8_9, © Springer Science+Business Media, LLC 2013

main limitation is the lack of relationship with synaptic monoamines. In vivo techniques such as ventricular perfusion via indwelling cannulae (2), superfusion of discrete brain regions by means of push-pull cannulae (3–5) or cups (6), and voltammetry with carbon fiber microelectrodes (7, 8), developed between 1950s and early 1980s, overcame many of the limitations of previous techniques allowing the continuous measurement of some neurotransmitters present in the brain interstitial fluid. However, these techniques still lack the sensitivity and selectivity to detect endogenous extracellular levels of monoamines. The development of brain microdialysis determined a technical advance in the measurement of endogenous monoamine release. Ungerstedt and coworkers, in 1974, using loop-type probes showed that striatal (^{3}H)DA release in freely moving rats was enhanced by amphetamine (9). The development of analytical techniques such as high-performance liquid chromatography coupled to electrochemical detection (HPLC-ED) (10, 11), selective and sensitive enough to detect femtomole amounts of DA, allowed the detection of basal levels of endogenous DA present in dialysate collected from the rat striatum (12–14). In the 1980s, methods based on the use of HPLC-ED were developed for the analysis of NA (15) and 5-HT (16–18) in dialysate and rapidly became the most popular analytical methods for the determination of monoamines in dialysate. These initial studies importantly contributed to the worldwide diffusion of the microdialysis technique for the measurement of monoamines in brain extracellular fluid. Search in PubMed (Feb 2011) shows that since 1970, over 13,500 microdialysis studies have been published. A large majority (9,770) are brain microdialysis studies mainly performed in rats (3,934) and, to a lesser extent, in mice (685). Brain microdialysis has been largely used to monitor extracellular levels of dopamine (1,682), 5-HT (951), and noradrenaline (457) in the rat brain. The largest number of microdialysis studies on rat brain monoamines was published in the last two decades (Fig. 1). However, their number tended to decrease in the last 10 years. By contrast, microdialysis studies in the mouse brain were markedly increased in the last decade (Fig. 2), likely because of the increased interest in transgenic mouse models of neuropathology.

This chapter describes in detail the microdialysis technique used in this author's laboratory to monitor the in vivo extracellular concentration of NA, DA, and 5-HT in discrete brain regions of the awake rat over 20 years of activity in the field (19–43).

Each section of the chapter is subdivided into two main subheadings to cover all aspects of the (1) microdialysis procedure and (2) analysis of samples. Particular emphasis is given to practical aspects of the microdialysis technique and indications to help improve the successfulness of the technique, rarely described in the scientific literature. Basic knowledge of neurophysiology and neurochemistry as well as basic principles of microdialysis and liquid

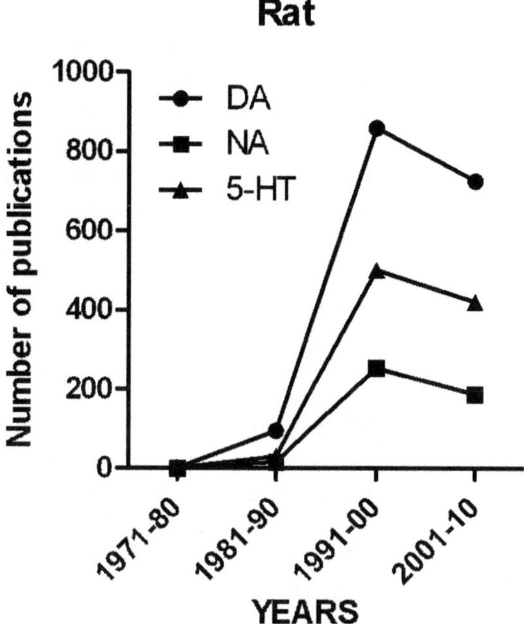

Fig. 1. The *curves* indicate the number of papers retrieved from PubMed in which microdialysis was applied to the analysis of *rat* brain monoamines. PubMed searches were done on Feb 2011 using "microdialysis brain rat and dopamine (or noradrenaline or 5-HT)" as key words. *DA* dopamine, *NA* noradrenaline, *5-HT* serotonin.

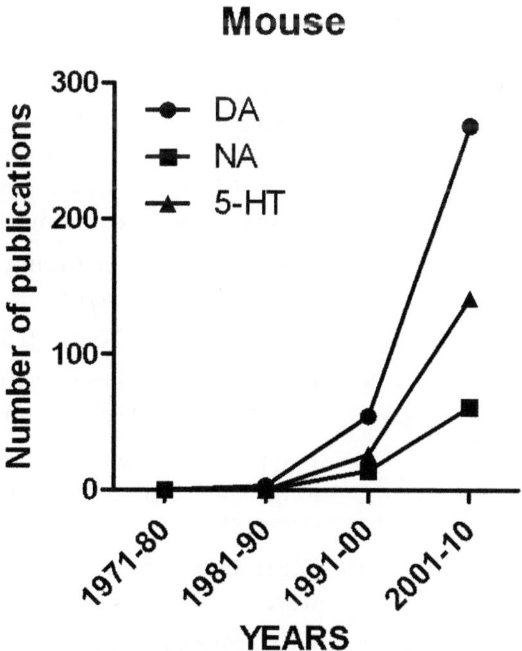

Fig. 2. The *curves* indicate the number of papers retrieved from PubMed in which microdialysis was applied to the analysis of *mouse* brain monoamines. PubMed searches were done on Feb 2011 using "microdialysis brain rat and dopamine (or noradrenaline or 5-HT)" as key words. *DA* dopamine, *NA* noradrenaline, *5-HT* serotonin.

chromatography are taken for granted and are not covered in this chapter. Excellent descriptions of the general aspects of the technique can be found elsewhere (44–46). Previous publications dealing with specific aspects of monoamine microdialysis are also available (47, 48). It is clear that any written description cannot substitute for hands-on experience in a laboratory in which microdialysis is a routine technique.

2. Materials

2.1. Brain Microdialysis

2.1.1. Construction of a Concentric Probe

There are several types of microdialysis probes: trans-cerebral or horizontal, loop-shaped, side-by-side, and concentric probes, also named I-shaped or vertical probe (see ref. (49) for schematic representation of the different probe types). Concentric probes are currently the most used because they are easy to construct and implant into the brain of laboratory animals. Construction and implantation of trans-cerebral and loop-type probes are not described here but details of their preparation can be found elsewhere (45).

The essential materials for constructing a microdialysis probe are the following:

- Small-diameter semipermeable dialysis hollow fibers easily obtained from hemodialysis filters used for patient with renal failure. Different materials and pore size (molecular weight cutoff) are available. Cellulose-derived membranes such as Cuprophan (216 μm outer diameter; 3,000 Da cutoff; Sorin Biomedica, Italy) and synthetic, polymer-derived membranes such as the acrylonitrile and sodium methallyl sulfonate copolymer (310 μm outer diameter; 44,000 Da cutoff; AN69, Hospal, Italy) are among the most common fiber materials.
- 26G stainless-steel tubing for Cuprophan membranes (23G for AN69 membrane) (Coopers Needle Works, UK)
- Tungsten wires, 76 mm × 80 μm (Clark Electromedical, UK)
- Fused silica capillaries, 150 μm outer and 75 μm inner diameters (Polymicro Technologies, USA)
- Polythene tubing for probe inlet (0.4 mm inner diameter; 3 cm long) and outlet (0.28 mm inner diameter; 4.5 cm long) (Portex)
- Water-resistant, two-component epoxy glue (most hardware stores)
- Dissecting stereomicroscope (Leica, Italy) or magnifying lens

2.1.2. Perfusion Medium Composition

Artificial cerebrospinal fluid (aCSF) containing 145 mM NaCl, 1.26 mM $CaCl_2$, 1 mM $MgCl_2$, 3.0 mM KCl, 1.4 mM Na_2HPO_4

is used as perfusion medium. aCSF is prepared by dissolving the following salts in ultrapure water:

- 8.47 g NaCl
- 224 mg KCl
- 203 mg $MgCl_2 \cdot 6H_2O$
- 250 mg $Na_2HPO_4 \cdot 2H_2O$
- 185 mg $CaCl_2 \cdot 2H_2O$

Salts are dissolved in 900 mL ultrapure water (Millipore Rios 50). $CaCl_2 \cdot 2H_2O$ is slowly added at the end to avoid precipitation. pH is adjusted to 7.4 with few drops of 0.6 M NaH_2PO_4 (prepared by dissolving 8.28 g in 100 mL water). Finally, volume is adjusted to 1 L with water. aCSF is filtered through a 0.22 μm filter under vacuum on a Millipore-like filtration system and stored at 4°C for 1 week.

2.1.3. Implantation and Tethering of Microdialysis Probe

Microdialysis probes are implanted in the target brain region using conventional stereotaxic techniques:

- Rat (or mouse)
- Microdialysis probe
- aCSF
- Microinfusion pump (CMA/100 or CMA/400, CMA Microdialysis, Sweden; Harvard 22, Harvard Apparatus, USA)
- Stereotaxic frame model 900 with one or two manipulators equipped with probe holder (David Kopf Instruments, USA). Mouse adapter is recommended if this species is used
- Stereotaxic atlas of the rat (mouse) brain
- Surgery lamp or fiber-optic cold light to illuminate surgery field
- Stereomicroscope (preferred), magnifying lens, or glasses
- 500 mL Equithesin containing 4.8 g sodium pentobarbital, 21.2 g chloral hydrate, 10.6 g $MgSO_4 \cdot 7H_2O$ 198 mL propylene glycol, and 50 mL absolute ethanol. Equithesin mixture was made by dissolving chloral hydrate and $MgSO_4$ in sterile water, adding pentobarbital and ethanol. When salts are dissolved, add propylene glycol and adjust the final volume with sterile water. If this sequence is used, a clear solution is immediately obtained. Attempting to dissolve all ingredients together may require prolonged stirring. Keep at 4°C for no more than 2 weeks. This anesthetic is recommended for acute preparation; typically rats are sacrificed at the end of the microdialysis experiment, 1 day after surgery. For chronic preparations, other anesthetics are more appropriate (see Note 4.1)
- Ventilated surgical table to evacuate volatile solvents (acrylic) or inhalation anesthetics
- Shaver or scissors to shave rat head

- Dental drill
- Surgical instruments including scalpel, fine scissors, hemostats
- Cotton swabs
- Spatula, jeweler forceps, and screwdriver
- Stainless-steel bone screws
- Acrylic dental cement
- 1.5 mL Eppendorf tubes or similar

Surgical tools should be sterilized.

2.1.4. Sample Collection and Storage

- Microinfusion pump
- Liquid swivel, two-channel (Instech; Eicom)
- Microbore FEP tubing (inner diameter, 0.12 mm) for connecting probe outlet to swivel (30–40 cm) and swivel to fraction collector (70–80 cm) (Microbiotech/se AB, Sweden)
- Polythene tubing (inner diameter, 0.4 mm) 60 + 40 cm long for connecting pump to swivel and swivel to probe
- Metal spring to protect inlet and outlet lines
- Balancing arm
- 300 μL Collection vials (Beckman, USA)
- Refrigerated fraction collector (optional but strongly recommended; Univentor, Malta)
- Antioxidant mixture composed by 0.1 M acetic acid, 0.27 mM Na_2EDTA, 3.3 mM L-cysteine, and 0.5 mM ascorbic acid, pH 3.2. 1 μL antioxidant is added to each vial before sample collection.

2.2. Detection and Quantification of NA, DA, and 5-HT in Dialysate with HPLC-ED

Catecholamines (NA and DA) and 5-HT are determined in the same sample (split into two aliquots) using two different HPLC systems, which include the following parts:

- HPLC pump, model LC-20AD (Shimadzu, Japan), and pulse dampener (SSI-Lab Alliance, USA)
- Refrigerated autosampler, model MIDAS, equipped with column oven (Spark Holland, The Netherlands), for both systems
- Reverse-phase analytical columns: Capcell Pack C18MG S3, 75 × 4.6 mm, 3 μm particle size (Shiseido Fine Chemicals, Japan), for DA and NA; Supelcosil LC10-DB, 150 × 4.6 mm, 3 μm particle size (Supelco, USA), for 5-HT
- Hand-made guard columns 30 × 4.0 mm, packed with Pelliguard 30–40 μm (Supelco, USA), for both systems (change every 2 months)
- Coulochem II electrochemical detector equipped with 5011 electrochemical cell (ESA, USA), for both systems

- Azur 5.0 software for the acquisition and elaboration of chromatographic peaks (Datalys, France). A two-channel interface allows to collect data generated by two independent HPLC systems

2.2.1. Mobile Phase for NA and DA

- Ultrapure water produced by Millipore Rios 50 (Millipore, USA)
- 61 mM (5 g/L) Sodium acetate anhydrous (Merck, Germany; MW 82.03)
- 9.57 mM (2.3 g/L) Citric acid monohydrate (Carlo Erba Reagents, Italy; MW 210.14)
- 0.3 mM (112 mg/L) Na_2EDTA dihydrate (Merck, Germany; MW 372.24)
- 1.02 mM (239 mg/L) Sodium octane sulfonate monohydrate (Fluka, Germany; MW 234.29)
- 7.0% Methanol HPLC-grade (Merck, Germany)
- Filtration apparatus (e.g., Millipore) consisting of borosilicate glass vacuum flask (1 or 2 L), fritted filter support for 47 mm diameter filters, 500 mL glass, and spring clamp. The apparatus should be equipped with an online, one-way valve or solvent trap to avoid reflux of "dirty solutions" into the mobile phase
- Vacuum source
- Regenerated cellulose disc filters, 0.22 μm (preferred) or 0.45 μm pore-size (Millipore, USA, or Schleicher & Schuell, Germany).

2.2.2. HPLC Setting for NA and DA Measurement

- Mobile phase flow rate: 1 mL/min
- Column temperature: 41°C
- Detector mode: DC
- Electrochemical cell potentials: E_1 +200 mV; E_2 −300 mV, for 5011 cells, which has two in-series electrodes. NA and DA are read as E_2 output signal
- Electrochemical detector response time: 10 s
- Detector gain: 2 nA
- Autosampler tray temperature: 4–6°C
- Background current is usually less than −5 nA (note that negative current is due to the negative potential applied to E_2)

2.2.3. Mobile Phase for 5-HT

- Ultrapure water produced by Millipore Rios 50 (Millipore, USA)
- 48 mM (6.5 g/L) Sodium acetate trihydrate (Merck, Germany; MW 136.08)
- 9.52 mM (2.0 g/L) Citric acid monohydrate (Carlo Erba Reagents, Italy; MW 210.14)

- 0.1 mM (37.2 mg/L) Na_2EDTA dihydrate (Merck, Germany; MW 372.24)
- 0.01%v/v Triethylamine (Fluka, Germany; MW 234.29)
- 4.0% Acetonitrile, HPLC-grade (Carlo Erba Reagents, Italy)

2.2.4. HPLC Setting for 5-HT

- Mobile phase flow rate: 1 mL/min
- Column temperature: 30°C
- Detector mode: DC
- Electrochemical cell potentials: E_1 +50 mV; E_2 +180 mV, for 5011 cells, which has two in-series electrodes. 5-HT is read as E_2 output signal
- Electrochemical detector response time: 10 s
- Detector gain: 2 nA
- Autosampler tray temperature: 4–6°C
- Background current is usually less than 3 nA

3. Methods

3.1. Microdialysis

3.1.1. Construction of a Concentric Probe: Step-By-Step Procedure

The main steps in the construction of a probe for rats are schematically illustrated in Fig. 3. Probes for mice are prepared with the same procedure and materials, except probe shaft and exposed membrane, are shorter (about half those used in rats). Also, inlet and outlet tubing for mice have shorter length. Probes can be prepared using membranes made with different materials. It is important to point out that AN69 membrane (Hospal), which has higher in vitro recovery and is more biocompatible than Cuprophan, is not appropriate for the analysis of 5-HT (see Note 4.2).

Preliminary steps include cutting of dialysis fibers into short pieces 20 mm long and cutting 26G stainless steel (SS) tubing to be used for the probe shaft into 15–20 mm long pieces. Typically, 15 mm shafts are suited for brain regions such as cortex, hippocampus, and dorsal striatum while longer shafts are used for hypothalamus and ventral striatum.

1. A tungsten wire is inserted into a 20 mm long piece of dialysis membrane made of Cuprophan (216 μm outer diameter, 8 μm wall thickness, with 3,000 Da cutoff; Sorin Biomedica, Italy). The fiber with the tungsten wire inside is threaded into a 15–20 mm long 26G stainless steel tubing until the fiber enters for about 2 mm into the tubing. The fiber is glued to the SS shaft with epoxy glue and left to dry for at least 2 h.
2. The tungsten wire is retracted and, under a dissecting microscope, the fiber is trimmed with a scalpel blade to a length 0.5 mm longer than desired (for 2 mm probes the fiber should

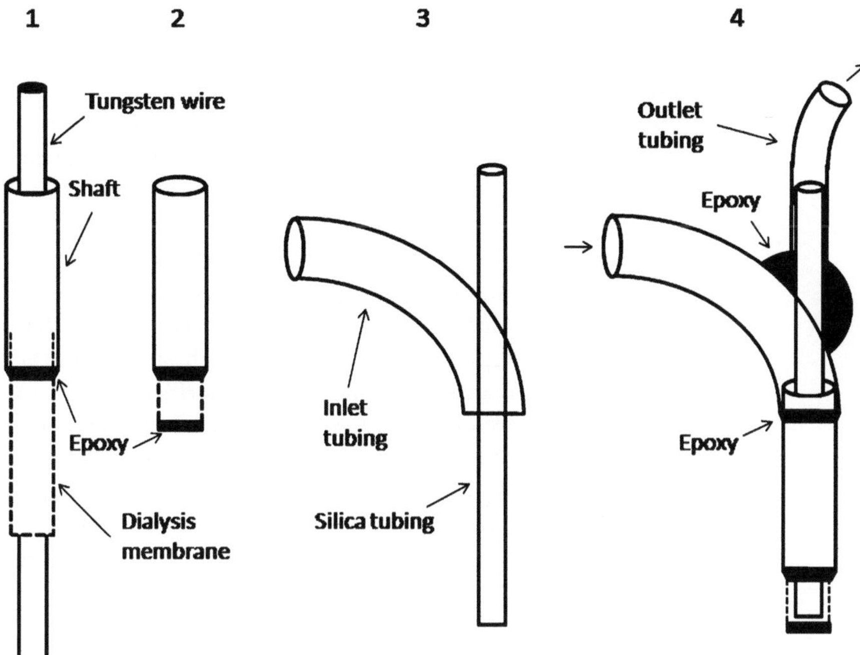

Fig. 3. Main steps in the construction of a concentric probe. (1) A tungsten wire is threaded into a piece (20 mm) of dialysis membrane, inserted into a stainless steel shaft, and glued to it with epoxy. (2) Tungsten wire is unthreaded, the fiber is trimmed to the desired length, and the distal end sealed with epoxy. (3) Silica tubing is threaded into a pierced 0.4 mm inner diameter PE tubing. (4) The silica-inlet tubing assembly is threaded into the shaft and the distance from the distal end of the silica tubing and glue cap is adjusted to 0.2–0.3 mm. The opposite end of the silica is threaded into a 0.28 mm inner diameter PE outlet tubing. The junctions between inlet, outlet, and stainless steel tubing are glued with epoxy and the assembled probe is left to dry for at least 24 h before implantation. Illustrations are not to scale.

be trimmed to 2.5 mm). The fiber tip is sealed with epoxy glue, which, by capillarity, goes up for 0.5 mm into the fiber lumen. Glue is left to dry overnight. Inspect the dialysis membrane under a dissecting microscope to ensure that the active portion of the membrane is free from epoxy.

3. The day after, a 0.4 mm inner diameter polythene (PE) tubing, 4.5 cm long, which serves as inlet, is pierced with a 30 G needle and the silica tubing (inner diameter 75 μm and outer diameter 150 μm; length 4.5 cm) threaded into this small hole. The silica-PE inlet tubing assembly is threaded into the SS shaft and the distance from the distal end of the silica tubing and glue cap is adjusted to 0.2–0.3 mm. The opposite end of the silica is threaded into a 0.28 mm inner diameter PE outlet tubing. The junctions between inlet, outlet, and stainless steel tubing are sealed with epoxy glue and the assembled probe is left to dry for at least 24 h before implantation.
4. It is recommended that probes are constructed at least one day and no more than 1 or 2 weeks before their usage. Prepare a number of probes exceeding those really needed as some may not work properly. Main reasons are leaking dialysis membrane, not properly sealed junctions, and clogged lines.

3.1.2. Perfusion Medium

Ionic composition of the perfusion medium should reflect ion concentrations in the extracellular fluid measured in vivo (50). However, a variety of different perfusion media has been used in microdialysis studies and different opinions as to the most appropriate composition still prevail. Calcium concentrations play a major role as relatively small changes of this ion in the perfusion medium may strongly affect the release of monoamines (51). "Physiological" Ca^{2+} concentrations range between 1 and 2 mM (52); 1.26 mM is the concentration measured in vivo in rat striatum extracellular fluid, with ion-selective electrodes (51, 53). It should be kept in mind that although small differences in the composition of extracellular fluid and perfusion medium are likely compensated by endogenous homeostatic mechanisms, large differences are not (50). For instance, increase of Ca^{2+} concentration much above physiological levels as in Ringer's solutions used in early studies alters basal and drug-induced changes of monoamine release (51, 54) and might contribute to discrepancies in results across laboratories.

Inclusion in the perfusion medium of monoamine reuptake inhibitors, such as citalopram for 5-HT and desipramine for NA (55, 56), facilitates their detection and may help to distinguish between uptake-dependent and release-dependent changes in extracellular monoamines (57–59). However, the presence of monoamine reuptake inhibitors in the perfusion media may prevent the response to certain drugs, such as amphetamine.

3.1.3. Implantation and Tethering of Microdialysis Probe

Using a brain stereotaxic atlas (60), determine the antero-posterior (AP), lateral (L) and dorso-ventral (DV) coordinates for the target brain region (Table 1). Bregma or interaural line can be used as reference points. Bregma, the midline intersection of the coronal sutures connecting frontal and parietal bones and sagittal sutures, is the preferred reference point as it is easily recognized on the exposed skull and close to the most frequently microdialyzed brain regions, such as striatum cortex and hippocampus. As the coordinates of bregma are taken after the rat is mounted in the apparatus, correct mounting of the rat is not as critical as when the interaural line is used as reference point. The latter is used as reference point, particularly for implanting probes into monoaminergic brain nuclei such as the mesencephalic dorsal raphè, where serotonergic neurons originate, the noradrenaline-containing neurons of the locus coeruleus, and the dopaminergic ventrotegmental area and substantia nigra, which all lie in close proximity of the interaural plane.

The procedure for mice is similar to that described for rats. Stereotaxic coordinates can be taken from the stereotaxic atlas specific for the mouse (61). It is recommended to use specific adapters available from stereotaxic producers for mounting mice on the stereotaxic frame.

Table 1
Dialysate monoamine levels in various regions of the rat, mouse, and gerbil brain

Monoamine	Brain region (probe length)	Dialysate concentration fmol/fraction (20 min)	Stereotaxic coordinates (in mm from bregma)	Perfusion flow rate; probe membrane	References
Rat (CD)					
NA	DH (2 mm)	6.8 ± 0.5 (n = 15)	AP -3.8, L 1.9, V-4.5	1 μl/min; cup	(37)
NA	PFC (4 mm)	10.4 ± 0.5 (n = 36)	AP 3.7, L 0.7, V-4.8	1 μl/min; AN69	(42)
NA	PFC (4 mm)	5.6 ± 0.6 (n = 13)	AP 4.2, L 0.7, V-5.5	1 μl/min; cup	(34)
DA	STR (2 mm)	70.4 ± 8.2 (n = 17)	AP 0.7 L 3.4, V-6.6	2 μl/min; AN69	(71)
DA	NAC (2 mm)	63.7 ± 7.7 (n = 18)	AP 1.9 L 1.6, V-8.2	2 μl/min; AN69	(71)
DA	PFC (4 mm)	20.2 ± 1.5 (n = 35)	AP 4.3 L 0.6, V-5.0	2 μl/min; AN69	(71)
DA	PFC (4 mm)	10.5 ± 0.5 (n = 83)	AP 3.7, L 0.7, V-4.8	1 μl/min; AN69	(41)
DA	PFC (4 mm)	8.3 ± 0.5 (n = 35)	AP 3.7, L 0.7, V-4.8	1 μl/min; AN69	(42)
5-HT	PFC (4 mm)	4.6 ± 0.4 (n = 10)	AP 3.7, L 0.6, V-5.4	1 μl/min; cup	(23)
5-HT	DH (2 mm)	3.1 ± 0.5 (n = 10)	AP -3.8, L 1.9, V-3.8	2 μl/min; cup	(31)
5-HT	STR (4 mm)	3.5 ± 0.7 (n = 17)	AP 0.2, L 3.0, V-7.5	2 μl/min; cup	(31)
5-HT	DR (2 mm)	4.5 ± 0.5 (n = 13)	AP -7.9, L 0.6[a], V-7.0	2 μl/min; CMA	(25)
Mouse (C57BL/6)					
5-HT	PFC (2 mm)	4.2 ± 0.2 (n = 19)	AP 2.1, L 0.3, V-2.5	1 μl/min; cup	(72)
5-HT	DH (1 mm)	3.2 ± 0.1 (n = 21)	AP -2.3, L 1.5, V-2.0	1 μl/min; cup	(72)
5HText (ZNF)	PFC	0.41 ± 0.06 nM	AP 2.1, L 0.3, V-2.5	1 μl/min; cup	(72)
5-HText (ZNF)	DH	0.29 ± 0.02 nM	AP -2.3, L 1.5, V-2.0	1 μl/min; cup	(72)
5-HT	DR (1 mm)	19.0 ± 0.9 (n = 60)	AP -4.4, L1.2[a], V-4.2	1 μl/min; cup	(22)

(continued)

Table 1 (continued)

Monoamine	Brain region (probe length)	Dialysate concentration fmol/fraction (20 min)	Stereotaxic coordinates (in mm from bregma)	Perfusion flow rate; probe membrane	References
Mouse (DBA/2)					
5-HT	PFC (2 mm)	2.7 ± 0.1 ($n=42$)	AP 2.1, L 0.3, V-2.5	1 μl/min; cup	(72)
5-HT	DH (1 mm)	2.5 ± 0.1 ($n=42$)	AP -2.3, L 1.5, V-2.0	1 μl/min; cup	(72)
5HText (ZNF)	PFC	0.21 ± 0.01 nM	AP 2.1, L 0.3, V-2.5	1 μl/min; cup	(72)
5-HText (ZNF)	DH	0.21 ± 0.02 nM	AP -2.3, L 1.5, V-2.0	1 μl/min; cup	(72)
5-HT	DR (1 mm)	6.6 ± 0.2 ($n=93$)	AP -4.4, L1.2[a], V-4.2	1 μl/min; cup	(22)
Gerbil					
NA	PFC (3 mm)	5.2 ± 0.4 ($n=36$)	AP 1.5, L 0.7, V-3.3	1 μl/min; AN69	(42)
DA	PFC (3 mm)	8.3 ± 0.7 ($n=36$)	AP 1.5, L 0.7, V-3.3	1 μl/min; AN69	(42)

Dialysate concentrations are not corrected for in vitro recovery of the probe

Abbreviations: *NA* noradrenaline, *DA* dopamine 5-HT serotonin, 5-HText extracellular 5-HT, *DH* dorsal hippocampus, *PFC* prefrontal cortex, *STR* striatum, *NAC* nucleus accumbens, *DR* dorsal raphè, *Cup* cuprophan, *AN69* polyacrylonitrile sodium methallyl sulfonate, *CMA* Carnegie microdialysis probe, *ZNF* zero-net-flux technique to estimate extracellular levels

[a]8° (rat) and 20° (mouse) angle to the sagittal plane to avoid damage to the sagittal sinus

The following steps are followed:

1. Anesthetize rat (mouse) with equithesin i.p. (3.0–3.5 mL/kg for both species). Check level of anesthesia by observing breathing, that should be abdominal, and pain reflexes (tail-pinch or foot-pinch with forceps). Pinna- and ocular-reflex should be absent. If rat still responds to pain stimuli or presents reflexes, inject a supplemental dose of anesthetic (0.1–0.2 mL). See Note 4.1 for alternative anesthetics.
2. Meanwhile, connect the microdialysis probe to the infusion pump, flow it with aCSF (1–2 μL/min), and clamp it to the stereotaxic manipulator. Inspect the probe for the presence of leaks or air bubbles. If necessary, increase the flow rate to 5–10 μL/min or invert the flow direction to remove bubbles.
3. Shave the rat head with a shaver or scissors before placing the rat on the stereotaxic apparatus.
4. Using blunt tip (45°) ear bars of the stereotaxic apparatus, adjust one of them at about 7 mm (interaural distance for a 250 g male rat is about 15 mm). Insert the ear bar into the auditory meatus and move the rat head forward and backward while gently pushing the ear against the ear bar to localize the correct timple in which ear bar should be inserted. Keeping the rat head in this position, insert the second ear bar in the opposite ear gently pushing it against the acoustic meatus and lock it. If ear bars are correctly positioned head movements in the lateral plane are not possible. Slightly loosen the screws locking both ear bars holding the bars in place and center the rat between them by locking both bars at the same distance from zero. Finally, insert the upper incisors over incisor bar set at −3.3 mm (flat skull position) and gently tighten nose clamp.
5. Disinfect the skin with Betadine or similar, cut it along the midline using a sterile scalpel, and retract skin with hemostats.
6. Scrape periostium with the scalpel and dry the bone with a cotton swab.
7. Identify bregma and position the tip of the probe as close as possible to this reference point.
8. Record AP and L coordinates and move the probe to the desired AP and L position according to stereotaxic atlas coordinates previously determined.
9. Drill the bone without damaging the dura mater and position the probe tip in contact with the dura surface. Record DV coordinate, remove the probe, and puncture the dura with a bent hypodermic needle tip.
10. Drill 2–3 additional holes for anchoring screws and insert screws into the holes using jeweler forceps and screwdriver.
11. Slowly insert probe while perfused with aCSF.

12. Once in place, apply dental cement to surround the probe shaft and cover the screws and the bones making sure that acrylic extends to the muscles.
13. Detach the cap and cut an Eppendorf tube to remove the conic bottom. Position the cut tube so that it surrounds the probe and protect it from bending and accidental damage. Fill the space between tube and probe with acrylic.
14. Remove hemostats and place skin over the acrylic which will adhere to it once dry. Usually, no sutures are necessary.
15. For anchoring a tether, embed the flat head of a slotted screw as far as possible from the probe.
16. Allow acrylic to dry and remove the holder.
17. Detach the rat from the perfusion pump and seal the inlet and outlet plastic tubes with a forceps heated on a lighter flame.
18. Free the rat from the stereotaxic apparatus and put it in a cage. Use a heating lamp or a blanket to maintain body temperature until awake.
19. After surgery, rat should be caged singly in a cage with flat cover to avoid damage to the implant until the beginning of probe perfusion, usually 20–22 h after probe insertion into the brain tissue. Food and water are freely available during post-surgery recovery.

3.1.4. Sample Collection and Storage

- Probe perfusion is started after about 20–22 h post-surgery recovery. Cut the sealed ends of probe inlet and outlet and connect them respectively to microinfusion pump (1 μL/min flow rate) and fraction collector via a two-channel liquid swivel mounted on a balancing arm (see Note 4.3 for swivel choice and maintenance). Polythene tubing (0.4 mm inner diameter) is used to connect the syringe mounted on the microinfusion pump to the side channel of the swivel and the swivel with the probe inlet. The connections between the probe outlet and central channel of the swivel and between swivel and fraction collector are made with microbore FEP tubing (0.12 mm inner diameter; inner volume 1.2 μL/10 cm) and tubing adapters. After about 90-min washout during which dialysate is discarded, subsequent 20–30-min fractions are collected by the refrigerated fraction collector. A metal spring anchored to the swivel and rat head is used to protect inlet and outlet tubing from biting.
- Refrigeration of samples during and after collection is necessary to warrant the stability of monoamines. For this purpose, a refrigerated fraction collector is strongly recommended. Addition of antioxidants to collected samples is also recommended, particularly if they should be stored before analysis. Fig. 4 compares the stability of NA, DA, and 5-HT in samples

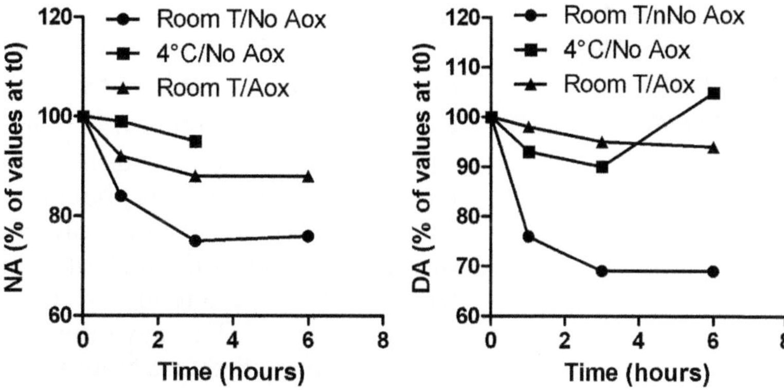

Fig. 4. Short-term stability of NA and DA at room temperature or at 4°C with or without the addition of 1 μL per 20 μL sample of the antioxidant (Aox) mixture (see Sect. 2.1.4). *Left* and *right panels* show, respectively, the stability of NA and DA (both 100 fmol/20 μL of aCSF) stored at room temperature (Room T) or at 4°C for 6 h. The addition of antioxidants prevents NA and DA degradation. After 18-h storage of samples with Aox at 4°C (not shown), most NA and 5-HT are not degraded (84% and 92% of the working standard, respectively) while only 77% of working DA standard is recovered. For longer storage periods it is recommended to add Aox and keep sample at −80°C. Data on 5-HT stability are not shown because standard prepared in aCSF without Aox is degraded by less than 10% after 6 h at room temperature.

stored at room temperature, in minced ice, with or without the addition of the antioxidant mixture (see Sect. 2.1.4). The addition of antioxidant mixture in collecting minivials is strongly advised to preserve monoamines. Add 1 μL of the antioxidant mixture described at the end of Sect. 2.1.4 to each collection minivial. The ratio 1:20 antioxidant:sample should be kept constant. If not injected immediately after collection, samples are stored at −80°C. It is recommended to avoid long-term storage. If not possible, store samples at −80°C (after the addition of antioxidant) and verify the stability of monoamines under storage conditions.

3.1.5. Calculation of Dead Volume

Dead volume introduces a time lag in sample collection that should be taken into account. Dead volume is the volume that analytes crossing the probe membrane from outside to inside should cover moving from the probe lumen to collection vial. Typically, this includes the inner volume of the probe and the volume of tubing used to connect probe to collection vial. The dead volume of the probe is calculated summating the inner volume of the silica tubing and that of the PE outlet tubing exceeding the silica (1.0–1.5 μL; see Fig. 3). The contribution of 0.12 mm inner diameter FEP tubing connecting probe to swivel and the swivel to collection vial is 1.2 μL for 10 cm length and the volume of central channel of the TCS2-23 swivel is 1.5 μL (Eicom, Japan). Assuming that 1 m FEP tubing is used, the resulting dead volume is 15 μL corresponding to 15-min delay in sample collection at 1 μL/min perfusion flow rate.

3.2. Detection and Quantification of NA, DA, and 5-HT in Dialysate with HPLC-EC

3.2.1. Preliminary Steps

- All solutions, particularly aCSF and mobile phase, should be prepared with HPLC-grade water and glassware that have been rinsed with HPLC-grade water.
- Once a new aCSF solution is prepared (usually every week) and before using it for perfusing implanted probes, it is recommended to check it for the presence of peaks that may interfere with the compounds to be analyzed. Inject 20 μL aCSF into the HPLC system at the highest sensitivity and check for interfering peaks at the retention times corresponding to those of NA, DA, and 5-HT or any other peak of interest that is measured and quantified (see Note 4.4 for troubleshooting).
- Dilute standards to appropriate concentrations and check for the optimal chromatographic separation of monoamines and sensitivity of the assay before samples are thawed.
- If optimal HPLC performance is achieved, thaw samples, shake them on a vortex for a few seconds, and store them in minced ice.
- As catecholamines and 5-HT are determined by separate HPLC systems, two series of autosampler vials should be prepared. Transfer samples into autosampler vials: half the sample volume (usually 10–15 μL) for catecholamines and the same volume for 5-HT. Put them in the refrigerated (4–6°C) autosampler tray.
- Program the injection sequence, which should include the blank (usually aCSF), the standard calibration curve prepared with at least three different concentrations of catecholamines or 5-HT bracketing those in the samples. The injection of a reference standard (usually the intermediate concentration of the calibration curve) every 4–5 samples to check for changes in sensitivity and retention times is highly recommended.

3.2.2. Preparation of the Mobile Phase for NA and DA

Chemicals should all be of reagent grade and should be added to water in the order listed in Sect. 2.2.1. Agitate with a magnetic stirrer for 15 min to dissolve salts. pH of the mobile phase before the addition of methanol is 4.83 (4.95 after methanol). pH should be checked for every new mobile phase but, usually, it is not necessary to adjust it if not exceeding ±0.03 pH units. Precise pH is critical for the separation of NA and DA (see Note 4.5). If needed, use concentrated sodium acetate or citric acid solution to adjust the pH. Add methanol and adjust to the final volume with water. After a 5-min stirring, filter under vacuum with a Millipore-like apparatus through regenerated cellulose disc filter. Transfer mobile phase in the HPLC reservoir and degas in ultrasonic bath for 5 min. Purge the pump with mobile phase and run it through the column for at least 20 column dead volumes before recycling and for at least 24–48 h before injecting samples. Mobile phase can be recycled when samples are not running. After a series of samples have been injected, discard at least 20 column dead volumes before recycling. To avoid bacteria growth, do not run the same mobile

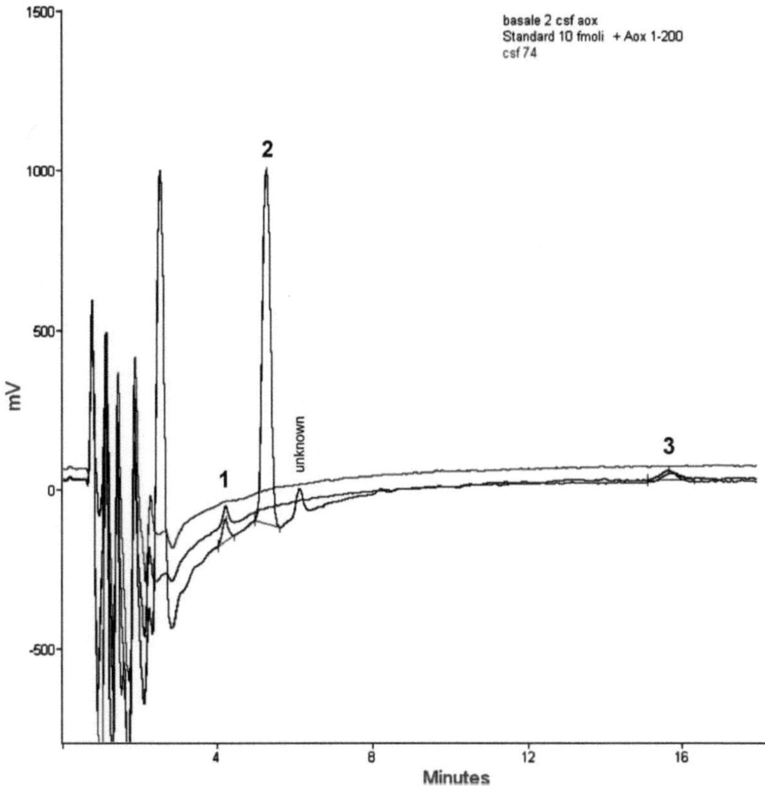

Fig. 5. Chromatograms showing the results of the HPLC-EC assay set up to detect NA and DA in the same sample of dialysate using reverse-phase-ion pair chromatography (see Sects. 2.2 and 3.2 for details). The three superimposed traces show the chromatograms obtained after injection of 20 μL aCSF (*upper trace*), standard solution containing 10 fmol/20 μL NA and DA (*middle trace*), and dialysate collected from the frontal cortex of an awake, freely moving rat (*bottom trace*). Peak 1 is NA (Rt, 4.2 min), peak 2 is 5-HIAA (Rt, 5.25 min), and peak 3 is DA (Rt, 15.65 min). Acid metabolites of DA elute in the solvent front.

phase for more than a week. Typical chromatographic separation of NA and DA is shown in Fig. 5. Total run time is about 30 min (see Note 4.8 for detection of late-eluting peaks).

3.2.3. Preparation of the Mobile Phase for 5-HT

"The simpler the easier" (and reliable) is a general rule that perfectly adapts to the mobile phase used for 5-HT determination and in general to the determination of 5-HT in dialysates with the current method used in this author's laboratory. As no ion-pair reagent is used, the time required for the equilibration of the column with mobile phase is relatively short. 1-h column equilibration with mobile phase is usually sufficient to obtain constant retention times, stable baseline, and detector response.

Preparing mobile phase is very simple. Just weigh appropriate amounts of citric acid and sodium acetate, Na_2EDTA, add to water in this order, and stir for 10–15 min to dissolve them. Then add triethylamine (not strictly necessary with modern silica-based HPLC column), acetonitrile, and water to the final volume. pH should be

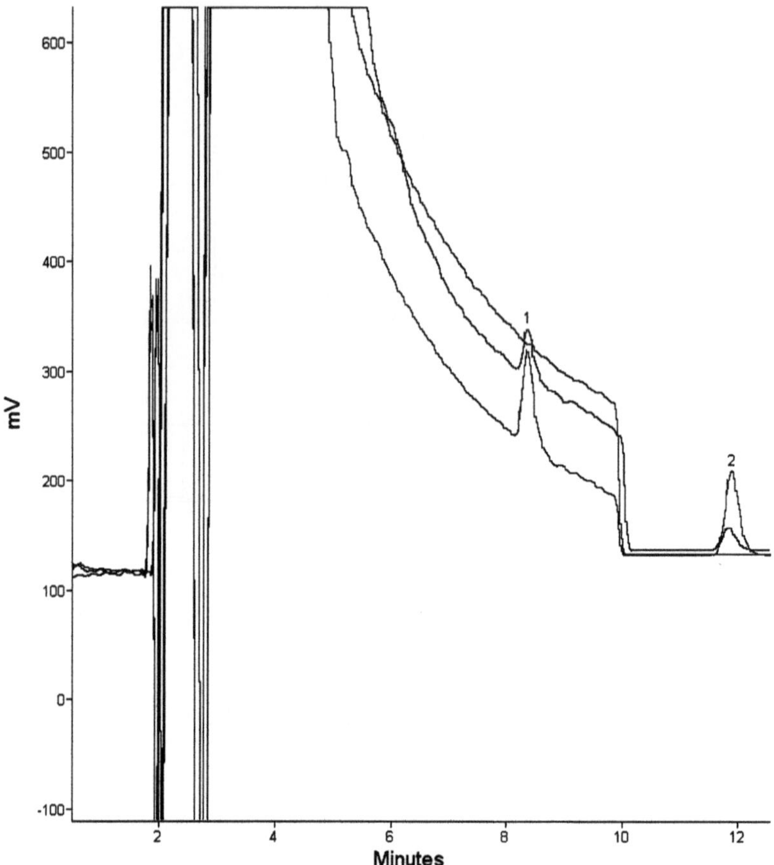

Fig. 6. Chromatograms showing the results of the HPLC-EC assay set up to detect 5-HT and its main metabolite 5-HIAA using reverse-phase chromatography (see Sects. 2.2 and 3.2 for details). The three superimposed traces show the chromatograms obtained after injection of 20 μL aCSF (*upper trace*), standard solution containing 10 fmol/20 μL 5-HT and 5 pmol/20 microL 5-HIAA (*bottom trace*), and 20 μL of dialysate collected from the frontal cortex of an awake, freely moving rat (*middle trace*). Peak 1 is 5-HT (Rt, 8.2 min), peak 2 is 5-HIAA (Rt, 11.9 min). NA, DA, and acidic DA metabolites elute close to or in the solvent front.

comprised between 4.9 and 5.1. If necessary adjust the pH to optimize separation of 5-HT from its main metabolite, 5-HIAA, adding drop by drop sodium acetate or citric acid concentrated solutions. After mixing, mobile phase is filtered through disc filters, transferred to the HPLC reservoir, and degassed in an ultrasonic bath for 5 min. Purge the pump with mobile phase and run it through the column for at least 20 column dead volumes before recycling and for at least 1 h before injecting samples. Mobile phase can be recycled when samples are not injected. After a series of samples have been injected, discard at least 20 column dead volumes before recycling. To avoid bacteria growth, do not run the same mobile phase for more than a week. Typical chromatographic separation of 5-HT is shown in Fig. 6. Total run time is 12–13 min but can be cut down by increasing the percentage of the organic modifier in the mobile phase.

Table 2
Commercial forms and molecular weight (MW) of monoamines and their main metabolites used to prepare stock solutions

Monoamines (salt) and metabolites	MW of the salt	MW of the free base	mg/100 mL
Noradrenaline (HCl)	205.64	169.18	20.56
Dopamine (HCl)	189.64	153.18	18.96
Adrenaline bitartrate	333.30	183.20	33.33
5-HT (creatinine sulfate monohydrate)	405.43	176.21	40.54
5-HIAA (free base)	–	191.19	19.12
DOPAC (free base)	–	168.15	16.82
HVA (free base)	–	182.18	18.22

Last column reports the amount of salt to be weighed to obtain 100 mL of 1 mM solution. Stock solutions are prepared in 0.1 M $HClO_4$ containing 0.05% $Na_2S_2O_5$ and 0.1% Na_2EDTA and stored at 4°C for 3 months

3.2.4. Standard Preparation and Storage

One mM stock solutions of NA, DA, and 5-HT are prepared by dissolving appropriate amounts of each compound in 0.1 M $HClO_4$ (3.9 mL $HClO_4$ 65% in 100 mL H_2O) containing 0.05% $Na_2S_2O_5$ and 0.1% Na_2EDTA as antioxidant agents and stored at 4°C for up to 3 months. Working solutions of 20 pmol/20 μL NA and DA or 5-HT are prepared on the day of assay by diluting stock solutions 1:1,000. Other concentrations of standard (the usual range is 3–100 fmol/20 μL) are prepared from working solution in 0.1 M $HClO_4$ containing antioxidants. Table 2 reports the monoamine salts used to prepare stock solutions, their molecular weight, and the amounts to be weighed to obtain 1 mM solutions. 5-HT creatinine sulfate is preferable because it is more stable than other 5-HT salts.

3.2.5. Limits of Sensitivity and Quantification

Excellent sensitivity of the assay should be warranted in order to measure the extremely small amounts of NA, DA, and 5-HT in samples, particularly in certain brain regions (see Table 1). Thus, it is important to establish the limits of sensitivity (LOS) and quantification (LOQ) of the assay. The two are not synonymous. LOS indicates the minimal amount or concentration of the analyte detectable by the assay. It can be defined as the amount of NA, DA, or 5-HT giving an oxidation (or reduction) signal in mV at least double of the noise (amplitude oscillations of the baseline) as established by injecting into the HPLC serial standard dilution approximating this limit.

LOQ can be defined as the minimal amount or concentration of NA, DA, or 5-HT falling into the linear portion of the respective

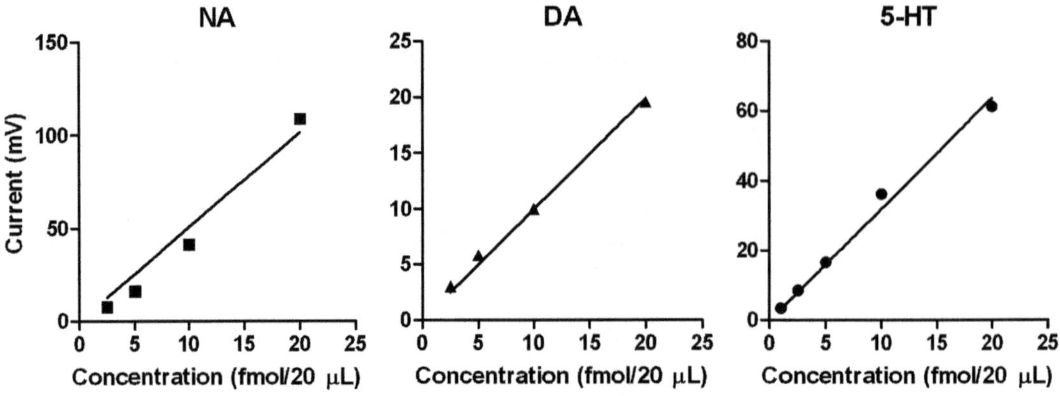

Fig. 7. Standard calibration curves for NA (2.5–20.0 fmol/20 μL), DA (2.5–20.0 fmol/20 μL), and 5-HT (1.0–20.0 fmol/20 μL).

Table 3
Calculation of LOD (*S*/*N* = 2) and LOQ based on data shown in Fig. 7

	NA	DA	5-HT
LOD (fmol/20 μL)	1.5	2.5	0.6
LOQ (fmol/20 μL)	2.0	3.0	1.0

calibration curves (see Fig. 7). It should be clear that precision of the assay in determining concentrations of analytes close to detection limit is far less than that for concentrations falling in the middle portion of the calibration curve.

LOQ for different monoamines obtained by this author's laboratory under the experimental conditions described in this chapter are reported in Table 3.

Useful advice on how to improve sensitivity of the assay is given in Notes 4.6 and 4.7.

3.3. Method Validation

3.3.1. Relationship Between Dialysate and Synaptic Levels of Monoamines

The main assumption on which microdialysis is based is that dialysate, extracellular, and synaptic concentrations of monoamines are strictly related. In other words, monoamine levels in dialysate reflect changes in synaptically available neurotransmitters. Although this cannot be directly proved, we can easily demonstrate whether dialysate levels of monoamines satisfy the basic criteria for exocytotic release such as the dependency on neuronal activity and calcium influx through the neuronal membrane (50).

Tetrodotoxin (TTX)-sensitivity and calcium-dependency of monoamine release are useful validation criteria to demonstrate that monoamines in brain dialysate arise from functional monoaminergic neurons rather than damaged cells or non-neuronal sources such as blood. Various studies have shown that basal efflux of NA, DA, and 5-HT is rapidly and markedly suppressed by perfusing

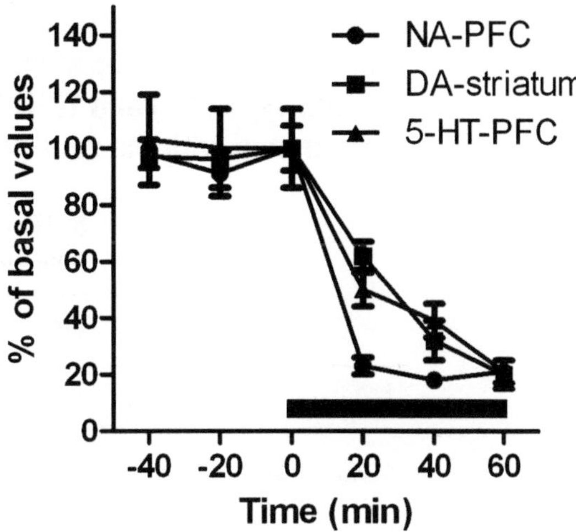

Fig. 8. The infusion of 1 μM tetrodotoxin (TTX) for 60 min (*horizontal bar*) reduced by about 80% basal NA and 5-HT efflux in the frontal cortex and DA in the striatum of the rat ($n = 4$–5 per group).

through the probe the fast voltage-dependent sodium channel blocker TTX or removing calcium from the perfusion medium (15, 18, 49, 62, 63). The effect of TTX on monoamine release is also illustrated in Fig. 8. Based on these findings, it has been concluded that the large majority of NA, DA, and 5-HT measured in dialysate is in equilibrium with that released in the synapse establishing an important relationship between changes in concentration of monoamines in dialysate and physiological activity of brain neurons. However, monoamine concentration in dialysate is influenced by another important mechanism: the reuptake of the neurotransmitter into presynaptic terminals. Therefore, dialysate monoamine levels reflect the balance between these two opposed processes and the contribution of either of them in determining extracellular levels cannot be easily established. This implies that although changes in monoamine release determine a corresponding change in dialysate levels, the opposite is not necessarily true.

The insertion of a microdialysis probe into the brain produces unavoidable damage and tissue reaction in the vicinity of implantation site. This is revealed by metabolic deactivation (reduced local blood flow and changes in glucose consumption in the tissue surrounding the probe) which depends on the probe type and the extent of initial damage and is usually normalized in about 24 h (50). As a consequence the release of neurotransmitters in the first period after probe implantation is mostly due to damaged neurons rather than exocytosis. Blood is another potential source of monoamines, particularly 5-HT, which is present in serum in high concentration (about 11 nmol/mL) compared to that in the extracellular fluid (about 0.5 pmol/mL). Most 5-HT in blood is stored in platelets and is released upon stimulation as following

Table 4
Extracellular concentrations of 5-HT in the frontal cortex of wild-type (WT) and TPH1 knockout (KO) mice determined with the zero-net-flux technique (68–70)

	5-HT concentrations in cortical extracellular fluid and serum of TPH1 KO mice	
	WT	TPH1 KO
Frontal cortex (fmol/20 μL)	10.5 ± 0.7 (n = 6)	9.3 ± 0.9 (n = 5)
Serum (μg/mL)	7.68 ± 0.96 (n = 5)	0.30 ± 0.09 (n = 5)

No significant differences in cortical 5-HT were found between WT and TPH1 KO mice. As expected, serum levels of 5-HT were dramatically reduced in TPH1 KO mice (>95% reduction compared to WT). The number of mice per group is indicated in parentheses. These data were generated by Dr. E. Calcagno in this author's laboratory. TPH1 KO mice were kindly donated by Dr. M. Bader (Berlin, Germany)

damage to blood vessels, as occurs during surgery and the few hours thereafter. In fact, a few hours after the implant of the probe in the striatum of rats in which 5-HT neurons were previously destroyed with a selective neurotoxin, basal levels of 5-HT are similar to those in the intact rats (18). Thus, blood 5-HT might importantly contribute to dialysate 5-HT under certain circumstances. However, 24 h after probe implantation, dialysate 5-HT in the brain is mostly derived from neuronal activity as shown by its large TTX-sensitivity (see Fig. 8). Consistently, unpublished findings from our laboratory (Calcagno and Invernizzi) show that knockout mice for the gene encoding TPH1, the enzyme responsible for the synthesis of 5-HT in nonneural tissues, have normal levels of cortical extracellular 5-HT (Table 4).

3.3.2. Protocols to Determine TTX-Sensitivity, Ca^{2+}-Dependency of Basal Monoamine Release

TTX-sensitivity

After stable baseline levels under perfusion with normal aCSF have been reached, three consecutive 20-min samples of dialysate are collected (basal levels). Switching to aCSF containing 1 or 2 μM TTX should determine a quick reduction of monoamine levels to 10–20% of basal levels. Typically, TTX reduces dialysate monoamines by about 80% of basal levels (Fig. 8).

Ca^{2+}-dependency

Prepare a first batch of aCSF containing 1.26 mM $CaCl_2$ as described in Sect. 3.1.2 and a second batch without $CaCl_2$. Minor osmotic changes due to the omission of $CaCl_2$ are compensated by a corresponding increase of $MgCl_2$ concentration in the aCSF, which reaches 2.26 mM.

Rat is perfused with normal aCSF for at least 2 h and 20-min baseline samples collected. On switching to the Ca^{2+}-free aCSF, concentrations of monoamines in dialysate should decline. The reduction should be rapid reaching at least 50–60% of basal levels measured under normal aCSF condition. Switching back to normal aCSF should restore basal monoamine concentrations.

Other validation criteria

As monoamine release is strictly controlled by inhibitory autoreceptors, another validation criterion consists of administering an appropriate dose of monoamine autoreceptor agonists in order to establish if the expected suppression of monoamine release occurs. Examples and doses can be found elsewhere.

Inhibition of monoamine synthesis and selective lesioning of DA, NA, or 5-HT containing neurons with specific neurotoxins should lead to a marked and selective decrease in dialysate levels of DA, NA, or 5-HT. Although these tools can contribute to establish the source and the identity of monoamine measured in dialysate, their limited selectivity or the possibility that spared neurons after lesioning restore normal extracellular monoamine levels (18) suggests that the validatory criteria based on TTX-sensitivity and Ca^{2+}-dependency should be preferred.

3.3.3. Identity of Chromatographic Peaks

Some of the criteria exposed in Sect. 3.3.2 such as marked reduction of monoamine peak height in response to TTX, Ca^{2+}-free medium, monoamine synthesis inhibitors, or lesioning of monoamines containing neurons may be used to validate the chromatographic identity of NA, DA, and 5-HT peaks. However, other criteria can be used to confirm the chromatographic identity of monoamine peaks. These include the measurement of NA, DA, and 5-HT concentrations in dialysate under different chromatographic conditions. Two protocols are suggested.

Protocol 1: Comparison between two different mobile phases.

Collect 3–4 samples of dialysate from a given brain region of the rat and pool them.

Prepare two mobile phases with the same composition except for the concentration of organic modifier: batch 1 is the currently used mobile phase; batch 2 is the same mobile phase with a lower concentration of organic modifier (30–40% less than the original mobile phase).

Take a volume of pooled sample and inject it into the HPLC in which batch 1 mobile phase is running. A known concentration of standard containing the monoamine under scrutiny should be injected as well.

Change to batch 2 mobile phase and once the system is equilibrated repeat the injections of standards and samples.

The ratio between the height of monoamine peaks in standard and sample under the two conditions should be similar. If important differences are noted, it is likely that the monoamine of interest is co-eluting with another, probably unknown, substance. In this case, chromatographic conditions should be changed to separate the co-eluting peaks and the method revalidated.

The presence of co-eluting peaks should be checked for each brain region analyzed.

Protocol 2: Comparing the response at two different oxidation potentials

The ratio between the heights of a given compound at two different oxidation potentials is constant. In a typical validation experiment, the maximum oxidation potential and the one at which about half of the maximum response is measured are used. Maximal and half-maximal oxidation potentials should be available from the tests initially done to establish the optimal oxidation potentials for monoamines of a given electrochemical cell.

Collect and pool 3–4 samples of dialysate.

Inject a portion of pooled sample and standard into the HPLC. The cell potential should be the one usually applied.

Reduce the oxidation potential of the electrochemical cell to that giving half-maximal response (to be established as described above) and wait for baseline and response stability.

Repeat the injection of standard and sample.

The ratio between the height of monoamine peaks in standard and sample at the two oxidation potentials should be similar. If important differences are noted, it is likely that the monoamine of interest is co-eluting with other, probably unknown, substance. In this case, chromatographic conditions should be changed to separate the co-eluting peaks and the method revalidated.

4. Notes

1. Although Equithesin is a relatively safe anesthetic, it may cause paralytic ileum that eventually may lead to death in a few days. As chloral hydrate is responsible for this major side effect, improvements may be obtained removing it from the mixture (64). In this author's experience mortality in the 24–30-h period after Equithesin is a very rare event. Alternatively, inhalation anesthetics such as isoflurane or parenteral anesthetics such as ketamine–xylazine mixture may be used (65).
2. Although the AN69 is an excellent membrane in terms of in vitro recovery, biocompatibility, robustness, and flexibility and quite widely used in microdialysis studies, its main drawback is that 5-HT binds to its negatively charged surface. The main consequence is that changes in 5-HT concentration in the extracellular medium are detected with delay. This is described in Fig. 9 showing that in vitro recovery of 5-HT of probes made with the AN69 membrane takes about 20 min to

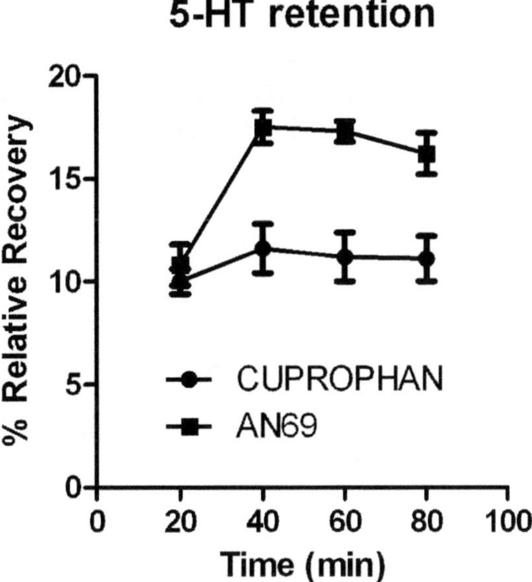

Fig. 9. Comparison between the in vitro recovery of concentric probes prepared with AN69 (outer diameter 340 μm) and Cuprophan (outer diameter 216 μm) membranes. Probe length was 2 mm. Probes (n=4–5) were perfused with aCSF at 2 μL/min and dropped into a plastic vial containing 5-HT standard solution (50 pg/20 μL) in aCSF. Collection of samples began immediately after the probe was dropped into the standard solution. Relative recovery was calculated for each fraction as follows: concentration of 5-HT in dialysate/concentration of 5-HT in standard solution × 100. Note that recovery is immediately stable with Cuprophan while there is a time lag of about 20 min with the AN69 membrane. The larger the surface of the active membrane the longer the time needed to reach stable recovery (about 80 min with the trans-cerebral probes made with AN69).

become constant while recovery with Cuprophan membranes is immediately stable. The larger the membrane surface the longer the time needed to reach a stable in vitro recovery. This implies that changes of extracellular 5-HT occurring in vivo measured with probes made with AN69 membranes are detected with delay. No delay is observed with NA or DA or the 5-HT metabolite 5-HIAA. This suggests that only 5-HT, among monoamines, selectively interacts with the AN69 membrane. Interestingly, the movements of potassium ions are also delayed across the AN69 membrane (66).

3. Although two-channel swivels are needed to run microdialysis in freely moving rats to avoid twisting and consequent stop flow of inlet and outlet tubing, they may also be the source of degradation of monoamines. Depending on the type and model, samples flowing into the channels of one- or two-channel swivels may contact metal parts. This should be avoided as metal (ferrous) ions released by stainless steel may catalyze the oxidative degradation of monoamines. In addition, leaking swivels due to worn seals cause rusting of the ball-bearings. Generated rust may contaminate swivel inner surfaces and

Table 5
Three swivels used regularly for at least 6 months were perfused with aCSF containing 200 fmol/20 μL NA, DA, and 5-HT (working standard)

	NA	DA
Swivel 1	65	18
Swivel 2	29	7
Swivel 3	2	0

Flow rate was 1 μL/min and was maintained with a microinfusion pump connected to the swivel via small bore polythene tubing. A microbore FEP tubing was used to connect swivel to the fraction collector. After 30-min washout, two consecutive samples were collected with a refrigerated (4°C) fraction collector and injected into the HPLC to determine the amount of NA, DA, and 5-HT after perfusion of the solution through the swivel. NA and DA concentration in collected samples (20 μL) was compared to that in the working standard. Data are expressed as percentages of the monoamine present in the working standard

strongly oxidize monoamines flowing into the swivel channels. Typically, this may result in almost total degradation of DA and partial destruction of NA (Table 5).

Data show that marked degradation of NA and, particularly, DA may occur because of swivels. This suggests that swivels should be checked regularly using the testing procedure described below. Swivels have negligible effect on 5-HT and data are not reported. No antioxidant mixture was present in the perfusion medium or in collection vials.

To avoid this problem, swivels with quartz- or Teflon-lined tubing should be preferred. However, even these swivels if not properly cleaned and dried may determine monoamine degradations. Inclusion of antioxidants in the collection vials or even in the perfusion medium is not very effective in preventing monoamine degradation due to rust contaminated swivels. Swivels should be rinsed and dried immediately after usage to prevent buildup of crystals and corrosion of metal surface. This is easily done by pulling 5 mL distilled water with a syringe connected to the swivel via PE tubing followed by 2 mL of 70% isopropanol in water and aspirating air (5–10 mL) to dry swivel channels. This procedure should be applied to each channel. It is preferable to pull the flushing solutions back rather than push liquid through the swivel, because excessive pressure may damage seals. It is recommended to check swivel "inertness" regularly by pumping a solution containing NA, DA, and 5-HT in aCSF through the swivel. Prepare a monoamines standard containing at least 200 fmol/20 μL NA, DA, and 5-HT in aCSF. Fill a syringe with the solution and connect it via PE tubing to the inlet of the central or side channel of the two-channel swivels. Mount the syringe on a microinfusion pump and pump the standard solution through the swivel at 1 μL/min (the same flow rate used for perfusing implanted probes). Connect microbore FEP tubing to the swivel outlet and collect two or three 20-min fractions in minivials with a refrigerated fraction collector. Inject collected samples into the HPLC to determine the concentration of monoamines. Compare the concentration of monoamines in collected samples with that in the original solution used to fill the syringe or even with that in the syringe itself if syringe-dependent degradation of monoamines is suspected. In fact, the stainless steel syringe needle, if not passivated, may also be source of metal ions and degradation of monoamines. The same goes for the stainless steel probe shaft. It is recommended to avoid contact between stainless steel surfaces and dialysate by replacing, whenever possible, stainless-steel with plastic or silica tubing. If swivel-dependent monoamine degradation is observed, stainless-steel surfaces of swivel channels may be passivated with HNO_3 (22%) followed by extensive washing with water. If not applicable (Teflon swivels) or not successful, inspect the inside of the swivel and check for leaking seals that may be replaced (Instech swivels) or replace the swivel with a new one.

4. Because even small amount of contaminants may strongly influence the correct detection of NA, DA, or 5-HT, all possible sources of contamination should be checked carefully. These include solutions, such as aCSF, mobile phase, water itself, salts, and glassware used to prepare solutions, and filter used to remove particulate from mobile phase and aCSF. Ghost peaks may also be generated by "dirty" injection needles and injection valves. Scratched rotor seals may retain/absorb electroactive compounds that are released once the valve is turned from load to inject position and detected together with the running sample.
5. For small changes in pH, the retention time of 5-HIAA (pK_a=4.7) may change a lot and cause NA to co-elute with 5-HIAA and DA to co-elute with an unknown peak. Increasing pH would shorten run time but cause NA to elute to close to the solvent front when baseline is still drifting, making it difficult to reliably determine NA peak height or area. With a new column, it may be necessary to adjust the pH between 4.85 and 5.0 and the concentration of methanol in the mobile phase (±1%).
6. Measurement of monoamines in dialysate samples presents many challenges. Small sample volumes are determined by the need to maximize recovery through the probe, which is inversely related with probe perfusion flow rate (50). Typically, 1–2 μL/

min are the most commonly used flow rates in monoamine microdialysis. This results in 10–30% relative recovery (depending on the membrane type and surface) but small-volume samples (20–40 μL for 20-min collection time). Samples cannot be reanalyzed because, usually, the entire volume collected is injected into the HPLC to meet the sensitivity requirements of the assay. If different HPLC methods are applied to measure catecholamines and 5-HT, as in the presently described method, samples should be split. This is only possible if sufficient sample volume is available and the amount of each monoamine in split sample is above the limits of quantification. This is usually achieved by increasing collection time from the usual 15–20 min to 30 min. In this latter case, neurochemical events occurring in 20 min or less cannot be picked up accurately. HPLC methods able to reliably measure all the monoamines in the same sample without splitting would improve the sensitivity of monoamine assay, halve the number of samples processed, and reduce the variability inherent to the manipulation of small sample volumes. Unfortunately, attempts to develop such a method have met with limited success. In this author's experience, it is not easy to optimize a reverse-phase HPLC-ED method for monoamines to detect NA, DA, and 5-HT in the same chromatographic run with the needed chromatographic selectivity and assay sensitivity. Quite a long retention time of 5-HT and poor separation of NA from solvent front, which respectively reduce the sensitivity for 5-HT and increase the risk of co-elution of NA with poorly retained peaks, are the major constraints of methods based on reverse-phase ion-pair chromatography. An interesting alternative is offered by strong cation exchange (SCX) HPLC columns. Although they do not allow the simultaneous detection of monoamine metabolites, separation of monoamines can be achieved in less than 20 min with good resolution from solvent front and from interfering peaks. Figure 10 shows the separation of monoamines standard on a cation exchange (SCX) column. Mobile phase composition was ammonium acetate 0.15 M, Na_2EDTA 0.14 mM, 15% CH_3OH, and 5% CH_3CN, pH 6.0 with acetic acid (before the addition of organic modifiers). A Coulochem II electrochemical detector equipped with a 5011A electrochemical cell (E_1 0 mV, E_2 +300 mV) was used to detect monoamines. Mass sensitivity of the assay can be further improved reducing the diameter of the column to 1–2 mm. In this case, a small-volume amperometric cell such as the 5041 (ESA, USA) should be preferred to avoid the loss of performance caused by the large volume of the coulometric cell.

7. Coulometric cells allow the detection of catecholamines and 5-HT at femtomoles level. The limits of detection (LOD) and quantification (LOQ) for NA, DA, and 5-HT using the

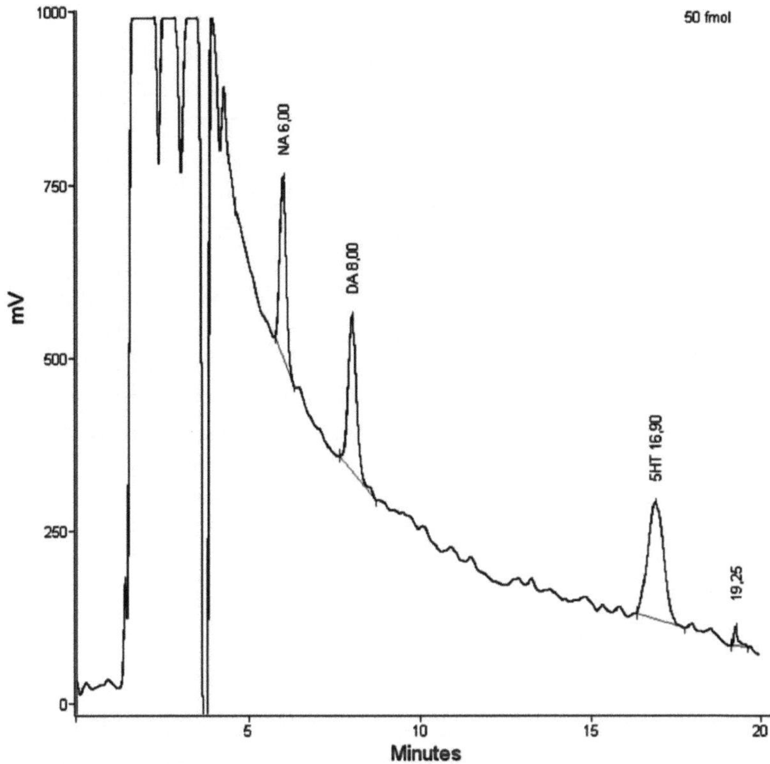

Fig. 10. Chromatogram showing the results of the HPLC-EC assay set up to detect NA, DA, and 5-HT in the same sample using a cationic exchange column (see Note 4.6). Standard concentration was 50 fmol/20 μL for each monoamine.

currently described procedures are shown in Table 3. As basal concentrations of NA, DA, and 5-HT in dialysate collected from the most commonly dialyzed regions of the rat brain (see Table 1) are sometimes close to the LOQ, the sensitivity of the assay should be the best possible. To maximize sensitivity, particular attention should be devoted to (1) eliminate air dissolved in the mobile phase by accurately degassing it under vacuum or by sonication; (2) optimize redox potentials at the two electrodes of the coulometric cell. Slight differences are observed depending on the electrochemical cell, age of the cell, and whether cell was reactivated by electrochemical or HNO_3 treatment. In this author's experience, the models 5011 and 5014 "microdialysis" electrochemical cells used in association with the Coulochem II detector have similar sensitivity for the analysis of NA, DA, and 5-HT although some cells show a better signal-to-noise ratio, regardless of the model; (3) using analytical columns of 1–2 mm diameter in association with an amperometric cell such as the model 5041 is also useful to improve mass sensitivity; (4) a pulse damper between the pump and injector further improves the signal-to-noise ratio by smoothing flow pulsations; (5) keeping organic modifier in the mobile phase to a minimum, the pH of the mobile phase as least acidic as possible, and the retention times of analytes of interest as short as possible. However, it should be kept in mind that changes to mobile phase composition to improve sensitivity of the assay should be compatible with an optimal chromatographic resolution of peaks of interest.

8. Although the retention time of DA, the last peak in the chromatogram, is about 18 min, total run time is 30 min. This is to avoid the overlay of late-eluting peaks, sometimes present in samples, with peaks of interest in subsequent injection(s). To determine the optimal run time, it is recommended to inject a sample and record the chromatogram for 60–120 min. Once the retention time of late-eluting peak(s) is identified, the optimal run time to avoid interference with peaks of interest is easily determined. Advice on the identification and elimination of late-eluting peaks can be found elsewhere (67).

Acknowledgements

I wish to thank all my coworkers that have contributed to set up, optimize, and validate the microdialysis technique applied to monoamine measurement in the rat, mouse, and gerbil brain as currently used in this author's laboratory. This Chapter is dedicated to the memory of my mentor, Dr. R. Samanin, who passed away in 2001.

Appendix

List of main suppliers and Web page address

Supplier	Product	Web page address
Beckman	Collection vials	www.beckmancoulter.com
Carlo Erba Reagenti	Chemicals	www.carloerbareagents.com
Clark Electromedical	Tungsten wires	www.warneronline.com
CMA/Microdialysis	Microinfusion pump	www.microdialysis.se
Coopers Needle Work	Stainless steel tubing	www.finestainlesstube.com
Datalys	Chromatographic software	www.datalys.net
David Kopf Instruments	Stereotaxic	www.kopfinstruments.com

(continued)

Supplier	Product	Web page address
Eicom	Swivels	www.eicomeurope.com
Eppendorf	Tubes	www.eppendorf.com
ESA	Electrochemical detector	www.esainc.com
Fluka	Monoamine standards	www.sigmaaldrich.com
Harvard Apparatus	Microinfusion pump	www.harvardapparatus.com
Hospal	Microdialysis fibers	www.hospal.it
Instech	Swivels	www.instechlab.com
Leica	Stereomicroscope	www.leicaitalia.it
Merck	Chemicals	www.merck-chemicals.com
Microbiotech	FEP tubing	www.microbiotech.se
Millipore	HPLC-grade water	www.millipore.com
Polymicro Technologies	Silica tubing	www.polymicro.com
Portex	PE tubing	www.smithsmedical.com
Schleicher & Schuell	Disc filters	www.schleicher-schuell.com
Shimadzu	HPLC pumps	www.shimadzu.it
Shiseido	HPLC columns	www.shiseido.co.jp
Sorin Biomedica	Microdialysis fibers	www.sorin.com
Spark Holland	Autosamplers	www.sparkholland.com
SSI-LabAlliance	Pulse dampener	www.laballiance.com
Supelco	HPLC columns	www.sigmaaldrich.com
Univentor	Fraction collectors	www.univentor.com

References

1. Dahlstroem A, Fuxe K (1964) Evidence for the existence of monoamine-containing neurons in the central nervous system. I. Demonstration of monoamines in the cell bodies of brain stem neurons. Acta Physiol Scand 62(p. SUPPL 232):1–55
2. Sarna GS, Hutson PH, Tricklebank MD, Curzon G (1983) Determination of brain 5-hydroxytryptamine turnover in freely moving rats using repeated sampling of cerebrovascular fluid. J Neurochem 40(2):383–388
3. Gaddum JH (1961) Push-pull cannulae. J Physiol 155:1–2P
4. Glowinski J (1981) In vivo release of transmitters in the cat basal ganglia. Fed Proc 40(2): 135–141
5. Philippu A (1984) Use of push-pull cannulae to determine the release of endogenous neurotransmitters in distinct brain areas of anaesthetized and freely moving animals. In: Marsden CA (ed) Measurement of neurotransmitters release in vivo. Wiley, Chichester, UK, pp 3–37
6. Besson M, Cheramy A, Feltz P, Glowinski J (1971) Dopamine: spontaneous and drug-induced release from the caudate nucleus in the cat. Brain Res 32(2):407–424
7. Gonon F, Buda M, Cespuglio R, Jouvet M, Pujol JF (1980) In vivo electrochemical detection of catechols in the neostriatum of anaesthetized rats: dopamine or DOPAC? Nature 286(5776):902–904

8. Stamford JA (1985) In vivo voltammetry: promise and perspective. Brain Res 357(2): 119–135
9. Ungerstedt U, Pycock C (1974) Functional correlates of dopamine neurotransmission. Bull Schweiz Akad Med Wiss 30(1–3):44–55
10. Kissinger PT, Refshuage CJ, Dreiling R, Blank L, Freeman R, Adams RN (1973) An electrochemical detector for liquid chromatography with picograms sensitivity. Analytical Letters 6:465–477
11. Mefford IN (1981) Application of high performance liquid chromatography with electrochemical detection to neurochemical analysis: measurement of catecholamines, serotonin and metabolites in rat brain. J Neurosci Methods 3(3):207–224
12. Imperato A, Di Chiara G (1984) Trans-striatal dialysis coupled to reverse phase high performance liquid chromatography with electrochemical detection: a new method for the study of the in vivo release of endogenous dopamine and metabolites. J Neurosci 4(4):966–977
13. Ungerstedt U, Herrera-Marschitz M, Zetterstrom T (1982) Dopamine neurotransmission and brain function. Prog Brain Res 55:41–49
14. Zetterstrom T, Sharp T, Marsden CA, Ungerstedt U (1983) In vivo measurement of dopamine and its metabolites by intracerebral dialysis: changes after d-amphetamine. J Neurochem 41(6):1769–1773
15. L'Heureux R, Dennis T, Curet O, Scatton B (1986) Measurement of endogenous noradrenaline release in the rat cerebral cortex in vivo by transcortical dialysis: effects of drugs affecting noradrenergic transmission. J Neurochem 46(6):1794–1801
16. Carboni E, Di Chiara G (1989) Serotonin release estimated by transcortical dialysis in freely-moving rats. Neuroscience 32(3):637–645
17. Hernandez L, Lee F, Hoebel BG (1987) Simultaneous microdialysis and amphetamine infusion in the nucleus accumbens and striatum of freely moving rats: increase in extracellular dopamine and serotonin. Brain Res Bull 19(6):623–628
18. Kalen P, Strecker RE, Rosengren E, Bjorklund A (1988) Endogenous release of neuronal serotonin and 5-hydroxyindoleacetic acid in the caudate-putamen of the rat as revealed by intracerebral dialysis coupled to high-performance liquid chromatography with fluorimetric detection. J Neurochem 51(5):1422–1435
19. Calcagno E, Carli M, Baviera M, Invernizzi RW (2009) Endogenous serotonin and serotonin2C receptors are involved in the ability of M100907 to suppress cortical glutamate release induced by NMDA receptor blockade. J Neurochem 108(2):521–532
20. Calcagno E, Carli M, Invernizzi RW (2006) The 5-HT(1A) receptor agonist 8-OH-DPAT prevents prefrontocortical glutamate and serotonin release in response to blockade of cortical NMDA receptors. J Neurochem 96(3):853–860
21. Calcagno E, Guzzetti S, Canetta A, Fracasso C, Caccia S, Cervo L, Invernizzi RW (2009) Enhancement of cortical extracellular 5-HT by 5-HT1A and 5-HT2C receptor blockade restores the antidepressant-like effect of citalopram in non-responder mice. Int J Neuropsychopharmacol 12:793–803
22. Calcagno E, Invernizzi RW (2010) Strain-dependent serotonin neuron feedback control: role of serotonin 2 C receptors. J Neurochem 114(6):1701–1710
23. Ceglia I, Acconcia S, Fracasso C, Colovic M, Caccia S, Invernizzi RW (2004) Effects of chronic treatment with escitalopram or citalopram on extracellular 5-HT in the prefrontal cortex of rats: role of 5-HT1A receptors. Br J Pharmacol 142(3):469–478
24. Ceglia I, Carli M, Baviera M, Renoldi G, Calcagno E, Invernizzi RW (2004) The 5-HT receptor antagonist M100,907 prevents extracellular glutamate rising in response to NMDA receptor blockade in the mPFC. J Neurochem 91(1):189–199
25. Invernizzi R, Belli S, Samanin R (1992) Citalopram's ability to increase the extracellular concentrations of serotonin in the dorsal raphe prevents the drug's effect in the frontal cortex. Brain Res 584(1–2):322–324
26. Invernizzi R, Bramante M, Samanin R (1994) Chronic treatment with citalopram facilitates the effect of a challenge dose on cortical serotonin output: role of presynaptic 5-HT1A receptors. Eur J Pharmacol 260(2–3):243–246
27. Invernizzi R, Bramante M, Samanin R (1995) Extracellular concentrations of serotonin in the dorsal hippocampus after acute and chronic treatment with citalopram. Brain Res 696(1–2):62–66
28. Invernizzi R, Bramante M, Samanin R (1996) Role of 5-HT1A receptors in the effects of acute chronic fluoxetine on extracellular serotonin in the frontal cortex. Pharmacol Biochem Behav 54(1):143–147
29. Invernizzi R, Morali F, Pozzi L, Samanin R (1990) Effects of acute and chronic clozapine on dopamine release and metabolism in the striatum and nucleus accumbens of conscious rats. Br J Pharmacol 100(4):774–778
30. Invernizzi R, Pozzi L, Samanin R (1995) Selective reduction of extracellular dopamine in the rat nucleus accumbens following chronic

treatment with DAU 6215, a 5-HT3 receptor antagonist. Neuropharmacology 34(2):211–215

31. Invernizzi R, Velasco C, Bramante M, Longo A, Samanin R (1997) Effect of 5-HT1A receptor antagonists on citalopram-induced increase in extracellular serotonin in the frontal cortex, striatum and dorsal hippocampus. Neuropharmacology 36(4–5):467–473
32. Invernizzi RW, Garattini S (2004) Role of presynaptic alpha2-adrenoceptors in antidepressant action: recent findings from microdialysis studies. Prog Neuropsychopharmacol Biol Psychiatry 28(5):819–827
33. Invernizzi RW, Garavaglia C, Samanin R (2003) The alpha 2-adrenoceptor antagonist idazoxan reverses catalepsy induced by haloperidol in rats independent of striatal dopamine release: role of serotonergic mechanisms. Neuropsychopharmacology 28(5):872–879
34. Invernizzi RW, Parini S, Sacchetti G, Fracasso C, Caccia S, Annoni K, Samanin R (2001) Chronic treatment with reboxetine by osmotic pumps facilitates its effect on extracellular noradrenaline and may desensitize alpha(2)- adrenoceptors in the prefrontal cortex. Br J Pharmacol 132(1):183–188
35. Invernizzi RW, Pierucci M, Calcagno E, Di Giovanni G, Di Matteo V, Benigno A, Esposito E (2007) Selective activation of 5-HT(2 C) receptors stimulates GABA-ergic function in the rat substantia nigra pars reticulata: a combined in vivo electrophysiological and neurochemical study. Neuroscience 144(4):1523–1535
36. Invernizzi RW, Sacchetti G, Parini S, Acconcia S, Samanin R (2003) Flibanserin, a potential antidepressant drug, lowers 5-HT and raises dopamine and noradrenaline in the rat prefrontal cortex dialysate: role of 5-HT(1A) receptors. Br J Pharmacol 139(7):1281–1288
37. Parini S, Renoldi G, Battaglia A, Invernizzi RW (2005) Chronic reboxetine desensitizes terminal but not somatodendritic alpha2-adrenoceptors controlling noradrenaline release in the rat dorsal hippocampus. Neuropsychopharmacology 30(6): 1048–1055
38. Pozzi L, Acconcia S, Ceglia I, Invernizzi RW, Samanin R (2002) Stimulation of 5-hydroxytryptamine (5-HT(2 C)) receptors in the ventrotegmental area inhibits stress-induced but not basal dopamine release in the rat prefrontal cortex. J Neurochem 82(1):93–100
39. Pozzi L, Invernizzi R, Cervo L, Vallebuona F, Samanin R (1994) Evidence that extracellular concentrations of dopamine are regulated by noradrenergic neurons in the frontal cortex of rats. J Neurochem 63(1):195–200
40. Pozzi L, Invernizzi R, Garavaglia C, Samanin R (1999) Fluoxetine increases extracellular dopamine in the prefrontal cortex by a mechanism not dependent on serotonin: a comparison with citalopram. J Neurochem 73(3):1051–1057
41. Renoldi G, Calcagno E, Borsini F, Invernizzi RW (2007) Stimulation of group I mGlu receptors in the ventrotegmental area enhances extracellular dopamine in the rat medial prefrontal cortex. J Neurochem 100(6):1658–1666
42. Renoldi G, Invernizzi RW (2006) Blockade of tachykinin NK1 receptors attenuates stress-induced rise of extracellular noradrenaline and dopamine in the rat and gerbil medial prefrontal cortex. J Neurosci Res 84(5):961–968
43. Sacchetti G, Bernini M, Bianchetti A, Parini S, Invernizzi RW, Samanin R (1999) Studies on the acute and chronic effects of reboxetine on extracellular noradrenaline and other monoamines in the rat brain. Br J Pharmacol 128(6): 1332–1338
44. Chefer VI, Thompson AC, Zapata A, Shippenberg TS (2009) Overview of brain microdialysis. Curr Protoc Neurosci Chapter 7: p. Unit 7 1
45. Zapata A, Chefer VI, Shippenberg TS (2009) Microdialysis in rodents. Curr Protoc Neurosci Chapter 7: p. Unit 7 2
46. Zapata A, Chefer VI, Shippenberg TS, Denoroy L (2009) Detection and quantification of neurotransmitters in dialysates. Curr Protoc Neurosci Chapter 7: p. Unit 7 4 1–30
47. Di Chiara G (1991) Brain dialysis of monoamines. In: Robinson TE, Justice JB Jr (eds) Microdialysis in the neurosciences. Elsevier, Amsterdam, pp 175–185
48. Sharp T, Zetterstrom T (1991) In vivo measurement of monoamine neurotransmitter release using brain microdialysis. In: Stamford JA (ed) Monitoring neuronal activity: a practical approach. IRL, Oxford, pp 147–179
49. Santiago M, Westerink BH (1990) Characterization of the in vivo release of dopamine as recorded by different types of intracerebral microdialysis probes. Naunyn Schmiedebergs Arch Pharmacol 342(4):407–414
50. Benveniste H, Huttemeier PC (1990) Microdialysis–theory and application. Prog Neurobiol 35(3):195–215
51. Moghaddam B, Bunney BS (1989) Ionic composition of microdialysis perfusing solution alters the pharmacological responsiveness and basal outflow of striatal dopamine. J Neurochem 53(2):652–654
52. Silver IA, Erecinska M (1990) Intracellular and extracellular changes of (Ca2+) in hypoxia and

ischemia in rat brain in vivo. J Gen Physiol 95(5):837–866

53. Nicholson C (1980) Modulation of extracellular calcium and its functional implications. Fed Proc 39(5):1519–1523
54. Invernizzi R, Pozzi L, Vallebuona F, Bonini I, Sacchetti G, Samanin R (1992) Effect of amineptine on regional extracellular concentrations of dopamine and noradrenaline in the rat brain. J Pharmacol Exp Ther 262(2):769–774
55. Lehmann J, Valentino R, Robine V (1992) Cortical norepinephrine release elicited in situ by N-methyl-D-aspartate (NMDA) receptor stimulation: a microdialysis study. Brain Res 599(1):171–174
56. Sharp T, Bramwell SR, Grahame-Smith DG (1989) 5-HT1 agonists reduce 5-hydroxytryptamine release in rat hippocampus in vivo as determined by brain microdialysis. Br J Pharmacol 96(2):283–290
57. Gundlah C, Martin KF, Heal DJ, Auerbach SB (1997) In vivo criteria to differentiate monoamine reuptake inhibitors from releasing agents: sibutramine is a reuptake inhibitor. J Pharmacol Exp Ther 283(2):581–591
58. Nakahara D, Ozaki N, Kapoor V, Nagatsu T (1989) The effect of uptake inhibition on dopamine release from the nucleus accumbens of rats during self- or forced stimulation of the medial forebrain bundle: a microdialysis study. Neurosci Lett 104(1–2):136–140
59. Romero L, Artigas F (1997) Preferential potentiation of the effects of serotonin uptake inhibitors by 5-HT1A receptor antagonists in the dorsal raphe pathway: role of somatodendritic autoreceptors. J Neurochem 68(6): 2593–2603
60. Paxinos G, Watson C (2005) The rat brain in stereotaxic coordinates. Elsevier Academic Press, Sidney
61. Franklin KBJ, Paxinos G (1997) The mouse brain in stereotaxic coordinates. Academic Press, San Diego
62. Auerbach SB, Minzenberg MJ, Wilkinson LO (1989) Extracellular serotonin and 5-hydroxyindoleacetic acid in hypothalamus of the unanesthetized rat measured by in vivo dialysis coupled to high-performance liquid chromatography with electrochemical detection: dialysate serotonin reflects neuronal release. Brain Res 499(2):281–290
63. Sharp T, Bramwell SR, Clark D, Grahame-Smith DG (1989) In vivo measurement of extracellular 5-hydroxytryptamine in hippocampus of the anaesthetized rat using microdialysis: changes in relation to 5-hydroxytryptaminergic neuronal activity. J Neurochem 53(1):234–240
64. Deacon RM, Rawlins JN (1996) Equithesin without chloral hydrate as an anaesthetic for rats. Psychopharmacology (Berl) 124(3):288–290
65. Waynforth HB, Flecknell PA (1992) Experimental and surgical techniques in the rat, 2nd edn. Academic, London
66. Tao R, Hjorth S (1992) Differences in the in vitro and in vivo 5-hydroxytryptamine extraction performance among three common microdialysis membranes. J Neurochem 59(5):1778–1785
67. Dolan JW (2010) Where did that peak come from? LC-GC, 23(7):358–361
68. Cosford RJ, Vinson AP, Kukoyi S, Justice JB Jr (1996) Quantitative microdialysis of serotonin and norepinephrine: pharmacological influences on in vivo extraction fraction. J Neurosci Methods 68(1):39–47
69. Parsons LH, Justice JB Jr (1994) Quantitative approaches to in vivo brain microdialysis. Crit Rev Neurobiol 8(3):189–220
70. Shippenberg TS, He M, Chefer V (1999) The use of microdialysis in the mouse: conventional versus quantitative techniques. Psychopharmacology (Berl) 147(1):33–34
71. Invernizzi R, Garavaglia C, Samanin R (2000) JL13, a pyridobenzoxazepine compound with potential atypical antipsychotic activity, increases extracellular dopamine in the prefrontal cortex, but not in the striatum and the nucleus accumbens of rats. Naunyn Schmiedebergs Arch Pharmacol 361(3): 298–302
72. Calcagno E, Canetta A, Guzzetti S, Cervo L, Invernizzi RW (2007) Strain differences in basal and post-citalopram extracellular 5-HT in the mouse medial prefrontal cortex and dorsal hippocampus: relation with tryptophan hydroxylase-2 activity. J Neurochem 103(3): 1111–1120

Chapter 10

Intracerebral Human Microdialysis in Parkinson's Disease

Salvatore Galati and Giuseppe Di Giovanni

Abstract

Microdialysis (MD) procedure is a versatile technique that allows the analysis of small molecular weight compounds from the interstitial space in different tissues. MD was used extensively in neuroscience animal studies at the end of last century, while only a single MD study in the human brain was performed at the end of 1980s. In contrast, over the last decade MD has been applied in humans mainly in neurological disorders, such as stroke, epilepsy, and Parkinson's disease (PD). In these disorders MD has allowed quantification of neurotransmitters, peptides, and hormones giving precious information in research as well as in clinical practice. Microdialysis in awake PD patients in conjunction with deep brain stimulation (DBS) surgery has received increasing interest, but is nonetheless still a novel approach. It is likely that MD will be instrumental as a research tool in clarifying important open issues relating to the basal ganglia functions and dysfunctions. The present chapter will summarize the principles of MD and its involvement in research studies in PD in humans, illuminating the limits and the future applications of this procedure.

Key words: Microdialysis, Deep brain stimulation, Parkinson's disease, Subthalamic nucleus, Substantia nigra, Basal ganglia

1. Introduction

The basis of microdialysis (MD) dates back to the 1970s, when it was used to measure dopamine in the striatum in awake animals (1). However, the first MD study in human brain was performed only after a decade (2) and still represents the cornerstone for the subsequent applications in clinical use of this methodology. Successively, MD potentiality was applied in the clinical assessment of head trauma (3), in stroke (4), epilepsy (5), and brain tumor (6, 7). So far as PD (8) is concerned, MD has shown considerable utility as a powerful investigative tool in experimental research either in animal (9) or in human studies (10–14). The main characteristic of this technique is the real-time sampling of endogenous neurotransmitters and other small molecules that undergo modification when

Giuseppe Di Giovanni and Vincenzo Di Matteo (eds.), *Microdialysis Techniques in Neuroscience*, Neuromethods, vol. 75,
DOI 10.1007/978-1-62703-173-8_10,

PD arises. Moreover, this neurodegenerative disease is characterized by rapid modifications of motor performance (commonly in response of drug therapy or during intense emotions) that underly corresponding changes in the amount of several neurotransmitters either in cortical and in subcortical regions. Among its pathophysiological characteristics, these distinctive features render MD rather suitable for deepening the knowledge of basal ganglia functioning and dysfunctioning (9).

1.1. Basal Ganglia Dysfunction in Parkinson's Disease

The neurochemical changes induced by the reduction of dopamine (DA) striatal level in consequence of substantia nigra pars compacta (SNc) neuronal loss have been recently and extensively reviewed by Di Giovanni and colleagues (9). As much as the electrophysiology is concerned, DA demise also triggers profound changes of the neuronal activity within the entire cortical-basal ganglia-thalamic loop (15–18). The deep brain nuclei are interconnected and in relation to the other brain regions (mainly cortex) in a hierarchical manner, this loop was simplified as a "box and arrow" circuit in the late eighties (9, 19). In the Parkinsonian state, as exemplified in Fig. 1, the strong inhibitory projection of the internal part of globus pallidus (GPi) output (Fig. 1a pathway 1) influences directly the ascending drive from motor thalamic nuclei to the cerebral cortex (Fig. 1a pathway 2) as consequence

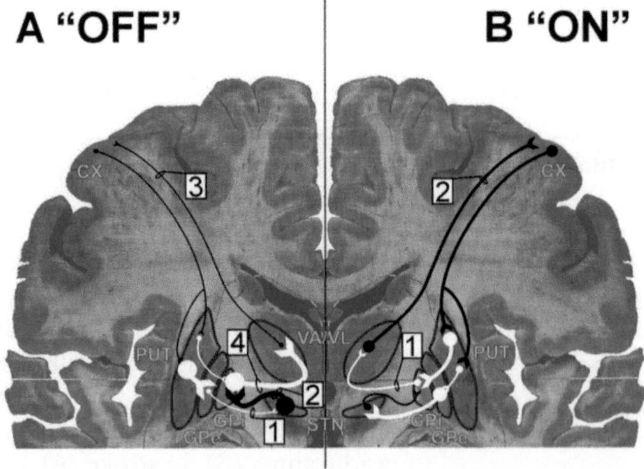

Fig. 1. The classical "rate model" representation of the cortico-basal ganglia loop in Parkinsonian "OFF" (*left panel*, (**a**)) and "ON" state (*right panel*, (**b**)). Excitatory (glutamate) pathways are in *black* whilst inhibitory ones are in *white* (GABA). The OFF state, i.e., the state of a PD patient without any therapy (**a**), is characterized by a decreased thalamo-cortical activity (pathway n° 2) as result of the lack of GABA-ergic inhibition from the GPe to the STN (n°4) and by the consequent overactivity of the excitatory STN-GPi pathway (n°3). In turn, GPi is overactive and, therefore, the GPi-VA/VL pathway (1) releases excessive GABA onto the VA/VL. The consequent VA/VL inhibition produces the decrease of the VA/VL inputs to cortex activity. After the drug challenge a rebalance of the activity of these pathways is observed producing e normalization of the thalamo-cortical drive.

of an excessively augmented glutamatergic tone to GPi driven by subthalamic nucleus (STN; Fig. 1b, pathway 3). This STN hyperactivity seems to be secondary to the loss of inhibitory GABAergic projections from the external segment of globus pallidus (GPe; Fig. 1b, pathway 4). As result, the ventrolateral thalamic (VL) and antero ventral thalamic nuclei (VA) are excessively inhibited working against motor initiative (Fig. 1b pathway 1) since both these thalamic nuclei are strictly connected to the motor and pre-motor cortex (Fig. 1b pathway 2) (20). The above mentioned changes of activity are largely inferred by rodent studies and confirmed in the MPTP primate Parkinsonian models (21).

1.2. Deep Brain Stimulation

On the basis of these notions, surgery was developed last century with the aim of rebalancing the inhibitory tone of GP upon thalamic relay to control symptoms of tremor and rigidity. These approaches consisted of surgically-created lesions of different nuclei (i.e., VA, STN, GPi). The deep brain stimulation (DBS) of different basal ganglia structures, mainly the STN and the GPi has nowadays replaced all earlier types of surgery, including pallidotomy. The effect of STN-DBS has received much interest, with several studies performed in animal models of PD, and to some extent also in human PD (for a review see ref. (22)). Either the clinical benefits of the lesion as well as the predicted augmented glutamatergic tone of the STN, suggested a DBS mediated inhibition of the STN (23–25). Indeed, findings obtained from 6-hydroxydopamine (6-OHDA)-denervated rats have shown that STN-DBS decreased the firing rate of substantia nigra pars reticolata (SNr), resulting in a disinhibition of thalamic motor nuclei (24). In contrast, other reports in Parkinsonian primate (26) reported an increase of GPi activity during HFS of STN in MPTP-treated monkeys. Trying to explain these discrepancies, a decoupling effect of high frequency stimulation (HFS) was posited, driving a decrease of firing rate of the cell soma with a coexisting increased of the output fibers (25, 27). These and further inconsistent data have weakened the functional classical explanation based on a mere play of activation and inhibition. A new hypothesis, supported by peculiar changes of the firing pattern rather than of firing frequency, was suggested (12, 24, 28, 29). In agreement, HFS suppresses STN spontaneous activity, generating a pattern of recurrent bursts of spike time locked to a stimulus pulse (30).

2. Microdialysis Data in Parkinsonian Patients

Microdialysis has easily bridged the gap between the animal models and man due to its simplicity and limited invasiveness. MD in human brain was performed for the first time during thalamotomy intended to relieve tremor in four patients with PD at the Department of Neurosurgery of Karolinska Hospital in Stockholm at the end of the 1980s by Meyerson and colleagues with the collaboration of Ungerstedt (31). The levels of dopamine (DA), dihydroxyphenylacetic acid (DOPAC), homovanillic acid (HVA), 5-hydroxyindolacetic acid (5-HIAA), hypoxanthine, inosine, guanosine, adenosine, gamma-aminobutyric acid (GABA), taurine, aspartate, and glutamate (GLU) were measured in the perfusate from the microdialysis probe inserted in the ventral intermediate nucleus (Vim) of the thalamus. The individual differences between the patients were most pronounced for DA, HVA, DOPAC, and HVA. Interestingly, the patient who received L-3,4-dihydroxyphenylalanine (L-DOPA) therapy had baseline levels of DA more than ten times higher than those of the other patients. The DA/DOPAC ratio in this study appeared to be approximately 1:10, whereas in the rat striatum the corresponding ratio is approximately 1:100. In these respects the human thalamus seems to be similar to the rat cerebral cortex. The new routine surgical implantation of neuroprosthetic devices for DBS in the treatment of PD has provided a window of opportunity for unprecedented monitoring of neurotransmission in the human brain using MD and recently fast-scan cyclic voltammetry (32). Although the clinical efficacy of STN DBS in PD has been well established, its mechanism of action is largely unknown and debated. Thus neurochemical data obtained by MD will be of paramount importance in the exploration of STN-DBS effects and in clarifying some aspects of pathophysiology of PD. Hitherto, MD in PD during DBS has been performed by standard single probe and dual probes approach. The latter allows simultaneous MD data from different structures measuring the inhibitory or excitatory neurotransmitter content during the clinical effective STN-DBS. This has been made possible by taking advantage of the "multi-target strategy" consisting of simultaneous and bilateral implantation of multiple basal ganglia nuclei (12, 33–35) with the aim of obtaining added clinical advantages from each target. In this chapter we will illustrate some of these advances in the knowledge of human PD physiopathology obtained by MD approach. We will also point out the limit of the intra-operative MD as well as the future directions of this procedure.

2.1. Microdialysis in DBS Surgery

The use of MD in conscious patients with PD undergoing DBS surgery in an acute setting is a relatively novel technique with several technical challenges but great possibilities. In vivo MD allows the analysis of the interstitial fluid that contains compounds released by the neuronal and glia cells directly exposed to it and also it is also suitable to study L-dopa pharmacokinetics in the brain of patients with PD (13). Nonetheless, MD is a sampling technique that needs an appropriate detecting device on which depends the analytical assay. Hitherto, the analysis of human dialysates has been performed with HPLC coupled to electrochemical and fluorometric detectors or by radio-immunoassay (36) for cGMP concentration determination (12).

After the pioneering MD data to test the reliability of the technique during surgery for thalamotomy in PD tremor patients (31), further results in the thalamus were achieved by Rada and colleagues (37) demonstrating a GABA reduction as consequence of HFS delivered before the thermolesion of the thalamus in tremor PD patients. In early 2000s, an MD study of the extracellular amino acid (aspartate, GLU, glycine, and GABA) levels measured in the GPi and in the STN of nine PD patients before and after an acute challenge of apomorphine demonstrated any significant changes of these compounds, despite the drug-induced clinical amelioration (10, 11). The same group, capitalizing from the so-called multi-target strategy by which both STN and GPi (35, 38) were bilaterally implanted in a cohort of advanced PD patients, performed MD before, during, and after DBS. This protocol had important impacts upon the understanding pathophysiology of PD and at the same time the DBS mechanisms of action. The combined target strategies fulfill its potential by the demonstration of elevated cGMP extracellular concentrations (expression of an augmented glutamatergic tone) in the GPi, despite negligible changes in GLU levels, during a clinically effective STN-DBS (39). In agreement, STN-DBS had promoted changes in the levels cGMP within the substantia nigra pars reticulata (SNr). In these patients, MD was performed from SNr before, during, and after STN-DBS, producing a significant increase of cGMP concentration in SNr up to +400%. These results suggested on the one hand that a clinical benefit is not necessarily accompanied by a GPi activity, while on the other that DBS effect is not mediated by an inhibition but rather by promoting excitation. The electrophysiological findings were compatible with the MD data showing the SNr mean firing rate significantly increased time locked with the clinically effective STN-DBS (12). Firing analysis allowed recognizing a pattern of SNr neurons from an irregular pattern into a clustered discharge. Although considered initially paradoxical, these data gave further fundamentals on the debate upon the increasingly disputed classical model of basal ganglia operation.

Nevertheless, by taking advantage of the neurosurgery trajectory to STN and GPi passing respectively trough VA and striatum, a biochemical study of the changes occurring simultaneously in these two crucial basal ganglia structures was achievable. Similar to that observed in GPi and SNr, in the striatum, extracellular cyclic GMP was also significantly augmented during clinically effective STN-DBS. This result was in line with an increased glutamatergic tone from the cortex. The biochemical result in VA, on the other hand, was rather bizarre in the light of the GPi data. Since the VA conveys pallido-fugal information to cortex, representing the target of the GPi, an increased GABA content after the STN-DBS induced cGMP increase was predicted. Surprisingly, the extracellular GABA content in the VA was decreased. Altogether the SNr/GPi and thalamic finding suggests a decoupling effect of STN-DBS. One explanation was centered on an opposite effect of electrical field on the GPi efferent fibers that are grouped in the ansa and the fasciculus lenticularis crossing the dorsal area of STN. Of note, one surgical procedure for advanced PD patients, in the pre-levodopa era, was the lesion of the pallid-fugal at this level with clear benefits on tremor as well as on rigidity (40). Thus, if on one hand it promotes the activity of the efferent STN fibers, on the other the STN-DBS electrical field causes a reduction of the pallido-fugal ones. This theory was essentially based on the different susceptibility of the fibers to the electric field on the basis of the diameter. Thus, the clinically effective STN-DBS parameter preferentially inhibits the large (1.5–5.0 μm) of the GPi fibers (41) directed to the thalamus through the dorsal boundaries of the STN region. Of note, the active contacts of the electrode are often located in this region, likely inhibiting the action potentials propagation at this level by producing a functional block every pulse with a duration of few milliseconds. At 130–180 Hz, the most effective frequency, there is a repetitive block that hampered any signaling transmission of action potentials. Therefore, the electric stimulation might work like a low-pass filter by blocking action potentials propagation along the pallid-thalamus pathway. Conversely, the thinner fibers direct to the GPi are excited by the electric field. In other words, on the basis of the electrode geometry and the intensity of the stimulation, the electrical field could activate the thinner STN axons directed to GPi while at the same time filtering the action potentials along the larger-diameter GPi-VA/VL fibers (see Fig. 3). As mentioned above, the overlying STN area has been recognized as effective in term of clinical benefit in Parkinsonian monkey (42).

In another series of patients who had undergone DBS, MD was performed from striatum that receives a large GLU inputs from the cortex, as well as from the centromedian (CM)-parafascicular (43) thalamic complex and from the VA/VL complex itself (19, 44). Stimulation of both these thalamic stations induces excitatory postsynaptic potentials in the striatum (45). Thus, it is likely that the observed cGMP increase in the PUT is related both to the larger cortical and the thalamic inputs. Indeed, both these possibilities are in line with the classical view of the basal ganglia circuitry (see Fig. 1).

Considering that CM/Pf-PUT pathway is under the direct control of GPi as the VA/VL, where GABA decrease is observed very rapidly after commencing DBS, a cGMP increase in the PUT related to the activation of this pathway would be expected to be within seconds. On the contrary putaminal cGMP increased only

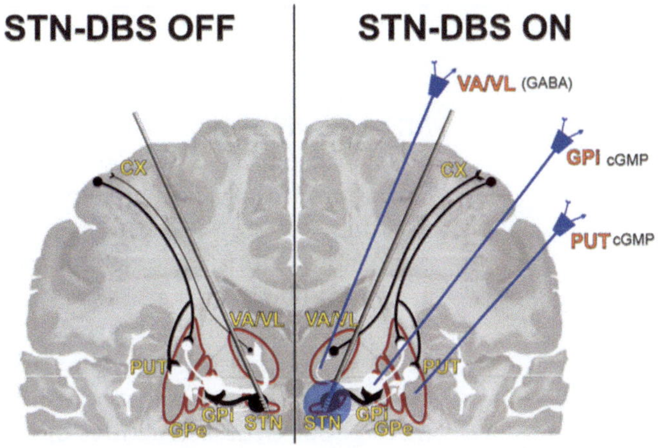

Fig. 2. Schematic representation of the structures (GPi, Putamen, and thalamus) investigated by mean microdialysis before, during, and after STN-DBS. cGMP was assessed in the structures receiving glutamatergic inputs such as GPi and Putamen respectively from STN and cortex/thalamus. On the other hand the content of GABA in the VA/VL that receive the GABAergic afferent from GPi was measured.

after several minutes from the decrease of thalamic GABA (see Fig. 3). This finding seems to suggest that a more complex loop involving the cortex, may be more correct.

These results also support a thalamic disinhibition that, in turn, reestablishes a more physiological level of PUT activity, as the main factor responsible for the clinical effect of STN-DBS in Parkinsonian patients. In a surgical and microdialysis session, STN and GPi target areas were identified preoperatively by means of ventriculography and intra-operatively by means of single unit recordings on two different trajectories, each performed with a multielectrode holder, one aimed at the STN and the second aimed at the GPi (35, 46) (see Fig. 2). Recently another two studies have investigated the use of MD in conjunction with DBS surgery (13, 14). Kilpatrick and colleagues (14) performed MD in STN prior to electrode implantation. They monitored GLU, GABA, and DA levels, showing that they were reliably identified in dialysate samples using HPLC obtaining a steady state baseline within approximately 30 min. In the study by Zsigmond and et al. (13) the use of a custom made probe for stereotactic MD was reported by which the authors studied the interstitial concentrations in the GPi of patients suffering from PD during intra-operative treatment, first with STN stimulation and then with intravenous (i.v.) L-DOPA infusion. During STN stimulation they observed increased L-DOPA levels in Gpi in some of the patients; however, the increase did not persist during the second fraction. Unfortunately, they were not able to dose DA for technical limitation and the concentration of noradrenaline, adrenaline, and serotonin were evaluated only during the steady state (13).

3. Methods

MD applied to human surgery for deep brain stimulation consists of several steps. We will describe the method used by Fedele and colleagues (10, 12). First step is the recognition of the target(s). For these, scope functional and stereotactic neurosurgery procedures are performed, based on planning system that includes an atlas and digital navigator system. The latter equipment is able to match the brain MRI with the CT scan acquired after the stereotactic device application on the patient skull. The procedure allows obtaining target coordinates with the useful trajectory not passing throughout critical structures of the brain or ventricles.

After the stereotactic localization, target areas are identified by means of single unit recordings. In case of multi-trajectory approach, two different trajectories aimed to STN and GPi can be obtained by means of a multi-electrode holder (35, 46) (see Fig. 2). With this procedure, it is possible to identify electrophysiologically the motor thalamus (VA), placed along the STN trajectory and the PUT along the GPi one. Subsequent to the electrophysiological identification of the targets, the recording electrodes are substituted by microdialysis probes. The real position of the probes and the stimulating electrodes is verified intra-operatively by X-ray in antero-posterior and latero-lateral projection. We used custom made probes (CMA, Microdialysis, Sweden) infused with an artificial cerebrospinal fluid (aCSF) containing 145 mM NaCl, 1.26 mM $CaCl_2$, 1 mM $MgCl_2$, 3.0 mM KCl, 1.4 mM Na_2HPO_4 (PBS) with a rate of 5 μl/min. For the perfusion we used a micro-infusion pump (CMA/400, CMA Microdialysis, Sweden) connected to the probes by means of tubes of 0.12 mm inner diameter (internal volume of 1.2 μl/100 mm length) of 50 cm length. This tube is connected to the inlet steel shaft of the probe. On the other hand the outlet steel shaft is connected to the sampling vials (300 μl collection vials Beckman, USA) by tubing adaptors. Each sample is collected every 10 min and on the basis of the perfusion rate (5 μl/min) with a volume of 50 μl. Probes were infused (5 μl/min) for stabilization (90 min) (10, 11). Then, basal microdialysis and clinical data are collected for 50 min. In order to avoid sample degradation, a mixture of antioxidant needs to be added before putting the vials into the fraction collector array. The solution is composed by 0.1 M acetic acid, 0.27 mM Na_2EDTA, 3.3 mM L-cysteine and 0.5 mM ascorbic acid, pH 3.2. Of particular importance for catecholamine detection, along with the antioxidant, each vial has to be protected by light and promptly closed on dry ice for the subsequent anlaysis or stored in a –80°C fridge.

After the first period of stabilization of 90 min, microdialysis and clinical data are collected for 50 min. Afterwards, STN-DBS is switched on for 60 min. Then, 60 min of recovery is performed (for

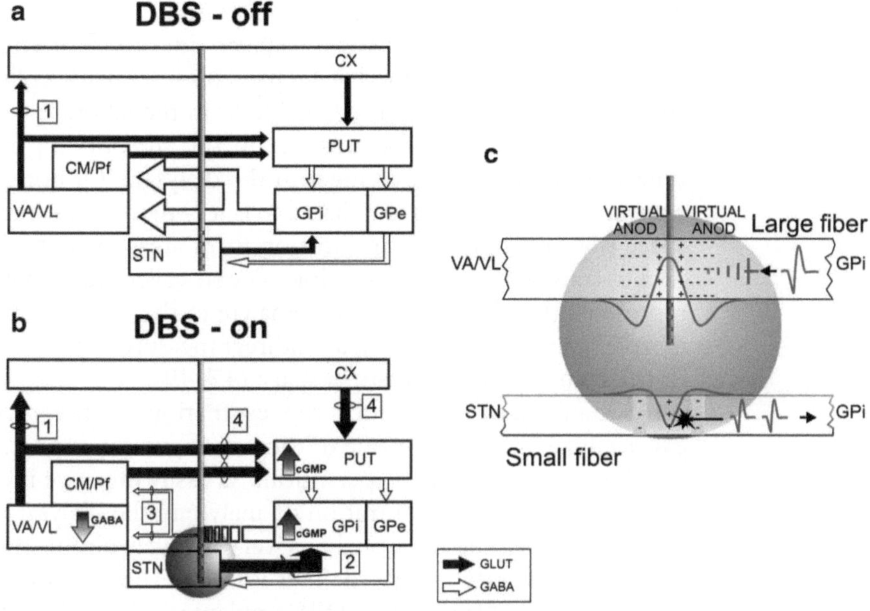

Fig. 3. Hypothetic mechanism of action of STN-DBS based on microdialytic data. (**a**) STN-DBS off. In the *upper part* of the figure is highlighted the classical alteration of basal ganglia activity as already mentioned in Fig. 1. (**b**) STN-DBS on. During the STN-DBS a clear cut decrease of GABA within the VA/VL likely consequent to the decrease of GPi afferents (n° 3) is observed. The reduction of GABA content causes an activation of the VA/VL-cortical pathway (n° 1). cGMP concentration in the GPi is increased according to a possible increased activity of the STN-GPi pathway (*arrow* marked by n° 2). Thus, the STN-DBS increases the activity of the STN-GPi pathway while decreasing the activity of the GPi-VA/VL pathway at the same time. The latter data was quite impressive and paradoxical in respect to the classical circuit. This paradox in (**c**) can be explained pointing out to the effect of an extracellular electrical field on large and small fibers. (**c**) The decupling effect between the large GPi-VA fibers and small STN-GPi fibers. In the proximity of the stimulating electrode the intracellular potential will be strongly depolarizing while at distances progressively farther from the stimulated point, the intracellular potential will became negative ("virtual anode"). According to the intensity of the electrical field, the "virtual anode" may block the action potential. The amplitude of the intracellular potential generated by an extracellular stimulating field is directly proportional to the diameter of the fiber, large fibers being excited more easily than small diameter fibers. On the basis of these considerations, and given that the contact normally activated in clinical use is one of the upper contact of Medtronic stimulating electrode, it can be inferred that the large fibers of the GPi-VA/VL pathway, positioned close to the stimulating electrode, will be blocked by stimulation ((**c**), *upper part*) while the small distant fibers of the STN-GPi pathway will be excited ((**c**), *lower part*).

details sees Fig. 3). All procedures are usually without additional risk for the patient, since guide tubes for micro-recordings will be already placed in each target. Clinical changes, contralateral upper limb rigidity, and akinesia needs to be continuously assessed by an expert neurologist utilizing selected items of UPDRS (Fahn et al., 1987) (rigidity 0–4, finger tapping 0–4, hand movement 0–4; total = 0 corresponds to "normal", 12 maximum score), while remaining blind to the stimulus intensity between 0 and 3 V. Cathecolamines, GABA, GLU concentration has been determined by HPLC with electrochemical detection (10, 13, 14, 39) and cGMP by radio-immuno assay respectively (12). The pefusate solution and the interstitial fluid are put in contact by means of an implanted probe characterized by a semipermeable polymeric membrane at the distal tip.

The specific pore size ranging from 5 to 20 kDa limits the diffusion of solutes from the brain's extracellular side to the perfusate (and vice versa) into the obtained dialysate. This is regularly collected and subsequently analyzed at the bedside or in the laboratory. Changes in the concentration of a substrate in the surrounding milieu are reflected by subsequent changes in the dialysate. Of note, the concentration of analyte in the dialysate is less than in the extracellular space obeying tight law of diffusion from areas of high concentration to those of low concentration. The so-called fractional recovery represents the ratio between the analyte in the tube and that in the extracellular space providing an indirect measure of the real content of the compound in interstitial space (47–49).

Several factors influence the estimation of the real analyte extracellular content but one of these is of greater importance, i.e., the flow rate of the probe perfusion. As easily intuitive it correlates inversely with the concentration of analyte in the dialysate samples. In other words, the fractional recovery parallels diminish when the perfusion flow rate increased. This is one of the major concerns to intra-operative MD during DBS implantation in PD patients. In order to obtain adequate analyte concentration and sufficient dialysate volume it is necessary to increase the sampling intervals and thus the length of surgery. One strategy to reduce the time of sampling could consist of an increase in the area of the interface membrane but this expedient reduces the anatomical accuracy (2). Less important factors such as the geometry of the probes and the volume of the extracellular fluid can also influence the MD findings. Worth mentioning is the probe placement time in the brain tissue. Long lasting experiments in animal studies have indeed shown edema and inflammatory cell infiltration of the membrane reducing variable extent the diffusion of the substances and ultimately the reliability of the results.

4. Limits of the Technique

MD clearly represents a formidable tool for research investigations in PD during surgery for DBS and for pharmacodynamics and pharmacokinetic studies. It allows detection of relatively rapid and long lasting changes that underlie clinical amelioration and worsening parallel to therapy. However, its potentiality is indeed hampered by the time of the procedure and by the limits of analytical methods. Moreover, MD requires a dedicated team that strictly collaborates with a neurosurgeon and that supports the patient during surgery monitoring the clinical state. Although not systematically weighted, a prolongation of surgery could theoretically represent an addition risk for patients during the procedure as well as in the post-operative recovery when the patient could be exposed, for instance, to infections of the wounds.

Undoubtedly MD will contribute to the evidence that clinical transitions in PD are linked to profound changes in modality rather than rate of activity of basal ganglia structures. The classical basal ganglia view schematically represented in a "box and arrow" diagram is definitely too simplistic. The MD technique per se is not free from limitations depending on the duration of probe placement. Indeed, the real-time measurement of endogenous substances might be unreliable due to prolonged probe placement causing brain tissue edema close to the membrane, altering the evaluation of the substances in the extracellular milieu. However, this drawback can be controlled by an appropriate time of probe wash after its implantation. Instead, changes of probe position with respect to the brain parenchyma should be avoided, since they might cause additional tissue damage, modifying the analysis of MD samples. In addition, the brain pulsation could theoretically distort analysis. In other words, since MD probe is anchored to the skull by the stereothactic frame with which it constitutes a unique body, on the other side, brain rhythmically moves parallel with the cardiac pulse. Although not yet demonstrated, we can figure out that the brain pulsing could generate an up and down of the membrane probe adding additional confounding factors.

As far as the cathecholamines are regarded, their detection needs scrupulous care in storage since it is easily degradable to the light and enzymatic exposure. This facet becomes critical when MD and sampling analysis cannot be performed at the same time, implying additional time for transportation, storage, and subsequent analysis. Unfortunately, the technique for sample analysis needs an apparatus that cannot be installed in the operating room and the same analysis should be done in a dedicated laboratory. Obviously, all the procedures, from the beginning steps to the sampling analysis should be planned by an experienced and skilled team working jointly to a functional surgeon.

5. Prospect for Future Studies

Besides the above limitations, MD has represented and surely will be a precious tool for researcher in movement disorders. MD could go further by biochemically recognizing the mark of akinesia (or other PD feature as freezing for instance) and activate the DBS, reducing the battery consumption as well as DBS side effects, when clinical performance does not need its activation. As we have recently demonstrated, thalamic GABA could be a good candidate as a biochemical marker of ON-OFF state (50, 51). Indeed, this theoretic MD application needs the development of miniaturized devices of GABA sampling and detection that will be simultaneously implanted with DBS apparatus.

New scenarios for MD future directions include also the *in loco* delivery of substances by the so-called reverse procedure. Together with detecting substances, we can speculate that MD might be used for chronic delivery of inhibitory neurotransmitters in structures known as hyperactive (i.e., STN) or growth factors able to reduce neurodegenerative process.

6. Conclusion

The development of MD as a general tool for studying brain chemistry has led to a truly translational venture as MD is brought into clinical use. Today there are MD catheters available for implantation in different tissues, including the brain, together with miniaturized pumps, microvials, and analysers suitable for intra operative as well as bedside use in the Intensive Care Unit. MD samples the neuronal environment that tells us whether the basal ganglia nuclei (1) are functioning abnormally; (2) neurotransmitter and second messenger concentrations are abnormal; (3) are receiving the expected drug concentration (e.g., L-DOPA). With ongoing research, MD will provide information about the clinical effect of DBS neurosurgery, improving the clinical outcome. Above all we will be able to monitor the impact of our interventions aiming to restore the normal physiology and chemistry of the BG circuitry. Therefore, further studies are warranted to correlate intracerebral MD with electrophysiology during evoked DBS therapy. The diffusion of MD technique also in medical practice will further the development of new analytical methods aimed at reducing the limit of sensitive chemical detectors and also overtake the limited temporal and spatial resolution of MD.

Acknowledgements

Authors are indebted to the EU COST Action CM1103 “Structure-based drug design for diagnosis and treatment of neurological diseases: dissecting and modulating complex function in the monoaminergic systems of the brain” for supporting our international collaboration. The research at the Ospedale Civico (GS) was supported by the Fondazione per lo Studio delle Malattie Neurodegenerative delle Persone Adulte e dell’Anziano, Lugano, Switzerland.

References

1. Ungerstedt U, Pycock C (1974) Functional correlates of dopamine neurotransmission. Bull Schweiz Akad Med Wiss 30(1–3):44–55
2. Lonnroth P, Smith U (1990) Microdialysis–a novel technique for clinical investigations. J Intern Med 227(5):295–300
3. Nilsson OG, Saveland H, Boris-Moller F, Brandt L, Wieloch T (1996) Increased levels of glutamate in patients with subarachnoid haemorrhage as measured by intracerebral microdialysis. Acta Neurochir Suppl 67:45–47
4. Meixensberger J, Kunze E, Barcsay E, Vaeth A, Roosen K (2001) Clinical cerebral microdialysis: brain metabolism and brain tissue oxygenation after acute brain injury. Neurol Res 23(8): 801–806
5. During MJ, Spencer DD (1993) Extracellular hippocampal glutamate and spontaneous seizure in the conscious human brain. Lancet 341(8861):1607–1610
6. De Micheli E, Alfieri A, Pinna G, Bianchi L, Colivicchi MA, Melani A, Pedata F, Della Corte L, Bricolo A (2000) Extracellular levels of taurine in tumoral, peritumoral and normal brain tissue in patients with malignant glioma: an intraoperative microdialysis study. Adv Exp Med Biol 483:621–625
7. Roslin M, Henriksson R, Bergstrom P, Ungerstedt U, Bergenheim AT (2003) Baseline levels of glucose metabolites, glutamate and glycerol in malignant glioma assessed by stereotactic microdialysis. J Neurooncol 61(2):151–160
8. Fahn S, Elton R (1987) Members of the UPDRS development committee. The Unified Parkinson's disease rating scale. In: Fahn S, Mardsen C, Goldstein M, Calne D (eds) Recent developments in Parkinson's disease. Mac Millan Healthcare Information, Florham Park, NY, pp 153–163
9. Di Giovanni G, Esposito E, Di Matteo V (2009) In vivo microdialysis in Parkinson's research. J Neural Transm Suppl 73:223–243
10. Fedele E, Mazzone P, Stefani A, Bassi A, Ansaldo MA, Raiteri M, Altibrandi MG, Pierantozzi M, Giacomini P, Bernardi G, Stanzione P (2001) Microdialysis in Parkinsonian patient basal ganglia: acute apomorphine-induced clinical and electrophysiological effects not paralleled by changes in the release of neuroactive amino acids. Exp Neurol 167(2):356–365
11. Fedele E, Stefani A, Bassi A, Pepicelli O, Altibrandi MG, Frasca S, Giacomini P, Stanzione P, Mazzone P (2001) Clinical and electrophysiological effects of apomorphine in Parkinson's disease patients are not paralleled by amino acid release changes: a microdialysis study. Funct Neurol 16(1):57–66
12. Galati S, Mazzone P, Fedele E, Pisani A, Peppe A, Pierantozzi M, Brusa L, Tropepi D, Moschella V, Raiteri M, Stanzione P, Bernardi G, Stefani A (2006) Biochemical and electrophysiological changes of substantia nigra pars reticulata driven by subthalamic stimulation in patients with Parkinson's disease. Eur J Neurosci 23(11):2923–2928
13. Zsigmond P, Nezirevic D, Kullman A, Augustinsson LE, Dizdar N (2012) Stereotactic microdialysis of the basal ganglia in Parkinson's disease. J Neurosci Methods 206(2):195–199
14. Kilpatrick M, Church E, Danish S, Stiefel M, Jaggi J, Halpern C, Kerr M, Maloney E, Robinson M, Lucki I, Krizman-Grenda E, Baltuch G (2010) Intracerebral microdialysis during deep brain stimulation surgery. J Neurosci Methods 190(1):106–111
15. Blandini F, Nappi G, Tassorelli C, Martignoni E (2000) Functional changes of the basal ganglia circuitry in Parkinson's disease. Prog Neurobiol 62(1):63–88
16. Alexander GE, Crutcher MD (1990) Functional architecture of basal ganglia circuits: neural substrates of parallel processing. Trends Neurosci 13(7):266–271
17. Guatteo E, Cucchiaroni ML, Mercuri NB (2009) Substantia nigra control of basal ganglia nuclei. J Neural Transm Suppl 73:91–101
18. Kravitz AV, Freeze BS, Parker PR, Kay K, Thwin MT, Deisseroth K, Kreitzer AC (2010) Regulation of parkinsonian motor behaviours by optogenetic control of basal ganglia circuitry. Nature 466(7306):622–626
19. Parent A, Hazrati LN (1995) Functional anatomy of the basal ganglia. II. The place of subthalamic nucleus and external pallidum in basal ganglia circuitry. Brain Res Brain Res Rev 20(1):128–154
20. Albin RL, Young AB, Penney JB (1995) The functional anatomy of disorders of the basal ganglia. Trends Neurosci 18(2):63–64
21. Bergman H, Wichmann T, Karmon B, DeLong MR (1994) The primate subthalamic nucleus. J Neurophysiol 72(2):507–520
22. Hammond C, Bergman H, Brown P (2007) Pathological synchronization in Parkinson's disease: networks, models and treatments. Trends Neurosci 30(7):357–364
23. Benazzouz A, Gao DM, Ni ZG, Piallat B, Bouali-Benazzouz R, Benabid AL (2000) Effect of high-frequency stimulation of the

subthalamic nucleus on the neuronal activities of the substantia nigra pars reticulata and ventrolateral nucleus of the thalamus in the rat. Neuroscience 99(2):289–295

24. Welter ML, Houeto JL, Bonnet AM, Bejjani PB, Mesnage V, Dormont D, Navarro S, Cornu P, Agid Y, Pidoux B (2004) Effects of high-frequency stimulation on subthalamic neuronal activity in parkinsonian patients. Arch Neurol 61(1):89–96
25. Filali M, Hutchison WD, Palter VN, Lozano AM, Dostrovsky JO (2004) Stimulation-induced inhibition of neuronal firing in human subthalamic nucleus. Exp Brain Res 156(3): 274–281
26. Hashimoto T, Elder CM, Okun MS, Patrick SK, Vitek JL (2003) Stimulation of the subthalamic nucleus changes the firing pattern of pallidal neurons. J Neurosci 23(5):1916–1923
27. McIntyre CC, Savasta M, Kerkerian-Le Goff L, Vitek JL (2004) Uncovering the mechanism(s) of action of deep brain stimulation: activation, inhibition, or both. Clin Neurophysiol 115(6): 1239–1248
28. Allers KA, Kreiss DS, Walters JR (2000) Multisecond oscillations in the subthalamic nucleus: effects of apomorphine and dopamine cell lesion. Synapse 38(1):38–50
29. Stefani A, Bassi A, Mazzone P, Pierantozzi M, Gattoni G, Altibrandi MG, Giacomini P, Peppe A, Bernardi G, Stanzione P (2002) Subdyskinetic apomorphine responses in globus pallidus and subthalamus of parkinsonian patients: lack of clear evidence for the 'indirect pathway'. Clin Neurophysiol 113(1):91–100
30. Garcia L, D'Alessandro G, Fernagut PO, Bioulac B, Hammond C (2005) Impact of high-frequency stimulation parameters on the pattern of discharge of subthalamic neurons. J Neurophysiol 94(6):3662–3669
31. Meyerson BA, Linderoth B, Karlsson H, Ungerstedt U (1990) Microdialysis in the human brain: extracellular measurements in the thalamus of parkinsonian patients. Life Sci 46(4):301–308
32. Kishida KT, Sandberg SG, Lohrenz T, Comair YG, Saez I, Phillips PE, Montague PR (2012) Sub-second dopamine detection in human striatum. PLoS One 6(8):e23291
33. Galati S, Scarnati E, Mazzone P, Stanzione P, Stefani A (2008) Deep brain stimulation promotes excitation and inhibition in subthalamic nucleus in Parkinson's disease. Neuroreport 19(6):661–666
34. Stefani A, Fedele E, Pierantozzi M, Galati S, Marzetti F, Peppe A, Pastore FS, Bernardi G, Stanzione P (2009) Reduced GABA content in the motor thalamus during effective deep brain stimulation of the subthalamic nucleus. Front Syst Neurosci 5:17
35. Peppe A, Pierantozzi M, Bassi A, Altibrandi MG, Brusa L, Stefani A, Stanzione P, Mazzone P (2004) Stimulation of the subthalamic nucleus compared with the globus pallidus internus in patients with Parkinson disease. J Neurosurg 101(2):195–200
36. McIntyre CC, Savasta M, Kerkerian-Le Goff L, Vitek JL (2004) Uncovering the mechanism(s) of action of deep brain stimulation: activation, inhibition, or both. Clin Neurophysiol 115(6):1239–1248
37. Rada P, Tucci S, Teneud L, Paez X, Perez J, Alba G, Garcia Y, Sacchettoni S, del Corral J, Hernandez L (1999) Monitoring gamma-aminobutyric acid in human brain and plasma microdialysates using micellar electrokinetic chromatography and laser-induced fluorescence detection. J Chromatogr B Biomed Sci Appl 735(1):1–10
38. Stefani A, Peppe A, Pierantozzi M, Galati S, Moschella V, Stanzione P, Mazzone P (2009) Multi-target strategy for Parkinsonian patients: the role of deep brain stimulation in the centro-median-parafascicularis complex. Brain Res Bull 78(2–3):113–118
39. Stefani A, Fedele E, Galati S, Pepicelli O, Frasca S, Pierantozzi M, Peppe A, Brusa L, Orlacchio A, Hainsworth AH, Gattoni G, Stanzione P, Bernardi G, Raiteri M, Mazzone P (2005) Subthalamic stimulation activates internal pallidus: evidence from cGMP microdialysis in PD patients. Ann Neurol 57(3):448–452
40. Meyers R (1955) Parkinsonism, athetosis and ballism; a report of progress in surgical therapy. Postgrad Med 17(5):369–381
41. Sidibe M, Bevan MD, Bolam JP, Smith Y (1997) Efferent connections of the internal globus pallidus in the squirrel monkey: I. Topography and synaptic organization of the pallidothalamic projection. J Comp Neurol 382(3):323–347
42. Benazzouz A, Tai CH, Meissner W, Bioulac B, Bezard E, Gross C (2004) High-frequency stimulation of both zona incerta and subthalamic nucleus induces a similar normalization of basal ganglia metabolic activity in experimental parkinsonism. FASEB J 18(3):528–530
43. Herzog J, Fietzek U, Hamel W, Morsnowski A, Steigerwald F, Schrader B, Weinert D, Pfister G, Müller D, Mehdorn HM, Deuschl G, Volkmann J (2004) Most effective stimulation site in subthalamic deep brain stimulation for Parkinson's disease. Mov Disord 19(9):1050–1054
44. Parent A, Hazrati LN (1995) Functional anatomy of the basal ganglia. I. The cortico-basal

ganglia-thalamo-cortical loop. Brain Res Brain Res Rev 20(1):91–127

45. Hull CD, Bernardi G, Price DD, Buchwald NA (1973) Intracellular responses of caudate neurons to temporally and spatially combined stimuli. Exp Neurol 38(2):324–336
46. Mazzone P, Brown P, Dilazzaro V, Stanzione P, Oliviero A, Peppe A, Santilli V, Insola A, Altibrandi M (2005) Bilateral implantation in globus pallidus internus and in subthalamic nucleus in Parkinson's disease. Neuromodulation 8(1):1–6
47. Bungay PM, Dedrick RL, Fox E, Balis FM (2001) Probe calibration in transient microdialysis in vivo. Pharm Res 18(3):361–366
48. Bungay PM, Newton-Vinson P, Isele W, Garris PA, Justice JB (2003) Microdialysis of dopamine interpreted with quantitative model incorporating probe implantation trauma. J Neurochem 86(4):932–946
49. Elmquist WF, Sawchuk RJ (1997) Application of microdialysis in pharmacokinetic studies. Pharm Res 14(3):267–288
50. Stefani A, Fedele E, Pierantozzi M, Galati S, Marzetti F, Peppe A, Pastore FS, Bernardi G, Stanzione P (2011) Reduced GABA content in the motor thalamus during effective deep brain stimulation of the subthalamic nucleus. Front Syst Neurosci 5:17
51. Stefani A, Fedele E, Vitek J, Pierantozzi M, Galati S, Marzetti F, Peppe A, Bassi MS, Bernardi G, Stanzione P (2011) The clinical efficacy of L-DOPA and STN-DBS share a common marker: reduced GABA content in the motor thalamus. Cell Death Dis 2:e154

Chapter 11

Monitoring Extracellular Amino Acid Neurotransmitters and hROS by In Vivo Microdialysis in Rats: A Practical Approach

Maria Alessandra Colivicchi, Chiara Stefanini, Wolfhardt Freinbichler, Chiara Ballini, Loria Bianchi, Keith F. Tipton, and Laura Della Corte

Abstract

The microdialysis technique has proved to be a powerful neuropharmacological tool for the measurement of extracellular levels of neurotransmitters in brains of freely moving animals. However, care in its application is essential if reliable results are to be obtained. More recent developments have made it possible to detect the release of highly reactive oxygen species (hROS) in the same systems as those used for determining amino acid neurotransmitter release. This chapter is intended to help in obtaining reliable and reproducible results by describing practical aspects of in vivo microdialysis in the rat, as applied to the determination of hROS and amino acid release in the same experiment. The presentation and validation of the results obtained are also discussed.

Key words: Microdialysis, High-performance liquid chromatography, Fluorescence detection, hROS, Amino-acid neurotransmitters, Aspartate, GABA, Glutamate, Taurine, In vivo release

Abbreviations

aCSF	Artificial cerebrospinal fluid
DNQX	6,7-Dinitroquinoxaline-2,3-dione
hROS	Highly reactive oxygen species
KA	Kainate
OH-TA	2-Hydroxyterephthalate
OPA	o-Phthalaldehyde
TA^{2-}	Terephthalic acid, terephthalate
RT	Retention time
TTX	Tetrodotoxin

Giuseppe Di Giovanni and Vincenzo Di Matteo (eds.), *Microdialysis Techniques in Neuroscience*, Neuromethods, vol. 75, DOI 10.1007/978-1-62703-173-8_11,

1. Introduction

1.1. Microdialysis Applications and Limitations

Microdialysis, which was originally developed by Ungerstedt and colleagues for determining monoamine release in brain (1), is used to sample the extracellular fluid (CSF) of living tissue by means of a small catheter (also referred to as microdialysis probe) inserted into the tissue of interest. The microdialysis probe is designed to mimic a blood capillary and consists of a concentric shaft with a semipermeable hollow fiber membrane at its tip (see Fig. 1), which is connected to inlet and outlet tubing. The probe is continuously perfused with an aqueous solution (perfusate/dialysate), which closely resembles the (ionic) composition of the surrounding tissue fluid, at a low flow rate of approximately 0.5–3.0 μl/min. Once inserted into the tissue or the fluid of interest, small solutes can cross the semipermeable membrane by passive diffusion. The direction of the analyte flow is determined by the respective concentration gradient, which allows microdialysis probes to function as delivery as well as sampling tools.

This semi-invasive technique is now in widespread use for monitoring neurotransmitter concentrations in the brain, and gradually spreading to many other research areas. In a recent review Chefer and colleagues (2) estimated that there were over 11,000 publications where microdialysis had been used to sample various substances from different tissues, with a majority still dedicated to monitoring extracellular monoamines and their metabolites in the CNS.

In this chapter we limit our description to the microdialysis procedures, which have been in use in our laboratory, exploiting the advantages of this technique and addressing the points that are critical to its successful use for monitoring the extracellular concentrations of the amino acid transmitters, aspartate, glutamate, taurine, and γ-aminobutyric acid (GABA), together with highly reactive oxygen species (hROS), in brain areas of the conscious, freely moving rat.

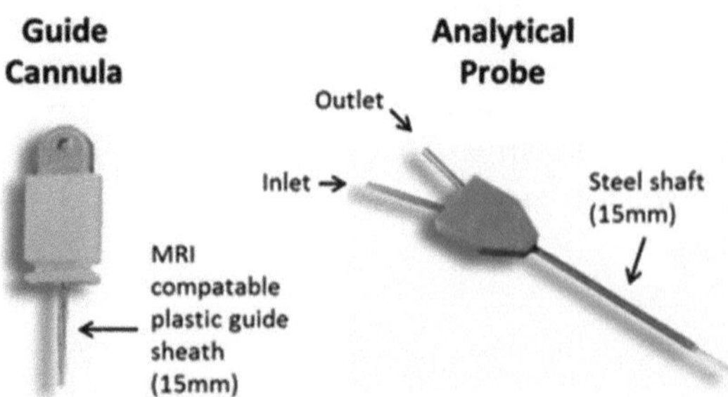

Fig. 1. Schematic representation of the guide cannula and microdialysis probe. Once the microdialysis probe (shown on the *right*) is inserted into its matching guide cannula (shown on the *left*), the membrane tip, which extends below the plastic guide sheath, is exposed to the extracellular space. Reproduced from CMA/Microdialysis Catalogue, with permission of CMA/Microdialysis AB, Stockholm, Sweden.

1.2. Amino Acid Neurotransmitters

The amino acids glutamate and aspartate mediate most of the excitatory synaptic transmission in the brain (3). However, the functions of the glutamatergic pathways are more diverse and complex than simply generating excitatory postsynaptic currents. Indeed, glutamatergic transmission plays a major role in neuroplasticity, which is implicated in memory formation, and its persistent activation may be responsible for excitotoxic neurodegeneration (3). The role of GABA as the main inhibitory neurotransmitter in the central nervous system, where it ameliorates excitatory transmission, is also well established. There has been less certainty about the functions of the β-amino acid taurine (2-aminoethanesulfonic acid), which is present at high concentrations in the brain. It has been shown to be released in response to glutamatergic stimulation and has been reported to act as an osmoregulator, a membrane-stabilizer, an indirect antioxidant, and an inhibitory transmitter acting in concert with GABA (4, 5).

Clearly, any full understanding of the interplay of these amino acids in the functions and dysfunctions of the CNS requires procedures for monitoring changes in their extracellular concentrations in response to stimulation. Here we describe a robust procedure developed in our laboratory (6) for doing this in the brains of freely moving rats that is based on fluorescence detection after reaction with *o*-phthalaldehyde (OPA) and HPLC.

1.3. Highly Reactive Oxygen Species

Reactive oxygen species (ROS) is a term that comprises oxygen-containing species that are more reactive than oxygen itself. These cover several different redox states of oxygen, ranging from the relatively unreactive superoxide ($O_2^{\bullet-}$) and hydrogen peroxide (H_2O_2), the non-radical species, singlet oxygen, and hROS, which may exist as free hydroxyl radicals ($OH^{\bullet-}$), as bound ("crypto") radicals, or as Fe(IV)-oxo (ferryl) species. hROS may be formed by reaction of hydrogen peroxide and the superoxide anion with transition metal complexes, particularly Fe(II), in the Fenton reaction (7). However, it is still uncertain whether the major product of this reaction is the free hydroxyl radical, a "crypto" radical, or a ferryl (Fe(IV)-oxo) species (8).

ROS have been implicated in the etiology of many diseases, including neurodegenerative diseases, ischemic or traumatic brain injuries, cancer, diabetes, and liver injury (9, 10) and the, direct or indirect, involvement of H_2O_2 and superoxide in a variety of cell-signaling processes is quite well established (11–13). However, the situation regarding hROS is less well defined (14) largely owing to difficulties in their detection, the lack of specificity of their formation, and their high nonspecific reactivities and short half-lives (15).

We have shown terephthalatic acid (TA^{2-}) to be effective as a trap to detect hROS release in microdialysis experiments (16). Reaction of TA^{2-}, which is essentially nonfluorescent, with hROS

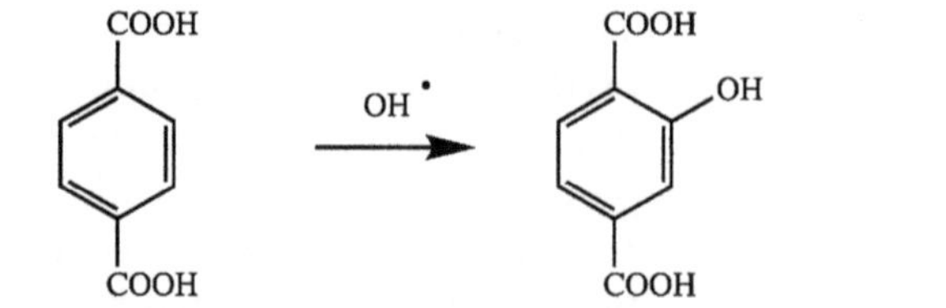

Terephthalic acid 2-Hydroxyterephthalic acid

Fig. 2. Reaction for fluorescent detection of hROS. Reaction of a hydroxyl radical with terephthalic acid to yield 2-hydroxy terephthalic acid.

produces a single, highly fluorescent product, 2-hydroxyterephthalic acid (OH-TA) (Fig. 2). This hydroxylation reaction is specific for hROS, as it requires the high redox potentials of OH-radicals or ferryl species (>1.6 V) (17) and there is no significant reactivity with other ROS, such as the superoxide radical, $ROO^{\bullet -}$ or H_2O_2. It is not cytotoxic and its presence in the microdialysis medium has been shown not to interfere with amino acid release or detection (16). In this account we describe the use of this, newly developed, HPLC method with fluorescence detection, for the determination of hROS and amino acid release in the same microdialysis experiments.

2. Materials

The materials described here are those that we have found to be appropriate, although those from other suppliers may also function satisfactorily.

2.1. Equipment for Brain Microdialysis

- *Stereotaxic apparatus.* Stellar stereotaxic frame (Stoelting Co., Wood Dale, IL, USA).
- *Microdialysis probes.* Concentric vertical probes, the most widely used in recent years, are the most appropriate for implantation in brain areas of laboratory animals, and have a relatively low mechanical damaging impact. They can be obtained commercially or constructed in-house (for probe construction, see, e.g., (18)). After a number of years using probes of our own construction, we found that purchasing them from commercially available sources resulted in much lower variability in the results obtained and was also less time consuming. Although they have a much higher cost than those constructed in-house, this is, to some extent, offset by the fact that, if properly cleaned and stored (see Sect. 3.1.2), they can be reused for up to 6–7 times. Commercially available probes can be obtained in different sizes, appropriate to the dimensions of the brain area of interest and of the size of

the laboratory animal used. Different membrane pore-sizes are also available to allow analytes of different molecular weights to be sampled. Furthermore, they can be obtained with complementary guide cannulae. Insertion of the guide cannula first, with the insertion of the actual probe, which extends below the guide (see Fig. 1), being delayed until the start of the microdialysis experiment, minimizes tissue damage at the sampling site and allows for more versatility in terms of delay in performing the microdialysis experiment, after the implantation surgery. For implantation in the rat striatum; probes of concentric design CMA 12 (4 mm exposed membrane surface, cutoff 20,000 Da) with their matching guide cannulae (from CMA/Microdialysis AB, Stockholm, Sweden) are appropriate.

- *Microdialysis, micro-infusion pump.* CMA/100 (from CMA/ Microdialysis AB, Stockholm, Sweden) mounted with a 1 ml syringe.
- *Polyethylene tubing.* Polyethylene tubing (i.d. 0.58 mm; o.d. 0.99 mm; Biological Instruments SNC, Varese, Italy) to connect the pump syringe to the probe inlet and the probe outlet to the collecting vessels.
- *Collecting vessels.* 250 μl Eppendorf centrifuge tubes (Starlab GMBH, Ahrensburg, Germany).
- *Disposable gloves.* Disposable latex gloves must be used in handling probes, tubing, and samples, to avoid bacterial contamination and that from amino acids present in sweat.
- *Cryostat.* CM 1800 (Leica, Milan, Italy).
- *Microscope.* Nikon ECLIPSE 80i (Nikon Instruments SpA, Calenzano, Firenze, Italy) equipped with a JVC camera.

2.2. HPLC Equipment

- *HPLC gradient system.* A variety of excellent HPLC systems are available. In our work we currently use the following. A reverse-phase Shimadzu HPLC gradient system (Shimadzu Italia Srl, Milano, Italy) consisting of two LC-10A$_{VP}$ pumps, a SIL-10AD$_{VP}$ refrigerated autoinjector, and an RF-551 fluorescent detector (λ_{ex} = 340 nm and λ_{em} = 455 nm, for the amino acid OPA-derivatives; λ_{ex} = 315 nm; λ_{em} = 435 nm, for OH-TA) and the manufacturer's software (class-VP™ 7.2.1 SPI Client/ Server Chromatography Data System) for controlling the system and for chromatographic peak recording and integration.
- *Chromatographic column.* The standard chromatographic column used is a 5 μm reverse-phase Nucleosil C18 column (250 × 4 mm; Machery-Nagel, Duren, Germany).
- *Sonicator.* Bandelin Electronic Sonorex TK 52 (Berlin, Germany).

2.3. Reagents and Media

Sources of the materials not given in the text are listed in the appendix.

- *Artificial cerebrospinal fluid (aCSF).* Glucose- and calcium-free aCSF is prepared by dissolving 8.17 g (140 mmol) NaCl, 223.7 mg (3 mmol) KCl, 203 mg (1 mmol) $MgCl_2$, 213 mg (1.2 mmol) Na_2HPO_4, and 37.4 mg (0.27 mmol) NaH_2PO_4 in 1 l double-distilled water (resistance: ≥18.2 MΩ). To avoid bacterial contamination and precipitation, glucose and $CaCl_2$ are dissolved, immediately before the experiment, into 100 ml of the glucose- and calcium-free aCSF, to a final concentration of 7.2 and 1.2 mM, respectively. The final pH of the solution is 7.2. When hROS are to be analyzed together with the amino acid neurotransmitters, the trapping reagent, 250 μM TA^{-2}, is also included in the aCSF. When treatments involve application of substances through the probe (retrodialysis), perfusion for one or more fractions is performed with aCSF prepared to include the following additions, singly or in combinations: 25 mM KCl, used as a depolarizing stimulus; 100 μM or 1 mM kainate (KA), a non-NMDA glutamatergic agonist used as a depolarizing or excitotoxic stimulus; 100 μM 6,7-dinitroquinoxaline-2,3-dione (DNQX), used as a specific non-NMDA glutamatergic antagonist; and 1 μM tetrodotoxin (TTX), to assess neuronal release. Other additions can be effected in a similar way.
- *OPA reagent.* The OPA-reagent for pre-column derivatization (6) is prepared by addition of 5 mg of OPA and 20 μl of methanol plus 5 μl of 2-mercaptoethanol to 5 ml of 0.5 M $NaHCO_3$, pH 9.5, to give a final OPA concentration of 1 mg/ml (7.5 mM).
- Chromatographic gradient elution
  - *Mobile phase.* All solvents, which must be of HPLC grade, are filtered through 0.22 μm Millipore filters and degassed by sonication for 5 min before use. The mobile phase comprises methanol and potassium acetate (0.1 M, pH adjusted to 5.48 with glacial acetic acid). The elution gradients are described in Sect. 3.2.2.
  - *Amino acid Standard solution.* A solution of the amino acids, each at 0.5 mM, in 0.1 N HCl is prepared by dilution of a 2.5 mM stock standard solution (Pierce, Rockford, Illinois, USA) to which 0.5 mM taurine and 0.5 mM GABA are added. Standard curves are then prepared by diluting the 0.5 mM standard solution. When appropriate, the external standard, 1.79 μM cysteic acid, is added to each dilution.
  - *Terephthalate* (*TA^{2-}*). Stock solutions of sodium terephthalate (3 mM) are prepared in aCSF or buffer and either used directly or after further dilution. If terephthalic acid is used, rather than its sodium salt, a 100 mM stock solution

is prepared by dissolving 2 g in 50 ml of 1 M sodium hydroxide (NaOH).

– *2-Hydroxyterephthalic acid (OH-TA)*. This is available from Giotto Biotech (Sesto Fiorentino, Firenze, Italy) but may be synthesized by a modification of previously published procedures (16). 2.465 g (10.0 mmol) of *ortho*-bromoterephthalic acid (95%, Sigma-Aldrich) and 0.91 g NaOH (20.1 mmol) are dissolved in 57 ml of water. For surface activation, 5 g of copper powder is treated with concentrated HCl, filtered, and washed neutral with H_2O; 12.9 mg (0.20 mmol) of the activated copper powder, together with 1.826 g (22.1 mmol) sodium acetate, is added to the solution. After the addition of a few drops of phenolphthalein the color of the solution turns purple. The mixture is then heated to reflux and stirred with a magnetic stirrer for 25 h. During the reaction the color quickly turns to yellow. At intervals, 10% (w/V) KOH is added, to keep the reaction mixture at a pH of ca. 8, since precipitation occurs if the solution becomes acidic. The solution is then allowed to cool to room temperature and filtered. The filtrate is acidified (pH of ca. 3) with 2 M HCl to precipitate the product. The mixture is then kept at 4°C in a refrigerator overnight to complete the crystallization. The cream colored product is filtered (frit pore size 3), dried in vacuo, and stored in a desiccator above phosphorous pentoxide. The final yield should be about 1.42 g (78%).

3. Methods

3.1. Brain Microdialysis

3.1.1. Animal Handing and Surgery

Male Wistar rats (220–250 g; Harlan Laboratories SrL, S. Pietro al Natisone, Udine, Italy) have generally been used in our work. All animal experiments should be performed according to national guidelines for Animal Care (D.L. 116/92) and international ones, such as the European Communities Council Directives (86/609/ECC), with all efforts to minimize animal sufferings. Only the number of animals necessary to collect reliable scientific data should be used.

Rats are anesthetized with chloral hydrate (400 mg/kg body weight i.p.) and placed in a stereotaxic frame. One or two microdialysis vertical guide cannulae may be implanted vertically in the selected brain regions and fixed to the skull with self-curing acrylic cement and the skin is then sutured. The stereotaxic coordinates are taken from the atlas of Paxinos and Watson (19). For example, the coordinates for the right neostriatum, relative to Bregma, are AP 0.7, L –3.2, and DV –5.5 mm.

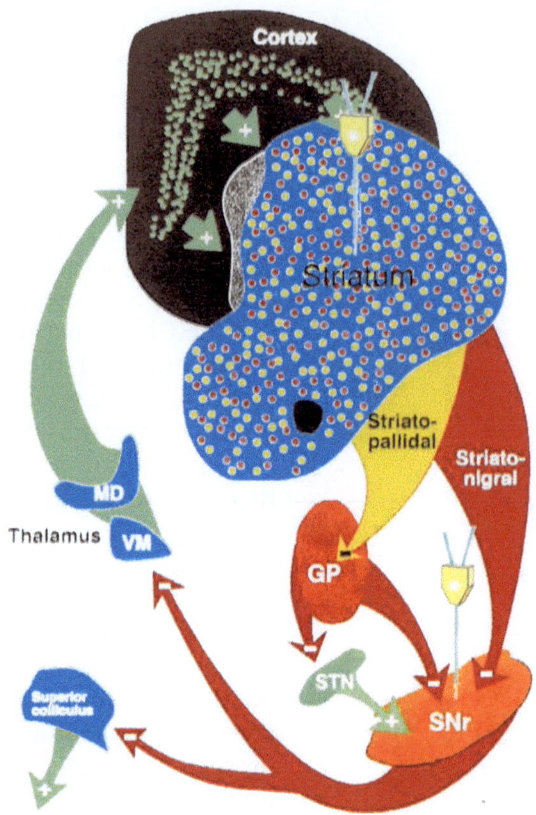

Fig. 3. Dual probe microdialysis in the striatonigral pathway. Diagrams of probe positions, one in the striatum (membrane tip of 4 mm) and one in the substantia nigra reticulata (membrane tip of 1 mm), are shown over a schematic representation of the striatonigral and striatopallidal GABAergic projections. In the case of the indirect striatonigral pathway, the second probe is located in the globus pallidus (membrane tip of 2 mm; not shown) rather than the substantia nigra. Modified from Gerfen (20).

As an example, diagrams of probe positions for direct striatonigral pathway are shown in Fig. 3 (20).

The animals are then returned to their cages and allowed to recover for *at least* 24 h before the microdialysis experiments are performed (see Notes 4.3 and 4.6.2.3).

3.1.2. Microdialysis Probe Handling and Care

All handling procedures require the use of disposable latex gloves to avoid sweat and bacterial contamination, which can easily lead to falsely high amino acid levels, particularly glutamate.

- *Washing procedure before use.* Before insertion into the guide cannula, the probe inlet is connected to the pump syringe through the polyethylene tubing, the syringe is filled with 70% ethanol, and the pump is set at 5 μl/min for 4–5 min. Then the syringe is filled with distilled water, and the probe is flushed for another 10 min. Finally, the syringe is filled with aCSF and the probe is flushed for a further 10 min. The probe, now ready

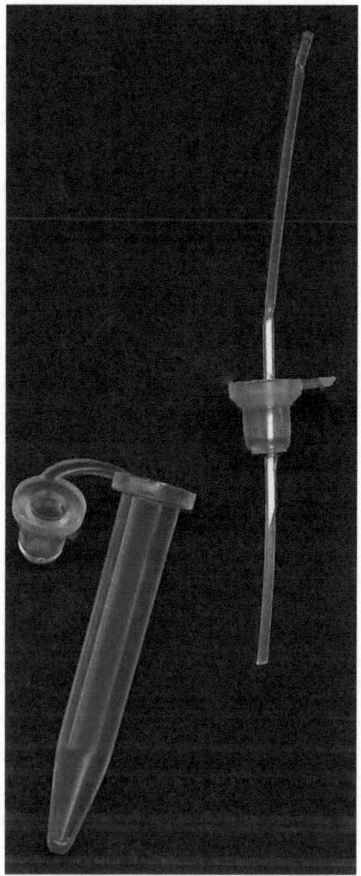

Fig. 4. Microdialysate collecting system. An Eppendorf centrifuge tube (shown on the *left* side) is connected to the outlet of the probe via a modification of the Eppendorf cap (shown on the *right* side), consisting in a 5 cm piece of polyethylene tubing onto a stainless steel tube cut from a syringe needle, which perforates an Eppendorf-tube cap, reaching close to the bottom of the collecting tube, via another piece of polyethylene tubing (2 cm).

for use, is inserted into the guide cannula and the outlet connected to a collecting Eppendorf centrifuge tube via a short (5 cm) piece of polyethylene tubing onto a stainless steel tube cut from a syringe needle (BD 25G1, Aldrich), which perforates an Eppendorf-tube cap, reaching close to the bottom of the collecting tube, via another piece of polyethylene tubing (2 cm) (see Fig. 4). The perfusion with aCSF can now start, setting the pump at the established perfusion rate, which in the case of striatal microdialysis is 3 μl/min, with a collecting time of 20 min for each fraction. In order to use the probe more than once, great care must be taken in the washing and storage procedures.

- *Washing procedure before storage for reuse.* At the end of the microdialysis experiment, the probe is removed and washed with 70% ethanol for 5 min followed by distilled water for 10 min, and then stored in a refrigerator at 4°C.

3.1.3. Probe Perfusion, Collection and Storage of Microdialysis Samples

After perfusing aCSF for a 90-min stabilization period, striatal dialysate samples may be collected every 20 min, unless shorter time periods are appropriate (see below). At the end of each collection period, the Eppendorf tube containing 60 μl of dialysate is quickly replaced, by hand, with a fresh tube, while the collected sample is capped and placed on ice for immediate analysis or frozen and stored at −80°C, where it is stable for up to 6 months (6). For determination of hROS, the 2-hydroxyterephthalate formed has also been shown to be stable for up to 6 months if stored in the same way before analysis.

Although it is possible to use an automatic fraction collector, the simple procedure described here performs well and is convenient if behavior is also to be monitored. In this way up to six rats, freely moving in their own cages, can be simultaneously monitored. The 20-min period allows sufficient material for determining amine neurotransmitters as well as the amino acids. If the dialysate samples are also to be analyzed for monoamines and their metabolites, they are collected into tubes containing 5 μL of 10 mM HCl, and then stored as above. A 5–10-min sample collection would be sufficient for the amino acids alone. We have found it unnecessary to cool the effluent collection tubes, each of sample is immediately cooled on ice or frozen for storage at the end of each collecting period. There is a clear advantage in delivering the microdialysate into the collecting vessel immediately as it comes out of the probe, reducing to a minimum the time before the microdialysate can reach refrigeration.

3.1.4. Protocol for Studying Evoked Release

After the 90-min stabilization, the first three samples are collected to evaluate basal levels. Whether a stimulus/treatment is applied systemically or locally, samples collected during and after the application represent the "stimulated" concentration–time curve. In the case of a depolarizing stimulus, aCSF containing 25 mM KCl is perfused for one 20-min fraction, collecting at least four more fractions before another stimulus is applied. In the case of KA stimulation, (1) if used as a non-NMDA agonist, aCSF containing 100 μM KA is perfused for one 20-min fraction; (2) if it is used as an excitotoxic stimulus, aCSF containing 1 mM KA is perfused for one 20-min fraction. The application of such a high concentration of KA results in epileptic episodes and must be done under anesthesia (chloral hydrate 400 mg/kg body weight given i.p. during the preceding 20 min). Another 4–6 fractions are then collected to monitor the stimulated concentration versus time curve. When appropriate, aCSF containing the antagonist DNQX (100 μM), to verify the selectivity of release evoked by a non-NMDA agonist, or TTX (1–3 μM), to verify the neuronal origin, or any other appropriate treatment, is applied throughout the experiment.

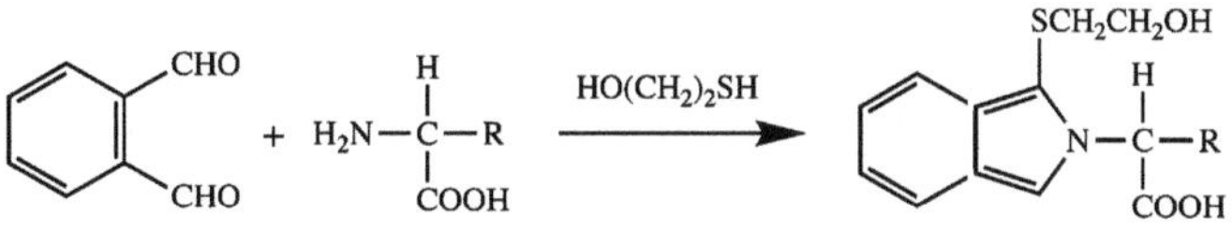

o-phthalaldehyde

Thioisoindole derivative

Fig. 5. Reaction for fluorescence detection of amino acids. Reaction of an amino acid and *o*-phthalaldehyde in the presence of 2-mercaptoethanol under alkaline conditions to give a highly fluorescent thioisoindole derivative.

3.1.5. Histological Probe Localization

At the end of the microdialysis experiment, the brains are removed and frozen in isopentane, to maintain the original volume and shape, at −45°C. Frozen brains are stored at −80°C before use. Coronal sections 30 μm thick are cut on a cryostat and mounted on slides and stained with 0.5% Cresyl fast Violet, to confirm the correct placement of the probe. Only data from rats with confirmed correct probe placement are used; those from a misplaced probe are discarded.

3.2. Determination of Aspartate, Glutamate, Taurine, and GABA in the Microdialysate

3.2.1. Derivatization

The amino acids are reacted with OPA (see Fig. 5) before application to the HPLC column. The reaction time, before on-column application, is quite critical (6). OPA-derivatives are obtained by adding an equal amount of the OPA-reagent, described in Sect. 2.3, to 5–10 μl of the sample and incubating at room temperature for precisely 1.5 min, after which time the entire incubation mixture is injected onto the HPLC column. The amino acid standard solution is treated in the same way. Automated pre-column derivatization may be used if high throughput is required (21).

3.2.2. Detection and Quantification of Amino Acids in Dialysates

HPLC is performed using the mobile phase system described in Sect. 2.2, according to Bianchi et al. (6) with minor modifications. During the first linear step (1.5 min), from 25 to 30% methanol, and the second linear step (15 min), from 30 to 70% methanol, it is possible to separate aspartate (RT = 9.6 min), asparagine (RT = 11.5 min), and glutamate (RT = 11.8 min), and then serine, histidine, glutamine, and OPA-reagent coelute in one peak (RT = 12.7 min), followed by glycine, arginine, and threonine co-elute in one peak (RT = 13.4 min) just before taurine (RT = 17.2 min), tyrosine (RT = 17.9 min), and alanine (RT = 18.5 min) followed by the well-separated peak of GABA (RT = 19.5 min) and that of tryptophan (RT = 20.2 min). Methionine, valine, and phenylalanine are eluted during the third step (2 min), during which methanol is increased from 70 to 90%, and during the subsequent isocratic hold at 90% methanol (1 min). Isoleucine, leucine, and lysine coelute during the subsequent steps, from 90% back to 25% methanol (2 min) and the isocratic hold at 25% methanol (3.5 min). After a total run time of 25 min, another run may be started immediately.

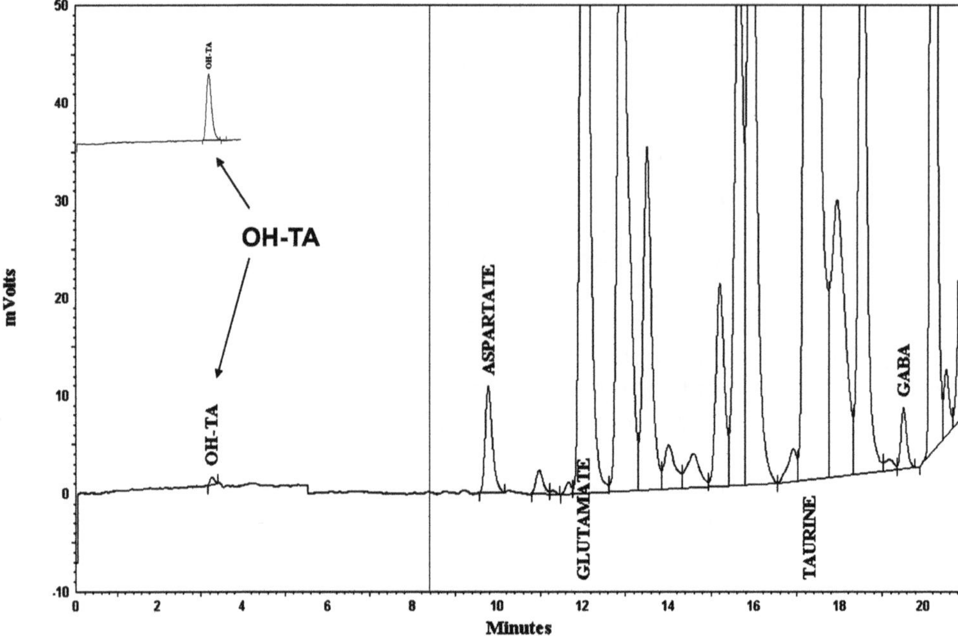

Fig. 6. Detection of OH-TA simultaneously to OPA-derivatized amino acids by HPLC and fluorescent detection. A typical chromatogram shows the peak of OH-TA detected at l_{ex} = 315 nm and l_{em} = 435 nm (expanded scale indicated by the *upper arrow*), followed by OPA-derivatized amino acids detected at l_{ex} = 340 nm and l_{em} = 455 nm. Reproduced from Freinbichler et al. (16), with permission of the International Society of Neurochemistry.

This elution system was optimized for the resolution and determination of aspartate, GABA, glutamate, and taurine and, as shown by Bianchi et al. (6), precise adjustment of the pH of the system is critical for effective separation.

3.3. Concurrent Determination of hROS and Amino Acid Release

Inclusion of 250 μM sodium terephthalate in the aCSF perfusion medium does not affect the yields of derivatized amino acids. The low retention time of OH-TA in this system (ca. 3.2 min) allows it to be determined, at excitation and emission wavelengths 315 and 435 nm, respectively, with no interference with the subsequent detection of amino acids (retention times >5 min), simply by changing the excitation and emission wavelengths to 340 and 455 nm, respectively, after its elution. When cysteic acid is included as a standard, the wavelengths are changed for OH-TA detection after its elution and then back to those for the OPA-derivatives, after emergence of the OH-TA peak. The external standard cysteic acid can be omitted, since detailed studies have shown the fluorescence yield of the amino acid derivatives to be stable under these conditions. 100 nM OH-TA may be added to samples as an internal standard and to identify the fluorescent peak position. A typical chromatogram is shown in Fig. 6.

3.4. Limits of Sensitivity and Quantification

There is a linear relationship between the amount of amino acids injected and the area of the corresponding peak ($r > 0.99$) from 0.05 to 2 pmol/μl of microdialysate. We have estimated the accuracy and precision of the method, by regression analysis according to the method of inverse prediction, from 9 to 15 replicate analyses of known aspartate, glutamate, taurine, and GABA concentrations added to the medium (6). In the above concentration range the accuracy, expressed as the mean percentage difference between the nominal and the estimated concentration, and the precision, expressed as the coefficient of variation (CV), were each below ±10%. Since blank values were not significantly different from zero, the detection limits for this method were defined for each compound as the amount showing 95% confidence interval (CI) that did not include 0 (zero). The detection limits ($n = 5$) of the four amino acids with their 95% CI shown in brackets were (fmol/μl microdialysate) 3.2 (2.6–3.6), 1.7 (1.0–2.3), 1.4 (1.1–1.7), and 2 (1.8–2.3) for aspartate, glutamate, taurine, and GABA, respectively (6).

Quantitation of hROS can be achieved by passing samples of OH-TA, prepared as described in Sect. 2.3, through the HPLC system under the same conditions. Our previous work (16) has shown the concentration of OH-TA formed, as measured by its fluorescence, to be linearly dependent on the concentration of hROS, generated from H_2O_2 in a Fenton reaction, in the range 0.5–1,000 nM. The detection limit is 0.5 fmol/μl microdialysate.

3.5. Statistical Evaluation of Results

The statistical analysis of amino acid and OH TA concentrations in microdialysate fractions is performed on the original values (nM). For graphical purposes only, concentrations can be expressed as percentage of their respective basal values, provided basal values are not significantly different from each other (see Sect. 4.6.2.4). The parameter used for statistical analysis of the evoked output is the area under the stimulated concentration–time curve (AUC_{stim}), normalized to the time unit corresponding to one 20-min fraction. Mean basal concentration values can be obtained from the AUC_{bas}, between -60 and -20 min normalized to the 20-min time unit, or taking the concentration value of the 20-min fraction collected immediately before the application of the stimulus (Bas). AUC_{bas} and Bas values should not be significantly different from each other. Thus mean values of the net stimulated kainate-induced output can be obtained from the stimulated AUC_{stim} value (nM/20 min) minus its respective basal value, i.e., the concentration in the 20-min pre-stimulation fraction. Confidence intervals (95% CI) of means and the one-sample test can be used for statistical significance of the evoked output over basal values. When appropriate, data are analyzed by ANOVA, followed by the Bonferroni test or another test for *post hoc* multiple comparisons, setting the probability level for statistical significance at $p < 0.05$. Data can be analyzed using the program Prism 5.0 for Mac OS X (GraphPad Software Inc., La Jolla, USA).

4. Notes

4.1. Behavior of the System: Basal Levels

The basal levels of the amino acids and hROS are determined in the microdialysate prior to stimulation. If release data are to be expressed as released minus of the basal value, it is important that the former remains constant and that the mean basal values, with errors, are presented together with the difference calculated.

In our studies low basal levels of OH-TA could be detected in all animals (mean value ± SEM 5.01 ± 0.38 nM ($n = 15$) (16). Some, but not all, studies that used different detection systems have reported higher basal levels. It has been suggested that hROS formation might be catalyzed by iron leaking from the stainless-steel probes that are used in many microdialysis studies (22). However, our own in vitro studies failed to observe any probe-catalyzed hROS formation. It has also been suggested that hROS formation might result from cell damage caused by the probe insertion (23). This would be an interesting field of study, but the use of adequate recovery times after probe insertion should minimize this response. The alternative possibility that the formation and release of hROS under basal conditions is a normal physiological process merits further investigation.

4.2. Behavior of the System: Evoked Release

The administration of 25 mM KCl by retrodialysis is frequently used as a depolarizing signal, whereas the non-NMDA agonist DNQX allows one to assess whether release evoked by kainate stimulation is solely a result of that agonist binding to receptors. Antagonists to other receptors may be used in a similar way. TTX, which blocks action potentials by binding to the voltage-gated, fast sodium channels in nerve cell membranes, preventing nerve cells from firing, is a standard way of determining how much of the evoked release is a result of nerve depolarization. Since vesicular release is calcium ion dependent an approach to determining the extent to which this contributes is to use calcium-free aCSF plus the specific chelator ethylene glycol tetraacetic acid (EGTA) (24, 25).

4.3. Behavior of the System: Recovery Period After Anesthesia

This may be critical since, as discussed below (Sect. 6.2.3), anesthetics affect amino acid release. A recovery time of *at least* 24 h after anesthesia with 400 mg/kg body weight (i.p.) chloral hydrate or ketamine/xylazine (80–120 mg/kg body weight, i.p.) has been found to be adequate, in terms of microdialysis studies on neurotransmitters. However, longer periods may be necessary if other anesthetics are used or when microdialysis is associated to behavioral studies. For example, we used a 7-day recovery period when rats had to be exposed to a step down inhibitory task (26).

4.4. Behavior of the System: Dual Probe Microdialysis

By positioning probes in two discrete brain regions it is possible to stimulate in one area and measure the evoked release at a distal site. Thus one can stimulate in a site containing cell bodies and determine both the local release and that occurring at sites enriched with terminals. Extensions to this allow the interactions of different neuronal pathways to be studied. Such an approach has, for example, been used to show that the taurine released in response to the local stimulus of the striatum by KA was from extraneuronal sources, whereas that evoked distally in the globus pallidus or substantia nigra was largely neuronal (27).

4.5. Behavior of the System: Behavioral Studies

It is, of course, possible to combine microdialysis, in the awake freely moving animal, with behavioral studies. However, as discussed above (Note 4.3), appropriately longer recovery times from anesthesia should be considered.

4.6. Advantages and Limitations in Detecting Amino Acid Neurotransmitters and hROS in Brain Microdialysis

The continued and developing use of microdialysis testifies to its value. However, there have been a number of criticisms of the approach that need to be considered. The main advantages can be summarized as follows.

4.6.1. Main Advantages

4.6.1.1. Microdialysis allows continuous monitoring (over several hours or days) of the concentrations of endogenous substances, drugs, or metabolites in the extracellular fluid of virtually any tissue.

4.6.1.2. The ability to administer compounds through the same probe as that used for sampling minimizes the risk of tissue damage and the pressure changes that can occur with direct injection into the brain.

4.6.1.3. The dialysis membrane acts as a filter against macromolecules (including enzymes), cells, or cellular debris present in the CSF.

4.6.1.4. Microdialysis represents an invaluable approach to study in vivo neurotransmitter interactions in the brain, as the regional chemistry of selected brain areas can be explored in the awake freely moving animal and associated with behavioral changes.

4.6.1.5. "Dual probe" microdialysis in the freely moving animal provides a means to monitor the effects of activation of a specific pathway, in terms of evoked neurotransmitter release, simultaneously detected at the site of neuronal cell bodies and at their distal, terminal sites.

4.6.1.6. Local tissue damage and correct probe positioning can be evaluated, *ex vivo*, at the end of the microdialysis experiment.

4.6.1.7. Microdialysis can be used for kinetic studies of drug distribution and clearance in different organs; this is of particular relevance to studies on the ability of drugs to penetrate through the blood–brain barrier.

4.6.2. Reservations and Possible Limitations

Intracerebral microdialysis is based on the assumption that the extracellular concentrations measured accurately reflect the concentration at the synapse. However, several workers have questioned the validity of this assumption, as discussed below. In addressing these points of criticism we performed a retrospective study of the literature, including our own work, of data obtained from the rat striatum, which is one of the most commonly used brain areas. We were able to analyze the basal unstimulated levels of the neurotransmitter amino acids, GABA, aspartate, and glutamate, reported in over 60 publications between the years 1986 and 2006.

The major points of criticism and responses to them are considered below.

4.6.2.1. The exact quantities of materials present/released in the extracellular fluid are not known

This will be dependent on the probe recovery rate, which may be affected by differences in several parameters, including diffusion velocity which will be governed by temperature, size, cutoff membrane area, as well as the flow rate concentration gradient and perfusate composition (28). These might be expected to result in variability of data obtained by different laboratories. Several authors have made the efforts to apply correction factors from probe recoveries, determined in vitro. However, such factors, even if determined in aCSF, will not necessarily reflect the actual situation in vivo. Comparison of data from the literature allows an evaluation of the effectiveness of this approach. Much to our surprise, analysis of the uncorrected basal GABA and aspartate levels reported in the literature revealed the data obtained in different laboratories to be in very close agreement. In Table 1 uncorrected data of GABA and aspartate basal values from the literature are reported, in comparison to our own data (29). In contrast, those studies where some recovery corrections were applied reported widely disparate values.

The uncorrected extracellular basal levels of GABA and aspartate were surprisingly close, despite differences in the length of the membrane (2–4 mm), rates of perfusion (1–4 μl/min), in vitro recovery rates (6–51%), as well as the analytical methods used. None of these differences or any other factors appear to affect the determination of GABA and aspartate. We could conclude that basal levels obtained in different laboratories can be compared *only if they are not corrected* for any one of these factors and different HPLC conditions used in different laboratories did not appear to be critical for detecting GABA and aspartate. In contrast, van der Zeyden et al. (30) claimed that the chromatographic conditions for the separation of GABA from other amino-acid derivatives were critical. This was because, in contrast with their earlier findings, they found that under new chromatographic conditions extracellular GABA levels were significantly decreased by TTX infusion, consistent with a neuronal origin. They ascribed the failure of their earlier work and that of several other groups to observe this to be

Table 1
Summary of values for the striatal basal levels of GABA and aspartate obtained from a literature survey

	Uncorrected basal levels (nM) in striatum	
	Mean ± SEM	**CV**
GABA		
Data from the literature		
Total (*n* = 42)	21.0 ± 1.6	0.488
Anesthetized (*n* = 22)	18.5 ± 1.5	0.367
Awake (*n* = 20)	23.4 ± 2.6	0.527
Our data (28)		
Awake (*n* = 15)	23.0 ± 3.0	0.500
ASPARTATE		
Data from the literature		
Total (*n* = 27)	202 ± 36	0.926
Anesthetized (*n* = 16)	177 ± 47	1.097
Awake (*n* = 11)	240 ± 54	0.751
Our data (28)		
Awake (*n* = 15)	232 ± 32	0.534

CV coefficient of variation; *n* number of values taken from different publications, with the exception of our data, where *n* represents the number of observations taken from (28). In the awake animals, GABA and aspartate data obtained from different laboratories are comparable, in terms of uncorrected mean values and CV, to those obtained from a single one

a result of inadequate chromatographic procedures. However, studies reporting the insensitivity of GABA levels to TTX (31–38) were generally performed under anesthesia, whereas those reporting GABA microdialysate levels to be TTX-sensitive were performed in the awake freely moving animal, indicating that detection of TTX sensitivity depends on the absence of anesthesia rather than on the chromatographic conditions (see Notes 4.6.2.2, 4.6.2.3).

In contrast, reported values for glutamate, summarized in Table 2, were highly variable, ranging from 160 nM to over 12,000 nM. However, they did not correlate with different membrane lengths (2–4 mm), perfusion rates (1–4 μl/min), or in vitro recovery values (6–51%) used in different laboratories. Incomplete separation of asparagine and/or glutamine, which elute quite closely to glutamate in many HPLC procedures (6), may be a reason for this discrepancy. HPLC conditions are indeed critical for the detection of glutamate by microdialysis. Bacterial contamination, reported to interfere with the determination of glutamate, may also have contributed to this variability. Under the conditions described here we have found glutamate levels to be comparable between experiments.

Table 2
Summary of values for the basal levels of glutamate obtained from a literature survey

	Uncorrected basal levels (nM) in striatum	
	Mean ± SEM	**CV**
Glutamate		
Data from the literature		
Total (*n* = 39)	1,387 ± 371	1.670
Anesthetized (*n* = 22)	1,207 ± 356	1.383
Awake (*n* = 17)	1,622 ± 728	1.850
<600 nM (*n* = 22)	319 ± 26	0.391
>600 nM (*n* = 17)	2,771 ± 733	1.091
Our data (28)		
Awake (*n* = 15)	396 ± 48	0.469

CV coefficient of variation; *n* number of values taken from different publications, with the exception of our data, where *n* represents the number of observations taken from (28)

4.6.2.2. The site from which release occurs (e.g., nerve terminal or astrocyte) is not known

Since sensitivity to TTX, described above, is often used to distinguish the neuronal or non-neuronal component of the extracellular concentrations of neurotransmitters in vivo, the question is whether this is a valid criteria. When we examined the effect of TTX on basal release, TTX sensitivity appeared as a valid tool for distinguishing the proportion (20–70%) of GABA release that was of neuronal origin from the extraneuronal pool, provided that anesthetized animals are not used, as discussed above (Note 4.6.2.1) and below (Note 4.6.2.3). As indicated above, such results can be supplemented by the use of specific antagonists and calcium-free medium.

4.6.2.3. The use of anesthetized animals may affect release

A major advantage of microdialysis is its use in conscious freely moving animals. Therefore, this is really a question of the effects of anesthesia on neurotransmitter behavior. Osborne et al. (39) reported that TTX sensitivity was a valid criteria only when release is monitored in the awake freely moving animal. Our comparison of 16 studies that reported the TTX sensitivity of striatal extracellular basal GABA levels in freely moving or anesthetized rats (Table 3) confirms this. Evoked release is also affected by anesthesia. The data shown in Table 4 indicate that kainate-evoked release of striatal GABA and, to a minor extent, aspartate, but not glutamate, is abolished in anesthetized rats.

Table 3
Effect of anesthesia on the TTX sensitivity of basal GABA release from rat striatum

	Anesthetized (n)		
TTX sensitivity	Yes	No	Total (n)
Yes	1	7	8
No	6	2	8
Total	7	9	16

n number of publications. Chi Square = 9.1428, $p < 0.01$, indicates a statistically significant difference between anesthetized and non-anesthetized rats in terms of their sensitivity to TTX

Table 4
Effect of anesthesia on the KA (100 μM)-evoked release of GABA, aspartate, and glutamate

	Anesthesia	
$AUC_{stim} - AUC_{bas}$ (nM)	[a]No	[b]Yes
GABA	196 ± 26	0.5 ± 5
Aspartate	1,423 ± 317	129 ± 41
Glutamate	1,683 ± 379	1,582 ± 741

[a](28)
[b]Unpublished observation; data, expressed as $AUC_{stim} - AUC_{bas}$ (fmol/μL), represent the mean net output ± SEM, i.e., the area under the concentration–time curve between 0 and 120 min (stimulated output) with the area under the concentration–time curve between −40 and 0 min (basal output) subtracted and normalized to the time interval of 20 min; number of animals, $n = 8–10$

These results indicate the necessity of allowing sufficient time for complete recovery from anesthesia before conducting microdialysis experiments. However, in some cases, such as the use of high concentrations of KA, which causes convulsions, it is necessary to conduct the experiments on the anesthetized animals, in which case the limitations described above must be acknowledged. In general, behavioral tests should be performed 2–7 days after the anesthesia to give enough time for the animals to recover.

4.6.2.4. The results are frequently expressed and analyzed as percentages with no indications of absolute values

This is not a criticism of the microdialysis approach itself, but rather of the expression of the results obtained. However, since it is common to find results expressed in this way, we should consider the question: What is wrong with expression of released values as stimulated/basal ratios?

First, there is no theoretical ground to support the validity of using ratios (stimulated/basal concentration), rather than absolute values, since that assumes proportionality of stimulated values to basal levels. As basal release represents the leakage from all cellular sites, neuronal and non-neuronal, *one might expect, in most instances, the amount released following stimulation to be additive, rather than proportional to basal levels.*

Second, there are major difficulties in the use of ratios (stimulated/basal levels) for statistical analysis. Over the past 20 years, most authors have used ratios or percentage increases, not only for graphical purposes but also for statistical analysis, often not even reporting the original basal values for each treatment group. This leads to serious artifacts when stimulation effects are compared between different treatment groups having significantly different basal values. Differences in ratios, in fact, not only reflect differences in the numerator, which refers to stimulated conditions, but also differences in the denominator, which refers to basal conditions (Fig. 7).

4.6.2.5. The insertion of microdialysis probes must result in some tissue damage

This is an unpleasant but well-established truth (22, 40, 41), which is too often ignored. The insertion of any probe in brain tissue will inevitably result in some tissue damage and injury responses, such as gliosis, ischemia, and immune responses, although such damage may be relatively minor. The extent of damage will depend on the size of the probe, and the insertion of the guide cannula some time before the probe, which extends into the area to be sampled, should minimize damage and trauma at the site of sampling. The gliosis response appears to develop rather slowly and may only become problematical if studies are to be extended over several days (42).

5. Conclusions

The uses and applications of microdialysis are becoming more widespread, despite the limitations discussed above. The development of amperometric or voltammetric sensors (sometimes enzyme linked) may lead to viable alternatives. They have the advantage of more rapid response times and appear to cause less tissue damage (40), although the latter may be offset if the effects of compounds administered directly to the brain are also to be studied. They are, however, generally limited to specific neurotransmitters or groups of neurotransmitters, whereas microdialysis can be used for multiple compounds. It is important, however, to recognize the possible limitations of whatever method is used.

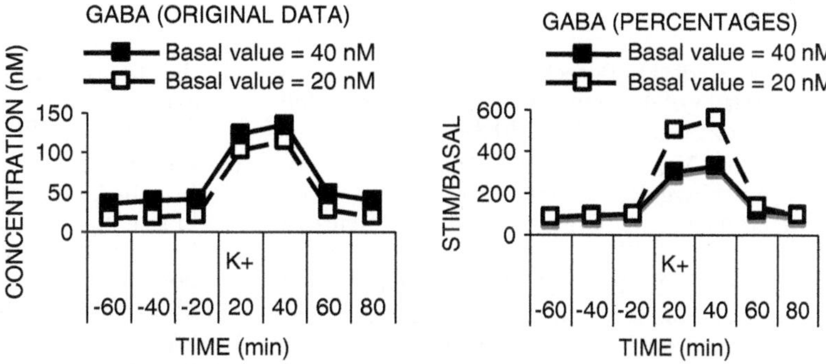

Fig. 7. Differences in basal values affect the evaluation of the magnitude of evoked release, when data are analyzed as ratios, rather than original values. In the *left panel*, the time courses of the original concentration values (*left panel*) or stimulated/basal ratios (*right panel*) of K^+-evoked GABA release curves, differing solely in terms of their basal concentration values, 20 nM (*open squares, dotted line*) and 40 nM (*closed squares, solid line*). Using original values, the amount released by stimulation is additive to basal levels; thus net K^+-stimulated AUC, i.e., $AUC_{stimulated} - AUC_{basal}$, equals 180 nM for both curves, independently of the difference in their basal concentration values. In contrast, when ratios/percentages are taken rather than original values, a proportionality between stimulated and basal values is assumed, i.e., the $AUC_{stimulated}/AUC_{basal}$ ratios for the two curves are 5.5 (550%) and 11 (1,100%), being affected by their different basal values of 40 and 20 nM, respectively.

Acknowledgements

We would like to thank Ente Cassa di Risparmio di Firenze (Firenze, Italy) for financial support, ERAB (Brussels, Belgium), and the EU COST action D34 and CM1103 for supporting our international cooperation.

Appendix

Sources of Materials

Most chemicals can be obtained from more than one supplier and should be of the highest purity available. Only those sources not given in the text are listed here.

- Acrylic cement. Self-curing acrylic cement was purchased from Kerr Italia (Salerno, Italy).
- Chloral hydrate, glacial acetic acid, methanol and potassium acetate, and all salts used to prepare aCSF were from Merck (Milano Italy).
- Cresyl fast Violet, DNQX, KA, 2-mercaptoethanol, *ortho*-bromoterephthalic acid, *o*-phthalaldehyde, terephthalic acid, and TTX were from Sigma Aldrich (Milano, Italy).

References

1. Ungerstedt U, Pycock C (1974) Functional correlates of dopamine neurotransmission. Bull Schweiz Akad Med Wiss 30:44–55
2. Chefer VI, Thompson AC, Zapata A, Shippenberg TS (2009) Overview of brain microdialysis. Curr Protoc Neurosci 47:7.1.1–7.1.28
3. Dingledine R, Bain CJ (199) Glutamate and aspartate. In: Siegel GJ (ed) Basic neurochemistry. Lippincott-Raven, Philadelphia, pp 315–333
4. Della Corte L, Crichton RR, Duburs G, Nolan K, Tipton KF, Tirzitis G, Ward RJ (2002) The use of taurine analogues to investigate taurine functions and their potential therapeutic applications. Amino Acids 23:367–379
5. Bianchi L, Colivicchi MA, Ballini C, Fattori M, Venturi C, Giovannini MG, Healy J, Tipton KF, Della Corte L (2006) Taurine, taurine analogues, and taurine functions: overview. Adv Exp Med Biol 583:443–448
6. Bianchi L, Della Corte L, Tipton KF (1999) Simultaneous determination of basal and evoked output levels of aspartate, glutamate, taurine and 4-aminobutyric acid during microdialysis and from superfused brain slices. J Chromatogr B Biomed Sci Appl 723:47–59
7. Pryor WA (1986) Oxy-radicals and related species: their formation, lifetimes, and reactions. Annu Rev Physiol 48:657–667
8. Freinbichler W, Tipton KF, Della Corte L, Linert W (2009) Mechanistic aspects of the Fenton reaction under conditions approximated to the extracellular fluid. J Inorg Biochem 103:28–34
9. Halliwell B (2001) Role of free radicals in the neurodegenerative diseases: therapeutic implications for antioxidant treatment. Drugs Aging 18:685–716
10. Moskovitz J, Yim MB, Chock PB (2002) Free radicals and disease. Arch Biochem Biophys 397:354–359
11. Stone JR, Yang S (2004) Hydrogen peroxide: a signaling messenger. Antioxid Redox Signal 8:243–270
12. D'Autréaux B, Toledano MB (2007) ROS as signalling molecules: mechanisms that generate specificity in ROS homeostasis. Nature Rev Mol Cell Biol 8:813–824
13. Forman HJ, Fukuto JM, Miller T, Zhang H, Rinna A, Levy S (2008) The chemistry of cell signaling by reactive oxygen and nitrogen species and 4-hydroxynonenal. Arch Biochem Biophys 477:183–195
14. Saran M, Michel C, Stettmaier K, Bors W (2000) Arguments against the significance of the fenton reaction contributing to signal pathways under *in vivo* conditions. Free Radic Res 33:567–579
15. Freinbichler W, Colivicchi MA, Stefanini C, Bianchi L, Ballini C, Misini B, Weinberger P, Linert W, Vareślija D, Tipton KF, Della Corte L (2011) Highly reactive oxygen species: detection, formation, and possible functions. Cell Mol Life Sci 68:2067–2079
16. Freinbichler W, Colivicchi MA, Fattori M, Ballini C, Tipton KF, Linert W, Della Corte L (2008) Validation of a robust and sensitive method for detecting hydroxyl radical formation together with evoked neurotransmitter release in brain microdialysis. J Neurochem 105:738–749
17. Koppenol WH, Liebman JF (1984) The oxidizing nature of the hydroxyl radical. A comparison with the ferryl ion (FeO_2^+). J Phys Chem 88:99–101
18. Steffes S, Sandstrom M (2008) Constructing inexpensive, flexible, and versatile microdialysis probes in an undergraduate microdialysis research lab. J Undergrad Neurosci Educ 7:A33–A47
19. Paxinos G, Watson C (1986) The rat brain in stereotaxic coordinates. Academic Press, Sydney
20. Gerfen CR (1992) The neostriatal mosaic: multiple levels of compartmental organization. Trends Neurosci 15:133–139
21. Fekkes D, van Dalen A, Edelman M, Voskuilen A (1995) Validation of the determination of amino acids in plasma by high-performance liquid chromatography using automated pre-column derivatization with o-phthaldialdehyde. J Chromatogr B Biomed Appl 669:177–186
22. Montgomery J, Ste-Marie L, Boismenu D, Vachon L (1995) Hydroxylation of aromatic compounds as indices of hydroxyl radical production: a cautionary note revisited. Free Radic Biol Med 19:927–933
23. Clapp-Lilly KL, Roberts RC, Duffy LK, Irons KP, Hu Y, Drew KL (1999) An ultrastructural analysis of tissue surrounding a microdialysis probe. J Neurosci Methods 90:129–142
24. Vezzani A, Ruiz R, Monno A, Rizzi M, Lindefors N, Samanin R, Brodin E (1993) Extracellular somatostatin measured by microdialysis in the hippocampus of freely moving rats: evidence for neuronal release. J Neurochem 60:671–677
25. Melendez RI, Vuthiganon J, Kalivas PW (2005) Regulation of extracellular glutamate in the prefrontal cortex: focus on the cystine glutamate exchanger and group I metabotropic glutamate receptors. J Pharmacol Exp Ther 314:139–147

26. Ballini C, Della Corte L, Pazzagli M, Colivicchi MA, Pepeu G, Tipton KF, Giovannini MG (2008) Extracellular levels of brain aspartate, glutamate and GABA during an inhibitory avoidance response in the rat. J Neurochem 106:1035–1043
27. Bianchi L, Colivicchi MA, Bolam JP, Della Corte L (1998) The release of amino acids from rat neostriatum and substantia nigra in vivo: a dual microdialysis probe analysis. Neuroscience 87:171–180
28. Plock N, Kloft C (2005) Microdialysis—theoretical background and recent implementation in applied life-sciences. Eur J Pharm Sci 25:1–24
29. Galeffi F, Bianchi L, Bolam JP, Della Corte L (2003) The effect of 6-hydroxydopamine lesions on the release of amino acids in the direct and indirect pathway of the basal ganglia: a dual microdialysis probe analysis. Eur J Neurosci 18:856–868
30. van der Zeyden M, Oldenziel WH, Rea K, Cremers TI, Westerink BH (2008) Microdialysis of GABA and glutamate: analysis, interpretation and comparison with microsensors. Pharmacol Biochem Behav 90:135–147
31. Girault JA, Barbeito L, Spampinato U, Gozlan H, Glowinski J, Besson MJ (1986) In vivo release of endogenous amino acids from the rat striatum: further evidence for a role of glutamate and aspartate in corticostriatal neurotransmission. J Neurochem 47:98–106
32. Drew KL, O'Connor WT, Kehr J, Ungerstedt U (1989) Characterization of gamma-aminobutyric acid and dopamine overflow following acute implantation of a microdialysis probe. Life Sci 45:1307–1317
33. Osborne PG, O'Connor WT, Drew KL, Ungerstedt U (1990) An in vivo microdialysis characterization of extracellular dopamine and GABA in dorsolateral striatum of awake freely moving and halothane anaesthetised rats. J Neurosci Methods 34:99–105
34. Campbell K, Kalén P, Wictorin K, Lundberg C, Mandel RJ, Björklund A (1993) Characterization of GABA release from intrastriatal striatal transplants: dependence on host-derived afferents. Neuroscience 53:403–415
35. Morari M, O'Connor WT, Ungerstedt U, Fuxe K (1993) N-methyl-D-aspartic acid differentially regulates extracellular dopamine, GABA, and glutamate levels in the dorsolateral neostriatum of the halothane-anesthetized rat: an in vivo microdialysis study. J Neurochem 60:1884–1893
36. Rimondini R, O'Connor WT, Ferré S, Sillard R, Agerberth B, Mutt V, Ungerstedt U, Fuxe K (1994) PEC-60 increases dopamine but not GABA release in the dorsolateral neostriatum of the halothane anaesthetized rat. An in vivo microdialysis study. Neurosci Lett 177:53–57
37. Ferraro L, O'Connor WT, Li X-M, Rimondini R, Beani L, Ungerstedt U, Fuxe K, Tanganelli S (1996) Evidence for differential cholecystokinin-B and A receptor regulation of GABA release in the rat nucleus accumbens mediated via dopaminergic and cholinergic mechanisms. Neuroscience 73:941–950
38. Ferraro L, O'Connor WT, Glennon J, Tomasini MC, Bebe BW, Tanganelli S, Antonelli T (2000) Evidence for a nucleus accumbens CCK2 receptor regulation of rat ventral pallidal GABA levels: a dual probe microdialysis study. Life Sci 68:483–496
39. Osborne PG, O'Connor WT, Kehr J, Ungerstedt U (1991) In vivo characterisation of extracellular dopamine, GABA and acetylcholine from the dorsolateral striatum of awake freely moving rats by chronic microdialysis. J Neurosci Methods 37:93–102
40. Shuaib A, Xu K, Crain B, Sirén AL, Feuerstein G, Hallenbeck J, Davis JN (1990) Assessment of damage from implantation of microdialysis probes in the rat hippocampus with silver degeneration staining. Neurosci Lett 112: 149–154
41. Jaquins-Gerstl A, Michael AC (2009) Comparison of the brain penetration injury associated with microdialysis and voltammetry. J Neurosci Methods 183:127–135
42. Hascup ER, af Bjerkén S, Hascup KN, Pomerleau F, Huettl P, Strömberg I, Gerhardt GA (2009) Histological studies of the effects of chronic implantation of ceramic-based microelectrode arrays and microdialysis probes in rat prefrontal cortex. Brain Res 1291:12–20

Chapter 12

Measurement of Neuropeptides in Dialysate by LC-MS

Omar S. Mabrouk and Robert T. Kennedy

Abstract

The development of analytical technologies must continue to improve to keep in step with new findings in the neurosciences. Neuropeptides serve a wide range of functions in the CNS and scientists study them intensively for a spectrum of neurological and psychiatric disorders. However, the measurement of neuropeptides in microdialysates has been particularly challenging based on their low basal concentrations in the brain. Here we describe an advanced capillary LC-MS/MS method for the detection of neuropeptides, specifically, methionine and leucine-enkephalin. This chapter is intended to describe a setup in our laboratory which does routine analyses of low picomolar level neuropeptides in dialysates. We describe a procedure for preparing capillary LC columns and how to interface this type of LC system to an ion trap mass spectrometer.

Key words: Mass spectrometry, Ion trap, Capillary liquid chromatography (LC), Neuropeptide, Enkephalin, Opioid, In vivo microdialysis

1. Introduction

Techniques applied to the analysis of microdialysates vary widely and are constantly being refined for more rapid and sensitive measurements. The most commonly used analytical methods for these measurements are HPLC coupled to fluorescence or electrochemical detection. In general these methods lack the sensitivity necessary to reliably detect neurotransmitters and neuromodulators that exist below the nanomolar threshold in the extracellular space.

The use of mass spectrometry (MS) for the analysis of neurotransmitters in microdialysates has gained considerable momentum in the past 10 years (1–4). Compared with conventional analytical methods coupled to microdialysis, MS can offer superior sensitivity and selectivity for some analytes. LC-MS is especially useful when measuring larger molecules such as neuropeptides because of good ionization efficiency and fragmentation, both of which aid in MS detection.

Giuseppe Di Giovanni and Vincenzo Di Matteo (eds.), *Microdialysis Techniques in Neuroscience*, Neuromethods, vol. 75, DOI 10.1007/978-1-62703-173-8_12,

Since their discovery, neuropeptides, such as enkephalins, vasopressin, and oxytocin, have been intensively studied based on their vast biological actions in the CNS and in the periphery. Yet their low basal concentrations in the brain (at picomolar levels) have posed a significant challenge to investigators interested in monitoring their release dynamics. Some investigators have relied on coupling microdialysis to peptide-specific radioimmunoassays (RIAs). However, RIAs require multiple sample preparation steps between sampling and analysis. Additionally, RIAs require large sample volumes (limiting temporal resolution) and only one analyte can be monitored at a given time (5). Finally, RIAs may not differentiate between peptides with similar amino acid compositions such as methionine- and leucine-enkephalin. This chapter describes a capillary LC-MS technique (including loading, desalting, and eluting phases) capable of measuring picomolar concentrations of endogenous opioid peptides (enkephalins) in dialysates.

Although mass spectrometers are cost prohibitive for many laboratories, they are becoming more available in core facilities operated at research institutions. Also, manufacturers of newer instruments are making strides in reliability, cost-effectiveness, and ease of use. When coupled to HPLC for separations, MS is a highly effective tool for neurobiologists who seek to uncover the function of the many neurochemicals currently undetected by standard analytical techniques.

1.1. Mass Spectrometry Overview

The most commonly used mass spectrometers for dialysate analyses are the triple quadrupole (QQQ) and the ion trap (1–4). QQQ instruments are generally considered most suitable for quantitative analyses of known analytes while ion traps are considered more suited towards qualitative analyses and the determination of unknown compounds. However, when configured properly, either type can be sufficiently quantitative and qualitative for reaching low detection limits needed for measuring neurotransmitter or neuropeptide content in microdialysates.

Analytes in samples (i.e., dialysates) are charged and introduced into the gas phase through various methods such as electrospray ionization (ESI) or matrix-assisted laser desorption ionization (MALDI). The high salt content of microdialysis samples interferes strongly with ESI and to a less degree, MALDI ionization; therefore HPLC is commonly used to desalt and separate analytes of interest prior to MS analysis. After samples are ionized, gas-phase charged analytes enter the ion optics of the mass spectrometer and are separated, filtered, or detected based on their mass-to-charge (m/z) ratio. High-accuracy m/z and/or tandem MS (MS/MS) measurements are usually required for confident identification of analytes, especially peptides, in complex samples such as microdialysate. For peptide identification, the structural information obtained by MS/MS is critical. In MS/MS, ions of a chosen m/z

are selected in the first stage of MS analysis for a second stage of MS analysis. Before the second stage, the selected ions are broken apart by a dissociation method, among which collision-activated dissociation (CAD) is the most widely employed. In CAD, collision of "parent ions" (molecular ions in most cases when ESI is used) with an inert gas causes the ions to fragment, generating so-called daughter ions. Fragmentation steps may be repeated to characterize the unique molecular characteristics of the parent ion in question. One advantage of using ion trap mass spectrometers is that multiple rounds of CAD can be performed supplying additional qualitative information on the parent ion while reducing noise or interference levels.

1.2. Capillary LC Overview

Irrespective of the type of mass spectrometer used, coupling the instrument to a capillary LC system (sometimes called nano-LC when using LC columns that have inner diameters ≤75 μm) is the preferred method for measuring low-concentration neuropeptides. Signal-to-noise ratio at the detector is dramatically improved in a capillary LC system since the dilution factor is less than a conventional LC system. Also, capillary LC uses lower flow rates which improves ESI efficiency, thereby improving detection limits. It has been found that MS response is inversely proportional to the flow rate at the emitter tip (6). Moreover, the use of capillary LC generally requires less sample allowing for lower microdialysis flow rates (more highly concentrated samples) or higher flow rates for improved temporal resolution (lower volumes). Dramatically reduced mobile phase solvent use is another beneficial aspect to using capillary LC compared to traditional HPLC methods. The method described here includes a pre-concentration step on the capillary analytical column (some configurations use two separate columns for this) to enhance sensitivity. Finally, capillary LC columns can be prepared in-house (as described in Sect. 3.1). This makes it possible to test various column configurations (i.e., particle material, size, column length, and inner diameter) in order to optimize chromatography for particular analytes, providing a more cost-effective alternative to purchasing commercial capillary LC columns.

2. Materials

1. Stainless steel high-pressure reservoir for packing capillary LC columns are custom-made—though they can be purchased commercially from Western Fluids (Wildomar, CA). Caution must be taken when using the high-pressure reservoir since a failure could result in explosive depressurization. This device must be built by competent machinists and must include a durable rubber seal and bolts to withstand high pressures (Fig. 1).

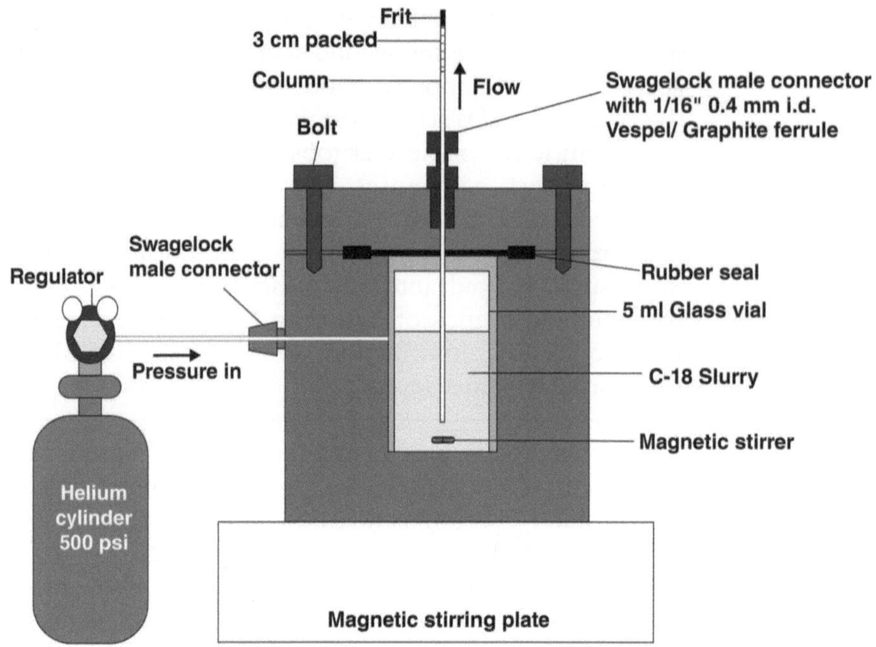

Fig. 1. Capillary LC columns are produced in-house using a high-pressure reservoir, frit material, and commercially available bulk reverse-phase C-18 particles. Building columns in-house allows for greater flexibility when testing different column conditions such as column length, column i.d., and particle diameter.

2. Solvent delivery pump is a two channel Eldex (Napa, CA) Micropro 2 ml syringe HPLC pump.
3. High pressure (HP) pump for loading/preconcentrating and desalting is a 25,000 psi (maximum) Haskel (Burbank, CA) DSFH-151 pneumatic liquid pump operated at 4,000 psi.
4. Autosampler is a Dionex (Sunnyvale, CA) WPS-3000TPL operating in full loop mode with a 5 μl sample loop. When analyzing samples for neuropeptides, it is best practice to keep autosampler cooled to ~6°C.
5. ESI source is a New Objective (Woburn, MA) PV-550 nanospray source.
6. Mass spectrometer is a Thermo (Waltham, MA) LTQ XL linear ion trap in positive mode using Thermo Xcalibur software for analysis.
7. Silica capillary material (20/360, 50/360, 75/360, 100/360 μm i.d./o.d.) for all LC connections, columns, and emitter tips are obtained from Polymicro Technologies (Phoenix, AZ).
8. Chromatographic columns (50/360 μm i.d./o.d.) and emitter tips (20/360 μm i.d./o.d.) are prepared in-house using a method described in Sect. 3.
9. Alltima C-18 reverse phase particles (5 μm diameter) are from Grace (Deerfield, IL).

10. Emitter tip puller is a P-2000 CO_2 laser puller from Sutter Instruments (Novato, CA).
11. Tee unions (UH-750) and inline filters (M-532) are purchased from Upchurch Scientific (Oak Harbor, WA).
12. Two C2 Cheminert 6-port valves and controllers are from Valco Instruments (Houston, TX).
13. All solvents used for standards, mobile phases, and capillary LC column preparations (including H_2O, acetonitrile, acetone, and methanol) are Honeywell Burdick and Jackson HPLC grade and purchased from VWR (Batavia, IL).
14. Methionine and leucine-enkephalin used for standards and instrument tuning are purchased from Phoenix Pharmaceuticals (Burlingame, CA). Peptide stock solutions are made up in 1 mM (in HPLC grade H_2O) solutions and kept in −20°C for a maximum of ~3 months.
15. Kasil 1624 and Kasil No. 1 silicate solutions for column frit are from PQ corporation (Malvern, PA). These materials are not commercially available from PQ corporation but can be requested for scientific purposes. Also, the material is good for ~1–2 years and should be stored in aliquots.
16. Formamide for frits and high purity 99%≥formic acid for mobile phases and acetic acid added to dialysate for peptide stability (7) are from Sigma Aldrich (St. Louis, MO).
17. 49% hydrofluoric (HF) acid is from Fisher Scientific (Waltham, MA) and is freshly added to a heavy plastic vial before use. HF is extremely corrosive and lethal. Extra care is taken by doubling nitrile gloves and wearing splash goggles.
18. Microdialysates analyzed are from brain areas known to contain enkephalinergic peptides such as the globus pallidus, ventral pallidum, and striatum. Acetic acid 0.5% (v/v) is added to fresh dialysate and standards to preserve peptide content (7). Dialysates can be stored this way for ~1 week at −80°C.

3. Methods

3.1. Preparation of Capillary LC Columns and Emitter Tips

1. Mix well 300 μl of Kasil 1624 and 100 μl of Kasil No. 1 potassium silicate solutions and then add 100 μl formamide to the mixture.
2. Vortex thoroughly for 2 min or until formamide is completely dissolved.
3. Dip the end of a freshly cut 20 cm section of 50/360 μm capillary into the frit mixture for 2 s. Frit solution should draw into the first 1–2 cm of capillary.

4. Place column in oven overnight at 100°C to polymerize frit material inside capillary.
5. After polymerizing the frit, rinse capillary with 75/25 H_2O/MeOH in a 5 ml glass vial for 1 min using a stainless steel reservoir pressurized with He at 500 psi (Fig. 1).
6. Make C-18 slurry by adding 20 mg of 5 μm particles into a 5 ml glass vial containing 75/25 Acetone/H_2O.
7. Vortex mixture and add stir bar to vial.
8. Slowly depressurize reservoir, then remove H_2O/MeOH from high-pressure reservoir and replace with C-18 slurry.
9. Set high-pressure reservoir on a magnetic stir plate to continuously mix slurry while inside the chamber.
10. Thread column through Swagelock (Solon, Ohio) connector fittings so that the end of column is submerged in C-18 slurry (Fig. 1) and tighten fitting with a 1/4 in. wrench.
11. Resume pressure (500 psi He) through high-pressure reservoir to begin column packing. Shine flashlight behind the capillary to visualize the packing progress.
12. Slowly reduce and then stop high pressure after the column has been packed to 3 cm.
13. Cut capillary with a ceramic capillary cutter leaving only ~1 mm of frit at the tip of the column (Fig. 1).

3.2. Capillary Emitter Tip

1. Burn a 20 cm piece of 20/360 μm capillary with a butane lighter to remove polyimide coating leaving a 2 cm clear window in the center of capillary.
2. Insert capillary into a P-2000 CO_2 laser puller to generate two fine emitter tips per capillary section.
3. Set laser puller to cycle once at line 1: Heat 300, velocity 30, delay 128, pull 0; line 2: heat 300, velocity 30, delay 128, and pull 125.
4. Submerge ends of tips in 49% hydrofluoric acid for 2 min to create sharp-edged electrospray emitters with ~3 μm i.d.
5. Join columns and tips joined in butt-connection by inserting them into a 2 cm piece of 1/16 in. × 0.10 in. PTFE tubing (Fig. 2).

3.3. Capillary LC Setup

1. Connect two channels of a binary LC pump (used for elution phase) to an Upchurch tee via 75/360 μm capillary.
2. Mobile phase A (MP A) for LC pump for channel 1 is HPLC grade H_2O with 0.1% formic acid (v/v).
3. Mobile phase B (MP B) for LC pump channel 2 is HPLC grade acetonitrile.

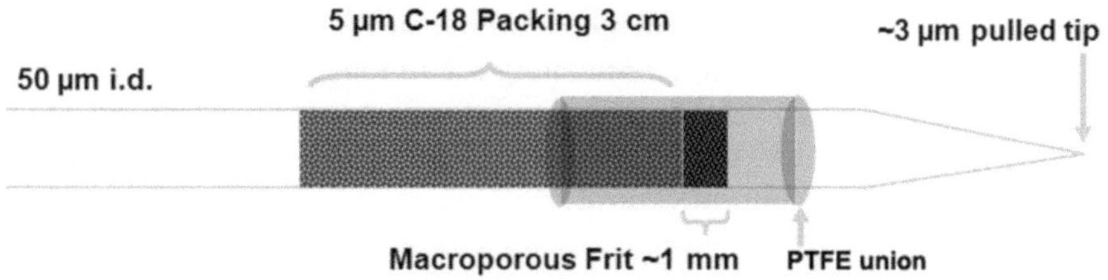

Fig. 2. Home-built emitter tip and capillary LC column are joined together by a 2 cm piece of 1/16 in. × 0.10 in. PTFE tubing.

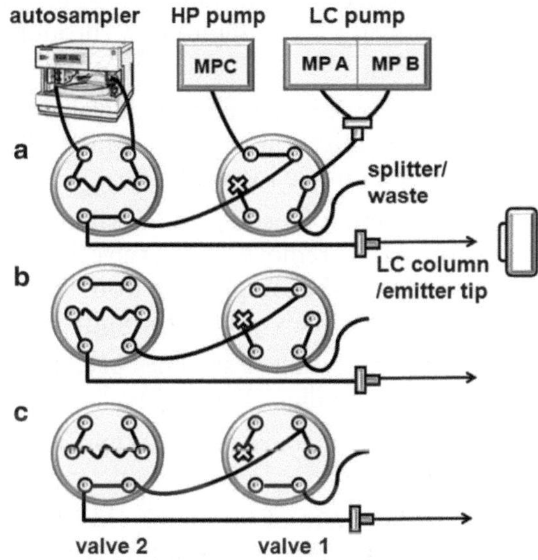

Fig. 3. When injection begins, valve 2 is in position A (load) to fill 5 μl sample loop (full loop injection mode) and valve 1 is set to position A putting HP pump on line (**a**). After sample is drawn into sample loop, valve 2 switches to position B (inject) allowing HP pump to flow through the sample loop to load sample onto capillary column (**b**). Meanwhile, LC pump is flowing to waste through the splitter. After sample is loaded onto column, HP pump continues to flow to desalt sample for 10 min total while valve 2 switches back to inject position, removing the sample loop from the flow path (**a**). Valve 1 then switches to position B putting the LC gradient pump on line for separation and elution (**c**). The "X" on valve 1 indicates that this channel is continuously blocked.

4. Connect high-pressure gas (N_2)-driven pneumatic pump (for loading and desalting) using 100/360 μm capillary.
5. Mobile phase C (MP C) for HP pump (operating at 3,000–4,000 psi) is HPLC grade H_2O with 0.1% formic acid (v/v).
6. Connect inline filters directly after the outlet of both LC pump and HP pumps.
7. Connect LC pump and HP pump to the same six port valve (see Fig. 3).
8. Connect sample loop to six port selection valve in autosampler.
9. Connect valve 1 and valve 2 in a series with 75/360 capillary so that switching between position A and B on valve #1 puts either HP pump (loading and desalting) or LC pump (elution) in-line with the column as described in Fig. 3.

10. Make a flow splitter to reduce flow rate through chromatographic column by attaching a 1 m long piece of 50/360 μm capillary to valve 1. This configuration puts LC pump through the splitter (to waste) while HP pump is loading and desalting through column (Fig. 3).

3.4. Detailed LC System Operation

1. Autosampler operating in full loop mode fills a 5 μl sample loop for injection.
2. Using Dionex Chromeleon software to control valve positions, at the start of a run, valve 1 automatically switches to position A, allowing HP pump to flow through column and emitter tip (Fig. 3a).
3. After sample loop is filled with dialysate sample, valve 2 switches back to injection position which allows HP pump to flow through sample loop (Fig. 3b).
4. After 8 min of loading and desalting (continuous aqueous flow from MP C removes salts), valve 2 returns to injection position closing off the sample loop from the flow path for 2 min while MP C continues to flow through column (Fig. 3a).
5. At 10 min, valve 1 switches back to original position putting LC pump online (Fig. 3c).
6. LC pump operates at 10 μl/min and generates ~450 psi after splitting.
7. Final flow rate of HP pump (MP C) is 2.5 μl/min while flow rate of LC pump after splitting is calculated to be 350 nl/min.
8. The gradient program for separation consists of an isocratic step of 10% B for 1 min, linear increase to 95% over 2.5 min, followed by isocratic at 95% B for 3 min, and then a ramp back down to 10% over 2 min.

3.5. Configuring MS to Detect a Known Neuropeptide

1. Prepare a 1 μM solution of neuropeptide of interest in 50/50 MeOH/H_2O with 0.1% formic acid (v/v).
2. Fill 50 μl syringe with peptide solution and connect to 100/360 capillary to emitter tip using a stainless steel union.
3. Use optimized emitter tip position (aligned ~3 mm from the center of the mass spectrometer inlet) and flow peptide solution through capillary and emitter tip at 0.2 μl/min.
4. Connect high voltage (1.5 kV) to the stainless steel union and then apply voltage.
5. In the Thermo LTQ Tune software begin monitoring mass spectra (MS^1).
6. Search spectra for the molecular ion (typically molecular weight + 1 when ion is singly charged).
7. Isolate parent ion within a mass window (±1 Da).

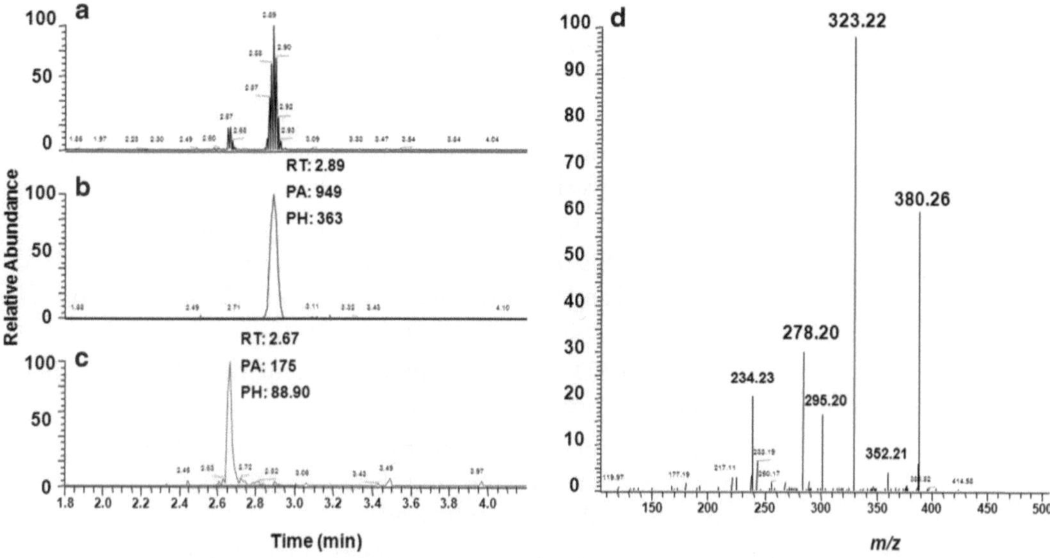

Fig. 4. Total ion chromatogram for a 5 pM standard of methionine- and leucine-enkephalin using the described method (**a**). Reconstructed ion chromatograms for leucine- (**b**) and methionine- (**c**) enkephalin. Mass spectra showing the selected granddaughter ions (MS^3) at *m*/*z*=380, 323, and 278 for methionine- and leucine-enkephalin. (**d**) RT=retention time, PA=peak area, PH=peak height.

8. Select the most abundant daughter ion using default collision energy.
9. Optimize collision energy so that fragmentation increases intensity of daughter ion response.
10. For MS^3 (note that this can only be done on ion trap-type instruments and not QQQ)—repeat procedure for granddaughter ions to be used for quantification (as in Fig. 4d).
11. Save parameters to be used in MS method.

3.6. Mass Spectrometer Setup

1. Couple the capillary LC system to a PV-550 nanospray ESI source (New Objective, Woburn, MA, USA) interfaced to a Thermo LTQ XL linear ion trap (LIT) mass spectrometer operating in positive mode.
2. Apply +2.5 kV potential to a stainless steel tee which connects LC system to capillary column for ESI.
3. Measurements are made with the following settings: Automatic gain control = on, collision-induced dissociation, $q = 0.25$, isolation width = 3 m/z, activation time = 0.25 ms, micro scan number = 1, and default target count values.
4. Collision energies are optimized for best sensitivity during constant infusion of each peptide (Sect. 3.5).

5. The MS^3 pathways for met-enkephalin and leu-enkephalin are (in m/z) 574→397→278, 323, 380 and 556→397→278, 323, 380, respectively (Fig. 4).
6. For improved quantification, use an internal standard. Past studies in our group have used deuterated leucine-enkephalin (with a mass shift of 4 Da) for internal standard.
7. Tuning of the mass spectrometer is performed monthly by directly infusing 1 μM met-enkephalin made up in 50/50 H_2O/methanol and 0.1% formic acid (v/v) at 0.2 μl/min using LTQ tune software in semi-automatic mode.
8. Typical RSD for methionine and leucine-enkephalin in our laboratory are around 5–10%.
9. Standard curves for both met-enkephalin and leu-enkephalin were generated by injecting 5 μl of 1, 5, 10, 50, and 100 pM in aCSF with 0.5% acetic acid.
10. Signal-to-noise ratios using MS^3 for 1 pM standards are generally around 5:1 for methionine-enkephalin and 10:1 for leucine-enkephalin.

4. Notes

1. One frequent issue that groups have with capillary LC systems is routine clogging, i.e., obstruction of liquid flow through narrow bore columns. We have found that by using 50 μm i.d. instead of 25 μm i.d. LC columns greatly improved the reliability of the system. Operators must still take great care to eliminate contaminants, however. For instance, filtering all aCSF with 0.2 μm syringe filters prior to use appears to help a great deal. Also, all mobile phases use in-bottle filters to minimize the introduction of any contaminants from solvents. In-line filters are used immediately after the LC and HP pumps, again, to minimize the introduction of any particulates that can obstruct flow through the capillary column and emitter tip.
2. Carryover on capillary LC-MS systems is also a frequent issue that some groups struggle with. We have found that by maintaining high acidity of mobile phases (~pH 2.5) and allowing for extended isocratic (i.e., 3.5 min) flow with high organic (95% or 100% acetonitrile) content can minimize carryover between runs. Also, not injecting high, non-physiological concentrations of peptide into the system (i.e., over 200 pM), can limit carryover. After autosampler injects, it is important that it does multiple wash cycles in mixed aqueous/organic wash solution.

3. As described in Sect. 3.5, direct infusion of peptides can be used for optimizing MS/MS detection. However, for some larger peptides such as β-endorphin or nociceptin/orphanin FQ, tryptic digestion may be necessary since smaller peptide fragments yield better signal for ESI and detection in some cases (7).

References

1. Haskins WE, Wang Z, Watson CJ, Rostand RR, Witowski SR, Powell DH, Kennedy RT (2001) Capillary LC-MS2 at the attomole level for monitoring and discovering endogenous peptides in microdialysis samples collected in vivo. Anal Chem 73:5005–5014
2. Zhang MY, Beyer CE (2006) Measurement of neurotransmitters from extracellular fluid in brain by in vivo microdialysis and chromatography-mass spectrometry. J Pharm Biomed Anal 40:492–499
3. Lanckmans K, Sarre S, Smolders I, Michotte Y (2008) Quantitative liquid chromatography/mass spectrometry for the analysis of microdialysates. Talanta 74:458–469
4. Mabrouk OS, Li Q, Song P, Kennedy RT (2011) Microdialysis and mass spectrometric monitoring of dopamine and enkephalins in the globus pallidus reveal reciprocal interactions that regulate movement. J Neurochem 118:24–33
5. Olive MF, Maidment NT (1998) Opioid regulation of pallidal enkephalin release: bimodal effects of locally administered mu and delta opioid agonists in freely moving rats. J Pharmacol Exp Ther 285:1310–1316
6. Shen Y, Moore RJ, Zhao R, Blonder J, Auberry DL, Masselon C, Pasa-Tolić L, Hixson KK, Auberry KJ, Smith RD (2003) High-efficiency on-line solid-phase extraction coupling to 15-150-microm-i.d. column liquid chromatography for proteomic analysis. Anal Chem 75:3596–3605
7. Li Q, Zubieta JK, Kennedy RT (2009) Practical aspects of in vivo detection of neuropeptides by microdialysis coupled off-line to capillary LC with multistage MS. Anal Chem 81: 2242–2250

Chapter 13

Achieving High Temporal Resolution for In Vivo Measurements by Microdialysis

Neil D. Hershey and Robert T. Kennedy

Abstract

Most methods for analyzing microdialysis samples require 1–20-min fraction collection times and off-line analysis of 1–50 min per sample; however, it is important to improve the temporal resolution to measure fast neurochemical events that go unnoticed with slower analyses. For example, many behavioral or pharmacological studies induce neurochemical release on the timescale of seconds or less. To better understand neurotransmitter release dynamics, there is a need to analyze microdialysis samples collected at shorter intervals. On-line analysis with fast analysis is a convenient way to avoid collecting many low-volume fractions. Capillary electrophoresis coupled with laser-induced fluorescence is a technique that allows for rapid measurements of neurochemicals. Separation times of 11 s can be achieved to measure GABA, taurine, serine, glycine, glutamate, and aspartate.

Key words: Capillary electrophoresis, LIF, Glutamate, GABA, *In vivo*, Temporal resolution, Microdialysis

1. Introduction

Measuring neurochemicals in the brain requires meeting several challenges. Sensitivity is important because neurotransmitter concentrations in the brain extracellular space vary from pM to μM. Temporal resolution is also key because neurochemical concentration changes can occur in a wide range of times. Microdialysis temporal resolution has typically been 1–20 min which is useful for pharmacologically induced changes, but behavioral activation can occur on a faster timescale.

In microdialysis, the temporal resolution is often dictated by the analytical method used to measure neurotransmitter concentrations. Microdialysis flow rates are from 300 nL/min to 5 μL/min so that when generating temporal resolution of less than 1 min, only nL of sample are available meaning that excellent

Giuseppe Di Giovanni and Vincenzo Di Matteo (eds.), *Microdialysis Techniques in Neuroscience*, Neuromethods, vol. 75,
DOI 10.1007/978-1-62703-173-8_13,

sensitivity is required. Further, the large number of samples that are generated (e.g., at 10-s temporal resolution, 360 samples are generated in 1 h) requires a high-throughput analysis method. If dialysate is analyzed on-line, that is, as it is being collected, then the method must also be rapid and automated.

Fluidic mechanics can also limit temporal resolution. Taylor dispersion, the broadening of a concentration zone due to flow and diffusion, occurs as dialysate is transported from the probe to the analytical instrument or the fraction collector (1). With a fast and sensitive analytical system, Taylor dispersion can limit the temporal resolution. Recent reports have shown methods of eliminating this effect, but they are still under development (2).

One method that has been shown to achieve good sensitivity and temporal resolution is on-line capillary electrophoresis coupled with laser-induced fluorescence (CE-LIF). CE is a miniaturized separation method in which sample loaded into a capillary tube (i.d. $<50\ \mu m$) is separated by applying an electric field along the length of the capillary. Molecules migrate through the capillary at a rate determined by their electrophoretic mobility which in turn is determined by the charge-to-size ratio. (Other factors can also affect the migration rate as discussed below.) The narrow channel allows excellent heat dissipation so that much higher fields, often over 1,000 V/cm, can be applied than traditional gel electrophoresis. The high fields enable rapid separation times, which is ideal for on-line microdialysis measurements. The narrow bore channels used also mean that only small samples are required (typically less than 1 nL is injected onto a capillary). Again, this property is well suited for microdialysis measurements where limited sample is available, especially for high temporal resolution measurements. On the other hand, the small samples require excellent sensitivity to measure the separated molecules.

Although analytes migrate by electrophoretic mobility, other factors can influence transport through the capillary. When high voltage is applied across a capillary that has a charged surface (fused silica typically has a negative surface charge because of deprotonation of the Si–OH groups on the capillary wall), electroosmotic flow (EOF) is produced as a result of movement of counterions of the charged surface dragging solvent with them. Thus, for fused silica with a negative voltage applied to the outlet, EOF will be towards the outlet. This flow causes all analytes to migrate in the same direction. A typical separation order is cations first, then neutrals, and finally anions. Other factors also affect the net migration of molecules. Most important is the pH which governs the ionization state of the molecule. The electrophoresis buffer may also contain molecules that bind to the analytes and affect their charge or size, thus affecting their mobility. Useful additives include cyclodextrins or micelles which can greatly aid in separation.

In addition to the separation, detection sensitivity is an important consideration. One detection method that has been shown to work reliably with high sensitivity is LIF. Because neurotransmitters are generally not naturally fluorescent, they must be derivatized to allow LIF detection. *O*-Phthalaldehyde (OPA) is a rapid reacting reagent, useful for on-line labeling, which has limits of detection of 10 nM. The sensitivity produced by LIF is well within the limit to measure multiple neuroactive amines.

Several reports of using CE-LIF for neurotransmitter separations have been made. Different fluorescent tags and buffer conditions have been used yielding a variety of assays. This chapter discusses a separation that allows separation and detection of six neuroactive amino acids: γ-aminobutyric acid (GABA), taurine, serine, glycine, glutamate, and aspartate in less than 11 s (3–6). The high speed, small sample requirements and automation allow this measurement to be used routinely for studying amino acid dynamics in the brain of behaving animals. This system has been used in behavioral studies, rapid pharmacological measurements, and study of brains altered by receptor overexpression (7, 8). Figure 1 illustrates an example where a glutamate response is detected during a fear response of rats to exposure to a fox odor (4, 9).

The system described here has three main parts: (1) microdialysis sampling; (2) on-line derivatization with interface to CE; and

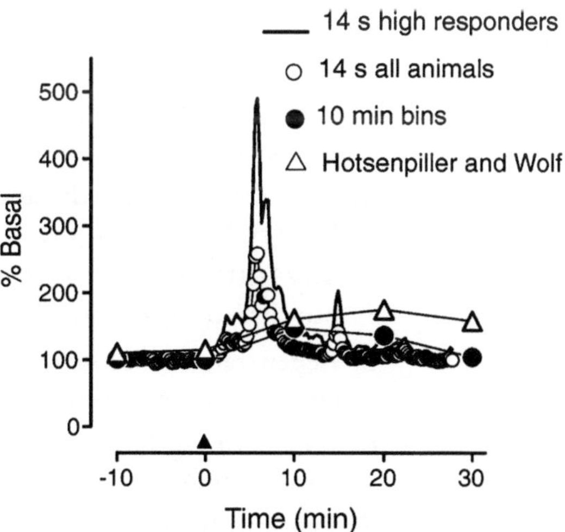

Fig. 1. Increased temporal resolution can elucidate neurochemical events. The graph shows the % basal levels of glutamate vs. time of rats that are presented with a fox odor. *Open triangle* shows glutamate levels measured by conventional HPLC. Animals that showed a high amount of activity to the fox odor were labeled high responders. The glutamate levels when measured by CE-LIF showed more dynamic release than previously measured (–). The overall average glutamate release (*open circle*) and the CE-LIF measurements with 10-min bins (*filled circle*) are also shown. Figure is supplied by John Wiley and Sons (4).

(3) microscope-based LIF detector. Several components are custom-made; however, commercial components can be substituted in some cases.

2. Materials

2.1. Chemicals and Buffers

1. Glutamate, aspartate, taurine, serine, and glycine are obtained from Sigma Aldrich (St. Louis, MO).
2. Artificial cerebral spinal fluid (aCSF) consists of 145 mM NaCl, 2.68 mM KCl, 1.0 mM $MgSO_4 \cdot 7\ H_2O$, 1.22 mM $CaCl_2$, 1.55 mM Na_2HPO_4, and 0.45 mM $NaH_2PO_4 \cdot H_2O$. The NaCl, KCl, and $CaCl_2$ are purchased from Fisher Scientific (Waltham, MA). The remainder of the salts is purchased from Sigma Aldrich (St. Louis, MO).
3. Ethylene glycol-bis(2-aminoethylether)-*N,N,N′,N′*-tetraacetic acid (EGTA), OPA, hydroxypropyl-β-cyclodextrin (HPbCD), β-mercaptoethanol (BME), and sodium tetraborate are obtained from Sigma Aldrich (St. Louis, MO).

2.2. Surgical Supplies and Microdialysis Probes

1. Ketamine (100 mg/mL), Dexdormitor (1 mg/mL), and soflurane are purchased from MWI Veterinary Supply (Boise, ID).
2. Cannulae (C312GP/O/SPC cut 8 mm below pedestal), dummy stylets which mate with the cannulae, push–pull connectors that mate with the cannulae, and mounting screws (2.4 mm length stainless steel) are purchased from Plastics One (Roanoke, VA). Push–pull equipment is used with dialysis probes threaded through one hole of the push–pull lines.
3. Dental cement powder and methyl methacrylate are purchased from A-M Systems, Inc (Sequim, WA).
4. 40/100 μm (i.d./o.d.) and 75/360 μm capillaries are purchased from Polymicro Technologies (Phoenix, AZ).
5. Regenerated cellulose membranes with an 18 kD cutoff are purchased from Spectrum Laboratories (Rancho Dominguez, CA).
6. Polyimide sealing resin is purchased from Fisher Scientific (Waltham, MA).
7. Tygon tubing (0.020″ inner diameter, 0.060″ outer diameter) was purchased from Fisher Scientific (Waltham, MA).

2.3. Syringes, Syringe Pumps, and Raturn Setup

1. 1 mL glass Hamilton syringes and 100 μL glass Hamilton syringe are purchased from Fisher Scientific (Waltham, MA).
2. Stainless steel unions (1/16″ tubing outer diameter, 0.15 mm bore) are purchased from Valco Instruments Co. Inc. (Houston, TX).

3. Syringe pumps (Fusion 200) are purchased from Chemyx (Stafford, TX).
4. A Raturn system is purchased from BASi (Lafayette, IN).

2.4. Reaction Capillary and Derivatization

1. A reaction tee (stainless steel, 0.15 mm bore, 1/16″ fitting) is purchased from Valco (Houston, TX).
2. 1/16″ PEEK reducing ferrules that reduce to 360 μm are purchased from Valco Instruments Co. Inc. (Houston, TX). Reducing ferrules are used in the reaction tee and unions.
3. The reaction capillary is a 150/360 μm capillary purchased from Polymicro Technologies (Phoenix, AZ).
4. Outflow from the on-line derivatization goes into a "flow-gate" interface that is used to control periodic injection of sample onto the separation capillary. The flow-gate is custom-made by the instrument shop at University of Michigan (Ann Arbor, MI) from a Plexiglass block. The flow-gate is symmetric and consists of a threaded portion that fits with Valco 1/16″ nuts, a 1/16″ bore that fits PTFE tubing, and a 360 μm bore to fit the capillaries. Refer to Fig. 2 for a diagram of the flow-gate (10).
5. The cross flow is pumped through the flow-gate interface with a K120 HPLC Pump (Knauer, North Potamac, MD).
6. Tubing used to transport the cross flow is tygon tubing (1/16″ i.d. 1/8″ o.d.) from Fisher Scientific (Waltham, MA).
7. Cheminert PEEK ferrules (1/16″) are used to make connec tions to the tees and flow-gates. They are obtained from Valco Instruments Co. Inc. (Houston, TX).
8. Two different types of PEEK nuts are obtained from Valco Instruments Co. Inc. (Houston, TX). One of the nuts has the dimensions 1/16″ hex, 0.45″ length, and the other is 1/16″ fingertight, 0.88″ length. Both types of nuts are used in the flow-gate.

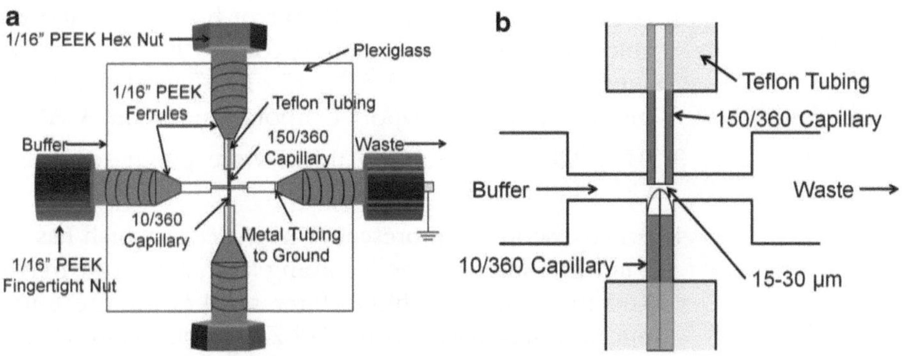

Fig. 2. Diagram of flow-gate interface. (**a**) Capillaries are secured by Teflon tubing and PEEK ferrules. (**b**) A diagram of the cross shows a honed separation capillary placed 15–30 μm from the reaction capillary.

9. 1/16″ stainless steel hex nuts are purchased from Valco Instruments Co. Inc. (Houston, TX).
10. Polytetrafluoroethylene (PTFE) tubing with an inner diameter of 0.006″ is purchased from Valco Instruments Co. Inc. (Houston, TX). This tubing will be a cap on the capillaries in the flow-gate to allow the ferrules to fit the capillary.
11. 350 μm drill bits are used to drill a hole in the PTFE tubing to allow the 360 μm capillary to fit. They are purchased from Small Parts (Lexington, KY).
12. A 3-way solenoid valve (1/16″ 12VDC), which toggles the cross-flow on and off, is purchased from Cole Palmer (Vernon Hills, IL).

2.5. Capillary Electrophoresis Equipment

1. The separation capillary is a 10/360 μm piece of capillary purchased from Polymicro Technologies (Phoenix, AZ).
2. The capillary tip is honed using a 300-1/24 Dremel tool fitted with a 425 polishing wheel purchased from McMaster Carr (Princeton, NJ).
3. To apply a voltage, a CZE1000R high-voltage power supply is purchased from Spellman High Voltage Co. (Hauppauge, NY).
4. The capillary is mounted on the microscope using a custom-made holder. The holder consists of a black plastic rectangle with three walls and a 360 μm channel drilled through the center. Each end of the channel has a wall at the end to keep out light. The third wall is parallel to the channel. A 360 μm hole needs to be drilled in the two walls at the end of the channel to thread the capillary through. A fourth removable wall is used constructed from the same material (refer to Fig. 3). A cover with a hole the diameter of the microscope objective blocks the light entering from the top.

2.6. Microscope-Based LIF Detector

1. A 355 nm, 100 mW laser is obtained from DPSS Lasers (San Jose, CA).
2. A 24″ × 36″ optical table, 1″ mirror holders, 4″ and 3″ adjustable vertical post holders, post collars, 90° angle brackets, and support posts (3″ × 1/2″, 4″ × 1/2″, and 12″ × 1/2″) are purchased from Newport Corporation (Irvine, CA).
3. 015/028 UV enhanced mirrors are purchased from Melles Griot Photonics (Carlsbad, CA).
4. An Axioskop 20 fluorescence microscope which has a 357 nm band-pass filter for the incoming laser, a dichroic mirror, and a 460 nm band-pass filter (filter set 42) for the fluorescence output is purchased from Carl Zeiss, Inc. (Hanover, MD). An adjustable iris is incorporated between the filter set and PMT to minimize background noise (refer to Fig. 3) (11).

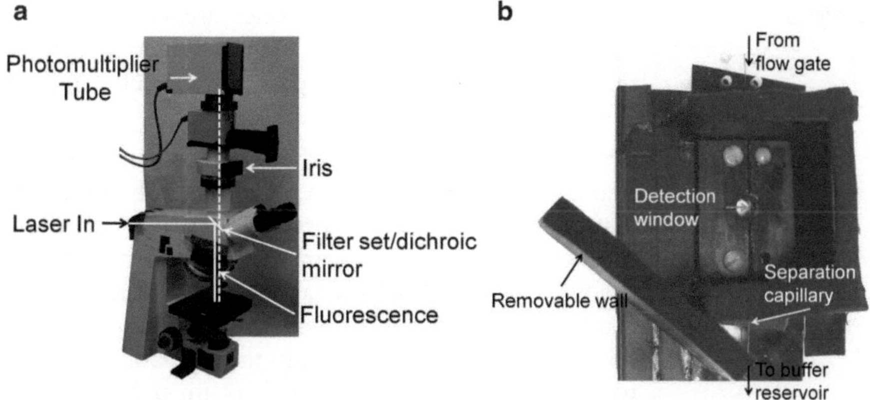

Fig. 3. (**a**) An Axioskop fluorescence microscope is used to collect the LIF. A dichroic mirror reflects the laser, but allows that fluorescence to pass through to the PMT where the signal is recorded. (**b**) Separation capillary holder that goes on the fluorescence microscope stage. Three of the four walls are fixed, while the fourth is removable so that the holder can fit past the microscope lens. The separation capillary is attached to the holder with superglue. The plastic is painted black to prevent background light.

5. The excitation laser is focused onto the separation capillary and fluorescence is collected by a 40×, 1.30 N.A. Fluar oil immersion lens (Carl Zeiss, Inc., Hanover, MD).
6. A 428 current amplifier can be purchased from Keithley Instruments (Cleveland, OH).
7. A photomultiplier tube (PMT, R1477; Hamamatsu, Bridgewater, NJ) is used to measure fluorescence.
8. Software for controlling the electrophoresis injections and data collection is a custom program written using Labview (National Instruments, Austin, TX).

3. Procedures

3.1. Surgery

1. Rats are anesthetized with ketamine (100 mg/mL) and dormitor (0.5 mg/mL) and boosters of 30 mg/mL ketamine and 0.25 mg/mL dormitor.
2. For awake rat experiments, the rat is placed in a stereotax and 4 holes are drilled into the skull, one for the cannula and three for anchoring skull screws.
3. Dental cement is used to fix the cannula in place.
4. For anesthetized rat experiments, only one hole needs to be drilled. The dialysis probe is lowered into the brain to the desired depth and anesthesia is maintained.

3.2. Microdialysis Probes

1. Side-by-side microdialysis probes are constructed by gluing two 40/100 μm (i.d./o.d) with the inlet capillary extending the desired active length past the outlet capillary (usually 1–4 mm) (12).
2. The capillaries are ensheathed with a semipermeable membrane that is sealed on both ends with polyimide sealing resin. The inactive area of the probe is coated with the polyimide sealing resin. A majority of the capillary between the brain and the reaction tee should be as narrow bore as possible without causing excessive backpressure and leaking to reduce Taylor dispersion. We use 40/100 μm tubing. Collars are constructed to couple the 100 μm o.d. to 360 μm o.d. tubing. The collar is made by gluing a 2.5 cm piece of 180/360 μm capillary to the ends of the probe.
3. A 350 μm bore is drilled into a 1 cm piece of the PTFE tubing. These pieces of PTFE tubing will act as a coupler to couple the probe capillaries to other capillaries.
4. 75/360 μm capillaries are used for the remainder of the tubing needs, which includes the tubing that goes from the syringe to the probe inlet and from the probe outlet to the reaction tee.

3.3. Reagent Mixing

1. A solution of 10 mM OPA, 40 mM BME, 9 mM sodium borate, 0.81 mM HPbCD, and 10 % methanol (v/v) is used as the fluorescent tag. The OPA solution is stored in the dark and is made fresh daily.
2. The sample syringe and the derivatization reagents are pumped into the reaction tee. Reducing ferrules and the stainless steel hex nuts are used to secure the capillaries into the tee.
3. A 150/360 μm capillary is coupled to the outlet of the tee which is known as the reaction capillary. The capillary is long enough that the OPA can react with the dialysate for 90 s at the flow rates used. The outlet of the reaction capillary will be inserted into the flow-gate.

3.4. Preparing the Separation Column and Flow-Gate Interface

1. A 350 μm bore is drilled into two more 1 cm pieces of PTFE tubing. The end of the separation capillary is inserted into the PTFE tubing. The second will go on the end of the reaction capillary. These PTFE sleeves will allow for the PEEK ferrules to clamp down onto the capillaries.
2. The end of the separation capillary is honed. A Dremel tool is secured on its side with a clamp. The 420 polishing wheel is attached to the Dremel and the speed is set to the lowest setting. Holding the separation capillary between the index finger and thumb, the capillary is rolled at a 45° on the right side of the polishing wheel. This is continued until a smooth, symmetric cone is obtained (refer to Fig. 2). This cone will allow for easier clearance of analyte from the capillary inlet during the injection.

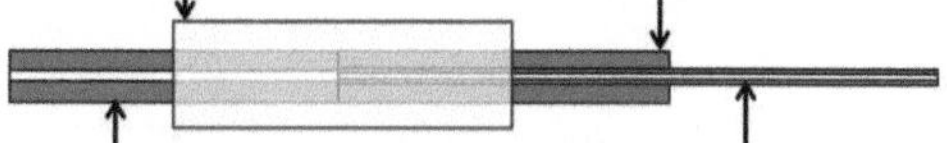

Fig. 4. Teflon tubing is used to couple two capillaries together. The couple can only connect 360 μm o.d. capillaries. When a 100 μm capillary is needed, a second capillary with an o.d. of μm is glued overtop of it (as shown).

3. If there is only air in the separation capillary, there can be no voltage applied across the capillary. To push liquid through the separation capillary, the capillary is coupled to a 100 μL Hamilton syringe in a syringe pump and a 50 % methanol in water solution is pumped through the capillary at 20 nL/min. Stainless steel Valco unions, stainless steel hex nuts, and reducing ferrules are used to couple the syringes to the capillary tubing. The methanol solution cleans the inside of the capillary. Once liquid is seen exiting the capillary, deionized water is pushed through the capillary. The syringe is coupled to a capillary using the Valco stainless steel unions. Capillaries are coupled to one another through PTFE tubing (refer to Fig. 4). The end of each capillary is pushed halfway through the couple so that the two ends meet in the middle of the PTFE tubing.
4. The reaction and separation capillaries are threaded through PTFE tubing and PEEK ferrules. The two capillaries are aligned in the flow-gate under a microscope so that they are 15–30 μm apart. The gap should be centered as much as possible. The inner diameter of the separation capillary is used to gauge the distance of the gap. PEEK hex nuts are used to secure the capillaries to the flow-gate.
5. A window is burned into the capillary using a butane lighter to burn the polyimide coating. The burned polyimide is removed by gently wiping the capillary with a paper towel soaked with methanol. The distance between the inlet of the separation capillary and window is 10 cm.

3.5. Microscope Setup

1. The separation capillary is threaded through the holes and channel of the separation capillary holder. The capillary is secured to the channel by placing drops of superglue intermittently across the capillary (refer to Fig. 3).
2. Using the mirrors, holder, and stands, the laser is aligned to go through the back of the fluorescence microscope.
3. The outlet of the separation capillary is placed in a 10 mL vial of cross flow buffer along with a platinum wire connected to the high-voltage power supply (refer to Fig. 5 for fluidic design).

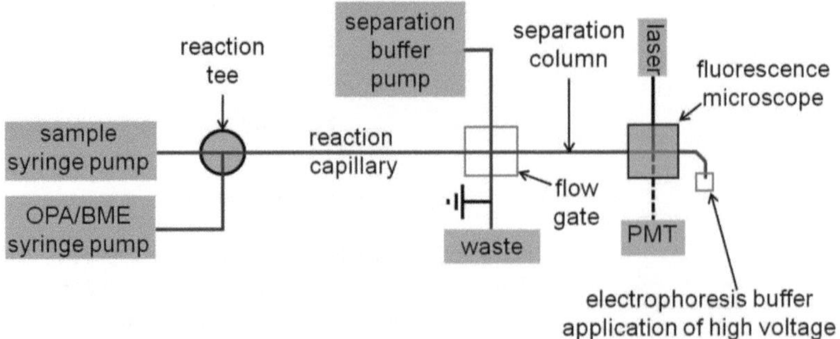

Fig. 5. Fluidic design of the CE-LIF setup. Sample is mixed with OPA in the reaction tee and allowed to react for 90 s. Discrete injections are made through the flow-gate. Fluorescence is recorded through a PMT.

3.6. Alignment

1. A 1 mL Hamilton syringe is filled with the OPA mixture and a second with ~7 μM glutamate, aspartate, GABA, taurine, serine, and glycine dissolved in aCSF. The two are pumped at the same flow rate as what will be used during the experiment, typically 0.3–1 μL/min.
2. The cross flow, which consists of 10 mM sodium borate and 0.9 mM HPbCD at pH 10.0, is pumped through the flow-gate interface. The pump used for the cross flow is an HPLC pump. The cross flow buffer should be pumped at a speed that allows for the clearance of analyte from the gap between the separation and reaction capillary. Typically, 300 μL/min will suffice (10).
3. A solenoid valve is used to switch the cross flow between going to the flow-gate and to waste. An in-house program using Labview software is used to control the solenoid valve and voltage. Pictures demonstrating successful flow-gate injection have been published previously (10). Using dye in the sample while observing the flow-gate allows similar tests of the injection.
4. The X–Y stage of the microscope is adjusted until the window in the separation capillary is centered in the field of view. A reticule can aid in this process. The microscope light source is then covered with a piece of cardboard wrapped in electrical tape so that ambient light will not contribute to the background.
5. With the cross flow diverted to waste, 10 kV is applied to the separation capillary. This setup causes continuous flow of derivatized analyte through the separation capillary. Using the adjustable mirrors, the laser is aligned so that it illuminates in the middle of the separation capillary and peak fluorescence is observed.
6. Once the alignment is completed, a finer adjustment is made by looking through the lens above the adjustable iris. The iris is set to allow the inner diameter of the capillary and the length of the detection window to be observed. The laser is aligned again until the peak amount of fluorescence is observed through the upper eyepiece.

3.7. Electrophoretic Injections

1. To initiate injection, the cross flow is diverted to waste for 500 ms (known as the pre-injection delay). This allows the analyte to diffuse across the gap in the flow-gate and reach the separation capillary.
2. The injection voltage (2 kV) is then applied to the separation capillary for 200 ms.
3. The voltage is then turned off and the cross flow is switched to the flow-gate for 1 s. The analytes at the separation capillary will be washed away and a plug of analytes will remain in the separation capillary.
4. The voltage is then ramped to 20 kV so that the analytes separate out. Figure 6 illustrates the timing diagram used to make a discrete injection.
5. The neurotransmitters will elute out in the following order: GABA, taurine, serine, glycine, aspartate, and glutamate. The separation is stopped 1 s after glycine elutes out. The next injection then begins. This will cause overlapping injections so that the aspartate and glutamate peaks will elute out in the beginning of the next run. Overlapping injections allow shorter separation times without sacrificing resolution.

3.8. In Vivo Neurotransmitter Measurements

1. Set the flow rate of the aCSF to be between 0.3 and 1 μL/min. Increasing the flow rate higher will cause a higher backpressure in the probe, which can cause the probe to leak or break.
2. For awake rat experiments, the rat is implanted the night before the experiments. The rat is lightly anesthetized with isoflurane and the probe is inserted into the brain while perfusing aCSF.

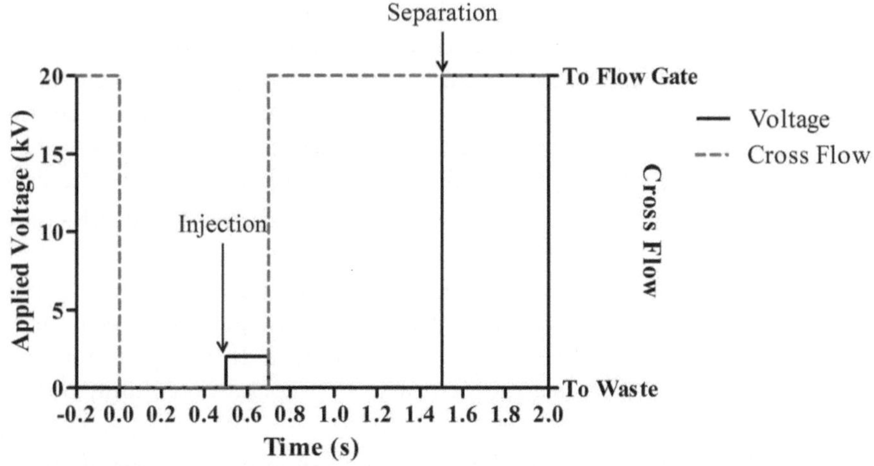

Fig. 6. Schematic for cross flow and voltage application is shown. The cross flow is diverted to waste prior to the injection. Flow is then resumed to the flow-gate for 1 s prior to separation. This timing allows for discrete injections of analyte to be injected.

3. For anesthetized rat experiments, the rat is implanted with the probe during surgery.
4. The outlet of the probe is connected to the reaction tee in the sample syringe. The OPA is pumped through to react with the dialysate.
5. The same injection/separation parameters from before are used for the dialysate. Electropherograms are collected for the duration of the experiment.
6. Data analysis is completed through Cutter, a custom-written Labview software which can be downloaded from http://www.umich.edu/~rtkgroup/NetworkMI.htm.
7. Once the experiment is complete, all the fluid lines that contained aCSF are flushed with water and the lines that contained OPA are flushed with methanol and then water. While flushing with water, the separation voltage is set to 10 kV in order to fill the separation column with water.

4. Notes

1. The largest source of problems occurs with clogs. The fluid lines are narrow and the flow-gate is small. To prevent clogs after each experiment, the capillaries are flushed with deionized water. When not in use, water is pumped at a slow rate through the capillaries and the flow-gate. All solutions used are filtered to prevent clogs.
2. A clog in the gap of the flow-gate is the most common clog. If the gap becomes clogged, analytes will either continuously leak into the separation column or no analytes will be injected. Removing the clog involves disassembling the flow-gate and running water through it. The clogs are usually visible under a microscope, so once the flow-gate appears clean, it can be reassembled. Before inserting the separation capillary, water is pushed through the capillary to dismantle potential clogs in the capillary.
3. If the signal is low, an injection is made with the pre-injection delay set to 10 s. If the peak heights appear normal, then there is a problem in the flow-gate. The flow-gate is either clogged or the gap is too large. The flow-gate is disassembled, cleaned, and the capillaries are realigned. If the signal does not change, there is a problem with the separation column and the separation column is replaced.
4. With care, the separation column can be maintained for many months. Once the resolution decreases to a point where neurotransmitters coelute, the column is replaced.

References

1. Norton LW, Yuan F, Reichert WM (2006) Glucose recovery with bare and hydrogel-coated microdialysis probes: experiment and simulation of temporal effects. Anal Chem 79:445–452
2. Wang M, Roman GT, Schultz K, Jennings C, Kennedy RT (2008) Improved temporal resolution for in vivo microdialysis by using segmented flow. Anal Chem 80:5607–5615
3. Lada MW, Vickroy TW, Kennedy RT (1998) Mediated regulation of extracellular glutamate and aspartate in rat striatum in vivo following electrical stimulation of the prefrontal cortex. J Neurochem 70:617–625
4. Venton BJ, Robinson TE, Kennedy RT (2006) Transient changes in nucleus accumbens amino acid concentrations correlate with individual responsivity to the predator fox odor 2,5-dihydro-2,4,5-trimethylthiazoline. J Neurochem 96:236–246
5. Schultz KN, von Esenwein SA, Hu M, Bennett AL, Kennedy RT, Musatov S, Toran-Allerand CD, Kaplitt MG, Young LJ, Becker JB (2009) Viral vector-mediated overexpression of estrogen receptor-{alpha} in striatum enhances the estradiol-induced motor activity in female rats and estradiol-modulated GABA release. J Neurosci 29:1897–1903
6. Bowser MT, Kennedy RT (2001) In vivo monitoring of amine neurotransmitters using microdialysis with on-line capillary electrophoresis. Electrophoresis 22:3668–3676
7. Lada MW, Vickroy TW, Kennedy RT (1997) High temporal resolution monitoring of glutamate and aspartate in vivo using microdialysis on-line with capillary electrophoresis with laser-induced fluorescence detection. Anal Chem 69:4560–4565
8. Venton BJ, Robinson TE, Kennedy RT, Maren S (2006) Dynamic amino acid increases in the basolateral amygdala during acquisition and expression of conditioned fear. Eur J Neurosci 23:3391–3398
9. Hotsenpiller G, Wolf ME (2003) Baclofen attenuates conditioned locomotion to cues associated with cocaine administration and stabilizes extracellular glutamate levels in rat nucleus accumbens. Neurosci 118:123–134
10. Hooker TF, Jorgenson JW (1997) A transparent flow gating interface for the coupling of microcolumn LC with CZE in a comprehensive two-dimensional system. Anal Chem 69:4134–4142
11. Shou M, Ferrario CR, Schultz KN, Robinson TE, Kennedy RT (2006) Monitoring dopamine in vivo by microdialysis sampling and on-line CE-laser-induced fluorescence. Anal Chem 78:6717–6725
12. Church WH, Justice JB (1987) Rapid sampling and analysis of extracellular dopamine *in vivo*. Anal Chem 59:712–716

Chapter 14

Indirect Analysis of Nitric Oxide and Quantitation of Selective Nitric Oxide Synthase Inhibitors in Microdialysate Samples

Gary M. Pollack

Abstract

While a longstanding body of evidence has suggested that nitric oxide plays a key role in maintaining vascular tone, work over the last decade has indicated that nitric oxide also is an important messenger in the central nervous system. Due to the extremely labile nature of this molecule, quantitation in bulk brain tissue is problematic. Indirect methods of analysis, therefore, have been developed, with the use of microdialysis technology for sample collection. To probe the physiologic and pharmacologic importance of nitric oxide, a variety of nitric oxide synthase inhibitors also have been identified and are amenable to quantitative analysis by standard chromatographic techniques in brain tissue microdialysate. The use of the microdialysis approach for the combined purpose of measuring concentrations of nitric oxide synthase inhibitors at the site of action and changes in nitric oxide content at that site illustrates the power of microdialysis in studies designed to elucidate the linkage between drug concentration (pharmacokinetics) and drug action (pharmacodynamics). This chapter is intended to describe the experimental design, procedural issues, and types of data associated with the use of brain microdialysis to explore the role of nitric oxide in central nervous system pharmacokinetics and pharmacodynamics.

Key words: Microdialysis, Nitric oxide, Nitric oxide synthase, Greiss assay, Pharmacokinetics, Pharmacodynamics

1. Introduction

A tremendous amount of effort has been expended since the early 1970s to fully develop and elaborate the complementary disciplines of pharmacokinetics and pharmacodynamics. Pharmacokinetics focuses on the processes that determine the time course of drug concentrations, in the systemic circulation and in selected regions (including target sites for therapeutic or toxic responses), following administration of a specified dose of the drug. Pharmacodynamics focuses on the processes that determine the magnitude and time

Giuseppe Di Giovanni and Vincenzo Di Matteo (eds.), *Microdialysis Techniques in Neuroscience*, Neuromethods, vol. 75, DOI 10.1007/978-1-62703-173-8_14,

course of a biologic response to a specified dose of a drug. Arguably, the most significant value of these two disciplines is at their intersection, and indeed the development of integrated pharmacokinetic–pharmacodynamic models has been an active area of research for many years.

The collection of data for pharmacokinetic analyses, while technically challenging at times, is nevertheless relatively straightforward. One simply develops an appropriate analytical technique to quantitate the drug in question (or derived metabolites if that is the primary focus) in the biologic matrices of interest. Assay specificity, sensitivity, and reproducibility all are factors that may limit the applicability of the method to the goals of a given pharmacokinetic experiment. However, there are well-accepted approaches to assay development and validation (1), and the quality of data produced by a well-validated assay is rarely a point of contention. In most pharmacokinetic applications, and the vast majority of experiments in human subjects, drug concentrations are measured only in blood despite the fact that the relevant site of action almost invariability is outside the systemic circulation. While it is well-recognized that measuring drug concentrations at the site of action is more relevant for understanding the pharmacokinetic–pharmacodynamic interface, pharmacokineticists often rely on mathematical models with "hypothetical" compartments in which the effect site resides (2). These mathematical constructs link drug concentrations in the systemic circulation to drug responses (3), often by presupposing certain types of pharmacokinetic behavior in an "effect compartment" or at the "receptor biophase."

The collection of data for pharmacodynamic analyses can, and often is, a much more complicated issue than for pharmacokinetics. Some drug-related responses (e.g., the degree to which blood pressure is reduced by an antihypertensive agent) are objective and can be measured with a high level of precision. However, many pharmacologically- relevant events are difficult, or perhaps impossible, to measure in any realistic way. This situation is especially acute for drugs that act in the central nervous system. How does one quantitate, for example, the degree to which an antidepressant ameliorates the depressive state in a patient? Consequently, there has been a high level of interest in identifying so-called biomarkers of drug response (4), which represent surrogate (usually biochemical) events for the therapeutically relevant (or toxic) response to the drug under consideration. If a predictable relationship exists between the pharmacologic outcome and the response of a surrogate marker to a drug, and if the surrogate marker can be quantitated with requisite accuracy and precision, then kinetic-dynamic models can be constructed, allowing prediction of overt pharmacologic response based on the dynamics of the surrogate.

The microdialysis approach is an effective way to address limitations associated with both pharmacokinetics (inaccessibility of the site at which drug response occurs) and pharmacodynamics (the need for identifying and quantitating biomarkers as a surrogate for the clinically relevant biologic response to a drug). Microdialysis probes can be implanted in tissue spaces that contain drug receptors relevant to a particular pharmacologic agent, and numerous studies have been published utilizing microdialysis in human tissues (5), including brain (6, 7). If a biomarker that is produced in close spatial proximity to the receptor can be identified, then microdialysis may represent an ideal solution to obtaining data for both the drug disposition and the drug action sides of the pharmacokinetic–pharmacodynamic linkage. The purpose of this chapter is to describe how microdialysis has been used to study nitric oxide, alone and in response to certain drugs including inhibitors of nitric oxide synthesis, as a case study for incorporating microdialysis methodology in the pharmacokinetic–pharmacodynamic arena.

Nitric oxide is a labile (the half-life for disappearance from biologic matrices is in the range of seconds or less) (8) physiologic and biochemical mediator that is produced from the amino acid L-arginine (9) through an enzyme-catalyzed reaction. Two isoforms of nitric oxide synthase (NOS) have been identified: an isoform expressed in vascular endothelium (eNOS) and an isoform expressed in brain tissue (nNOS). The endothelial form of the enzyme provides nitric oxide for the maintenance of vascular tone (10), and the nitric oxide system is the locus for the pharmacologic activity of nitrate-based vasodilators (11). Nitric oxide produced in the neuronal environment evidences some degree of regionality (12) and has been implicated in a variety of CNS events (13–23), including pain sensation (24), pain relief (25), and the development of tolerance to the antinociceptive effects of morphine (26, 27). The author's laboratory has, for many years, investigated a variety of aspects related to the pharmacokinetics (28–31) and pharmacodynamics (32–34) of morphine and other opioids, including the time course of loss of antinociceptive activity associated with the development of morphine tolerance (35, 36). A series of mechanistically based experiments indicated that exposure to morphine upregulates the production of nitric oxide in brain (37), that inhibiting the increase in nitric oxide production either chemically (37) or in animals that are nNOS-deficient (38) moderates the development of tolerance, and that administering L-arginine to increase nitric oxide production in morphine-naïve animals renders them functionally tolerant to morphine (37). Thus, increased production of nitric oxide in brain tissue appears to be both necessary and sufficient to decrease the antinociceptive effects of morphine. It is within this context that experimental procedures for evaluating the pharmacodynamics of nitric oxide with microdialysis techniques will be discussed in this chapter.

2. Materials

2.1. Reagents

2.1.1. Determination of Nitric Oxide by the Greiss Method (39)

The following reagents are required for the conduct of the indirect determination of nitric oxide (as a mixture of nitrate and nitrite) in microdialysate.

- NADPH
- Glucose-6-phosphate
- Glucose-6-phosphate dehydrogenase
- Nitrate reductase
- Sulfanilamide
- Hydrochloric acid
- *N*-(1-naphthyl)-ethylenediamine

2.1.2. Determination of 7-Nitroindazole by High-Performance Liquid Chromatography (40)

- *p*-Nitroaniline
- Acetonitrile
- HPLC-grade water
- Acetate buffer (20 mM, pH 4.0)
- Triethylamine

2.1.3. Medium for the Perfusion of Microdialysis Probes

Artificial cerebrospinal fluid is prepared with the following components in sterilized water:

- NaCl (147 mM)
- KCl (3.0 mM)
- $MgCl_2$ (1.0 mM)
- $CaCl_2$ (1.3 mM)
- NaH_2PO_4 (1.0 mM)

2.1.4. Chemicals for Animal Experimentation

- 7-Nitroindazole
- Morphine sulfate
- Naloxone hydrochloride
- Peanut oil
- Ketamine
- Xylazine
- Morphine
- Normal saline
- Penicillin G
- Betadine
- Heparinized normal saline
- Trypan blue

2.2. Consumable Supplies for Animal Experimentation

- Commercially available microdialysis probes and guide cannulae
- Bone wax
- Dental acrylic cement
- Syringes of various volumes (1–10 mL)
- Needles of various diameters (18–22 gauge)
- Silicone rubber tubing
- Polyethylene tubing
- Standard small-animal surgical tools
- Cotton swabs
- Gauze pads

2.3. Equipment

2.3.1. Equipment for Animal Experimentation

- Rodent stereotaxic frame
- Dental drill

2.3.2. Equipment for Nitric Oxide Quantitation

- UV/visible spectrophotometer (50 μL minimum cuvette; 540 nm) or
- Plate-reading spectrophotometer.
- Gas-permeable amperometric sensor (specifically, amiNO-700F-EH from Harvard Apparatus) with an external diameter of 0.6 mm.
- inNO-T recorder (Harvard Apparatus) to capture the signal generated by oxidation of nitric oxide at the electrode surface.
- inNO software (Harvard Apparatus) to record output signals.

2.3.3. Equipment for 7-Nitroindazole Quantitation

- HPLC system with isocratic mobile phase delivery.
- C8 column (5 μm stationary phase diameter).
- Ultraviolet absorbance detector (360 nm wavelength).

3. Methods

3.1. Pharmacokinetics and Pharmacodynamics of 7-Nitroindazole

3.1.1. Dosage Form

- 7-Nitoindazole was suspended in peanut oil to a final concentration of 33.3 mg/mL.

3.1.2. Animal Preparation

- Male Sprague-Dawley rats (275–350 g) were anesthetized with ketamine (50 mg/kg) and xylazine (10 mg/kg) by intraperitoneal injection, and received penicillin G (50,000 U/kg) by intramuscular injection 24 h before the experiment as a prophylaxis against infection. Surgical sites were shaved and cleaned with Betadine, and silicone rubber cannulae filled with heparinized saline (20 U/ml) were implanted in the right jugular and right femoral veins. The cannulae were exteriorized at the back of the neck.

- Polyethylene (PE60) tubing was heat-bent to match the curvature of the animal's abdomen, and a cuff was manufactured by heating one end of the cannula. The cuffed end was inserted through the abdominal wall into the peritoneal space with a 20-ga needle, sutured in place, and secured with surgical adhesive. The distal end of the cannula was tunneled subcutaneously and exteriorized at the back of the neck.
- Rats were placed in a stereotaxic frame to implant a microdialysis probe in the hippocampus. An incision was made in the scalp, connective tissue was removed with a cotton swab, and a hole was drilled through the skull at the following coordinates: Bregma −5.60, Lateral +5.00, DV −7.00 (41). A guide cannula was inserted 3 mm below the dura, the skull surrounding the guide cannula was sealed with bone wax, and the cannula was secured with dental cement. A 3-mm diameter microdialysis probe, which had been prepared by perfusing with ultra-pure H_2O, was inserted into the guide cannula. The inlet and outlet ports of the probe were sealed until the experiment to prevent drying of the probe. At the end of the experiment, the probe was perfused with 1% Trypan blue for 15 min, animals were sacrificed and examined for hippocampal staining to confirm probe location.

3.1.3. Pharmacologic Experimentation

- Rats ($n = 4$ per treatment group) were placed in a microdialysis apparatus to allow unrestricted locomotion.
- Microdialysis probes were equilibrated by perfusing with artificial cerebrospinal fluid (5 μL/min).
- Baseline nitric oxide (expressed as the sum of nitrate and nitrite) concentrations were determined over 2 h.
- Rats received either 7-nitroindazole through the peritoneal cannula (50 mg/kg initial dose, with 25 mg/kg every 2 h for 14 h) or peanut oil as a control.
- Normal saline (2 ml/h via the femoral vein) was infused to maintain hydration.
- Microdialysate was collected in toto every 20 min.
- Blood was collected through the jugular vein cannula at timed intervals and centrifuged for collection of serum.
- Serum and microdialysate samples were stored at −20 °C until chemical analysis.

3.1.4. Analysis of 7-Nitroindazole (40)

- Serum (100 μL) and microdialysate (80 μL) samples were thawed, and 30 μL of *p*-nitroaniline (PNA; 5 μg/mL in H_2O) was added as an internal standard.
- Sample volume was adjusted to 230 μL (serum) or 150 μL (microdialysate) with H_2O.
- For serum samples, ice-cold acetonitrile (460 μL) was added to precipitate proteins, and following centrifugation the supernatant was collected and evaporated to under dry nitrogen.
- Evaporated samples were reconstituted in 150 μl mobile phase.
- Aliquots (50 μl) of sample (reconstituted serum samples or native microdialysate) were injected on-column for analysis.
- Chromatographic separation was conducted on a C8 column (250 mm long, 4.6 mm diameter, 5 μm packing) with a mobile phase (pH 4.5; acetonitrile (18%), acetate buffer (82%; 20 mM, pH 4.0), and triethylamine (0.05%)) delivered at 1 mL/min. The internal standard and 7-nitroindazole eluted at 22 and 26 min, respectively.
- Absorbance of column effluent was monitored at 360 nm.
- Peak area ratios (7-nitroindazole to internal standard) were referenced to a standard curve constructed with known concentrations of analyte in naïve serum or microdialysate for absolute quantitation of 7-nitroindazole.

3.1.5. Determination of Efficiency of 7-Nitroindazole Recovery by the Microdialysis Probe (42)

- To estimate concentrations of 7-nitroindazole in hippocampal extracellular fluid, it was necessary to estimate the recovery of 7-nitroindazole in individual microdialysis probes.
- In vitro recoveries of 7-nitroindazole and the internal standard were not statistically different (35.6 ± 0.2 vs. 35.0 ± 0.4%, respectively) at physiologic pH, allowing the internal standard to be used to determine in vivo probe efficiency by retrodialysis.
- At the end of the experiment, *p*-nitroanaline in artificial CSF (1 μg/mL) was perfused through the microdialysis probe, effluent was collected, and *p*-nitroanaline was quantitated by HPLC.
- The fractional loss of *p*-nitroanaline across the probe was used an estimate of probe efficiency.
- Concentrations of 7-nitroindazole in microdialysate were corrected for probe efficiency to estimate the extracellular fluid concentration in vivo.

3.1.6. Indirect Analysis of Nitric Oxide in Microdialysate

- Aliquots (80 μL) of microdialysate were transferred to amber-colored polypropylene tubes.
- NADPH (10 μL; 1 μM) was added to each sample.

- A solution containing glucose-6-phosphate, glucose-6-phosphate dehydrogenase, and nitrate reductase (10 μL; final concentrations of 0.5 mM, 160 mU/ml, and 80 mU/ml, respectively) was added to each sample.
- Samples were incubated at 25°C for 1 h.
- Samples were placed on ice for 30 min.
- Sulfanilamide and HCl (17 μL each; 1 and 0.6 mM final concentration, respectively) were added and samples were incubated on ice for 15 min.
- *N*-(1-naphthyl)-ethylenediamine (NEDA) (17 μl; 1 mM) was added and samples were incubated at 25°C for 30 min.
- Sample absorbance was determined at 548 nm.
- Standard curves were constructed for quantitation of nitric oxide and to confirm that absorbance was linearly related to analyte concentration.

3.2. In Vitro Validation

In addition to standard methods of validating analytical assays (determination of linearity, sensitivity, and reproducibility relative to known standards), the indirect method of determining nitric oxide concentration can be validated relative to a direct, electrochemical approach. This validation step is helpful because, unlike analytical techniques (such as for 7-nitroindazole) that utilize chromatographic methods to separate the analyte from other components of a biologic matrix, it is difficult to assess assay specificity for approaches such as the Greiss method for quantitating nitric oxide based on total nitrate and nitrite (often symbolized as NO_x^-). The following steps were taken to assess the in vitro performance of the Greiss method relative to a direct electrochemical approach:

- Angeli's Salt was used to generate nitric oxide at 37 °C in 0.1 M phosphate buffer at pH 7.4 in vitro (43, 44).
- Angeli's Salt was dissolved in 0.01 M NaOH (final concentration 41 mM) and stored on ice.
- Sensors were polarized in aqueous solution for a minimum of 12 h before placement into nitric oxide generating solutions.
- The nitric oxide sensor was equilibrated in artificial CSF (10 mL) at 37 °C for 10 min to determine a baseline signal.
- Aliquots of the Angeli's Salt stock solution were added to produce concentrations ranging from 10 to 1,000 μM, and nitric oxide production was monitored.
- Alternatively, nitric oxide sensors can be calibrated with nitrite solutions after conversion of nitrite to nitric oxide with an acidic solution of potassium iodide.
- When nitric oxide concentrations were maximal, a 100-μL sample was collected for indirect analysis of nitric oxide as described previously.

3.3. Effect of 7-Nitroindazole on Morphine Tolerance and Brain Nitric Oxide Production

This experiment was performed to address the hypothesis that increased neuronal nitric oxide is necessary for development of tolerance to the antinociceptive effects of morphine. If increased production of nitric oxide in response to morphine administration is a requisite step in the production of tolerance to morphine, then decreasing nitric oxide formation with a selective nNOS inhibitor such as 7-nitroindazole should attenuate tolerance. To evaluate the effects of 7-nitroindazole administration on morphine tolerance, the following experiments groups were conducted.

- Rats ($n = 4$ per treatment group) were prepared surgically with venous and intraperitoneal cannulae as described above.
  - Group 1: 7-nitroindazole (50 mg/kg initial dose followed by 25 mg/kg every 2 h); a saline co-infusion (1 ml/h) was initiated in this group immediately prior to the second 25-mg/kg dose of 7-nitroindazole.
  - Group 2: 7-nitroindazole (50 mg/kg initial dose followed by 25 mg/kg every 2 h) plus a 12-h intravenous infusion of morphine (3 mg/kg/h; 1 ml/h) immediately prior to the second 25-mg/kg dose of 7-nitroindazole.
  - Group 3: peanut oil (1.5 ml/kg initial dose followed by 0.75 ml/kg every 2 h) plus an intravenous infusion of morphine (3 mg/kg/h; 1 ml/h) immediately prior to the second 0.75-ml/kg dose of peanut oil.
  - Group 4: a 12-h intravenous infusion of morphine (3 mg/kg/h).
  - Group 5: a 12-h intravenous infusion of saline (1 ml/h).
- Hot-water tail-withdrawal latency was determined before treatment to establish baseline nociception, and at timed intervals throughout the treatment period.
- To evaluate the effect of the selective neuronal NOS inhibitor 7-nitroindazole on hippocampal nitric oxide during morphine infusion, rats were prepared by implanting jugular vein, femoral vein, and intraperitoneal cannulae, together with a hippocampal microdialysis probe, as described above.
- Rats were placed in the microdialysis apparatus to allow free movement and the probe was equilibrated.
- Microdialysate was collected for 2 h to establish baseline nitric oxide content.
- Rats received 7-nitroindazole as a 50-mg/kg initial dose followed by 25 mg/kg every 2 h for 12 h.
- Morphine infusion (3 mg/kg/h; 1 ml/h, 12 h) was initiated immediately prior to the second 25-mg/kg dose of 7-nitroindazole.

- Positive and negative control groups received infusion of either morphine alone or saline.
- Microdialysate was collected every 20 min and analyzed as described previously.

3.4. Effect of Naloxone on Brain Nitric Oxide Production in Response to Morphine

Morphine mediates its central nervous system effects predominantly through interaction with μ-opioid receptors (45–48). To assess whether μ-receptor activation plays a role in modulating nNOS activity, this experiment was designed to determine if naloxone, which inhibits morphine binding to μ-receptors, affects hippocampal nitric oxide.

- A microdialysis probe was implanted in the hippocampus, and venous cannulae were placed for drug infusion, in rats ($n=4$ per group) as described above.
- Animals were placed in the microdialysis apparatus, the probe was equilibrated by perfusing with artificial CSF, and microdialysate samples were collected every 20 min to determine baseline nitric oxide.
- Naloxone (9 mg/kg/h) or saline was infused for 2 h.
- Morphine infusion was begun at either 9 or 3 mg/kg/h.
- Microdialysate samples (100 μL) were collected every 20 min throughout the infusion. Samples were analyzed as described above.

3.5. Effects of L-Arginine-Induced Increases in Brain Nitric Oxide on Morphine-Associated Antinociception

If increased production of nitric oxide in brain is sufficient to induce functional tolerance to morphine, administration of the nitric oxide precursor L-arginine would be expected to decrease the antinociceptive effects of morphine in morphine-naïve animals. To test this hypothesis, an initial experiment was performed to identify an infusion regimen for L-arginine that would produce nitric oxide concentrations similar to those observed during morphine infusion. Next, the impact of L-arginine-associated increases in nitric oxide on the antinociceptive effects of morphine was assessed.

- Rats ($n=4$ per group) were prepared surgically by implantation of venous cannulae and hippocampal microdialysis probes as described above.
- Animals were placed in the microdialysis apparatus and the probe was equilibrated by perfusion with artificial CSF for 2 h.
- Microdialysate was collected to determine baseline nitric oxide.
- L-Arginine was infused intravenously at a rate of 200 mg/kg/h, and the infusion rate was increased by 100 mg/kg/h at 4-h intervals for a total infusion duration 12 h.

- Microdialysate was collected every 20 min throughout the L-arginine infusion.
- Comparison groups of rats received infusions of morphine (3 mg/kg/h) or saline.
- Samples were analyzed as described above.

To determine the effect of elevated brain nitric oxide concentrations on morphine-associated antinociception:

- Rats ($n=4$ per group) were equipped with venous cannulae as described previously.
- L-Arginine was infused at a rate of 300 mg/kg/h for 12 h, to produce nitric oxide concentrations similar to those observed during morphine infusion, or saline as a control.
- After an 18-h post-infusion washout, rats received subcutaneous morphine (5 mg/kg).
- Nociceptive response was evaluated prior to, and at timed intervals following, morphine administration as described above.
- Morphine concentrations in serum were determined, when required, by high-performance liquid chromatography with fluorescence detection (35).

3.6. Data Analysis

The approach to expressing and analyzing data is a key element in pharmacokinetic and pharmacodynamic experimentation. Several different strategies may be used to express drug concentration or effect data. Depending on the goals of the particular study, the data collected in the experiments described above were manipulated as follows.

- Nitric oxide concentrations in microdialysate at discrete post-treatment intervals were expressed as a percentage of baseline nitric oxide for each animal to account for interanimal variability in basal nitric oxide production.
- The area under the nitric oxide concentration versus time profile (AUC) was calculated by the linear trapezoidal method to express the effect of each treatment on overall nNOS activity.
- The area under the antinociceptive effect versus time profile (AUE) was calculated in a similar manner and was used as an index of overall pharmacologic response.

4. Illustration of Microdialysis-Derived Data and Associated Analyses

4.1. Pharmacokinetics and Pharmacodynamics of 7-Nitroindazole

Concentration–time profiles for 7-nitroindazole in serum and microdialysate from the hippocampus are presented in Fig. 1. In this experiment, the nitric oxide synthase inhibitor was administered to rats as a series of multiple doses through an indwelling peritoneal cannula to minimize stress to the animal. In vivo microdialysis probe recoveries (mean ± SD) as determined by retrodialysis (49, 50) with the analytical internal standard *p*-nitroanaline were 25.5 ± 4.8% (range of 19.6–32.3%). Microdialysate concentrations of 7-nitroindazole were corrected for the efficiency of recovery calculated in each individual probe to provide estimates of 7-nitroindazole concentrations in hippocampal extracellular fluid.

7-Nitroindazole concentrations accumulated in both serum and hippocampus during repeated administration of 7-nitroindazole, as would be expected for a multiple-dose administration paradigm. Equilibration of 7-nitroindazole between hippocampal tissue and the systemic circulation was rapid, with a time-independent tissue-to-serum concentration ratio of 0.42 ± 0.02. Because 7-nitroindazole is approximately 50–60% bound to proteins in rat serum (40), the microdialysis data indicate that unbound concentrations in hippocampal extracellular fluid were essentially identical to unbound concentrations in serum, as would be expected for a substrate that distributes passively across the blood–brain barrier.

These results demonstrate the advantages of utilizing the microdialysis approach to explore CNS pharmacokinetics. This relatively simple experiment demonstrated that (a) 7-nitroindazole equilibrated rapidly across the blood–brain barrier and (b) at equilibrium,

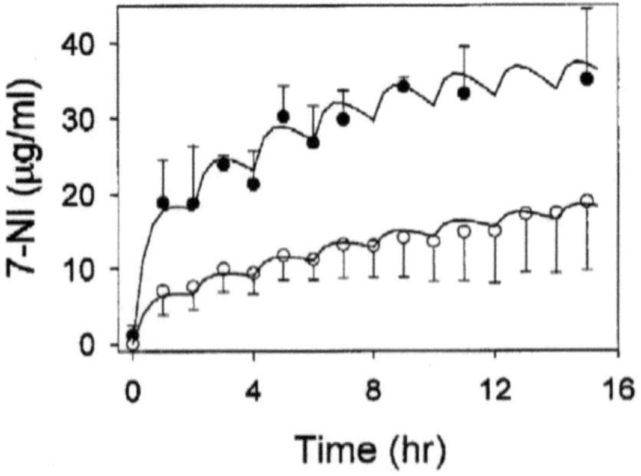

Fig. 1. Concentration–time profiles for 7-nitroindazole (7-NI) in serum (*closed symbols*) and hippocampal mircodialysate (*open symbols*) during intraperitoneal administration of a 50-mg/kg loading dose (time 0) plus 25 mg/kg every 2 h thereafter. Data are presented as mean ± SD for four rats; *lines* indicate the fit of the pharmacokinetic model to the hippocampal concentration–time data and model predictions of the serum concentration–time profile. From Bush and Pollack (53).

unbound concentrations on either side of the barrier were equivalent, indicating that asymmetric transport at the blood–brain interface likely does not occur for this molecule.

From the perspective of pharmacokinetics, the utility of the microdialysis approach is in providing relatively data-rich information for the construction of mathematical models. Such models may be used to assess the potential validity of mechanisms underlying drug disposition or action, to leverage informatics in order to assess treatment-related or other (e.g., disease state-associated) effects, or to make predictions regarding behavior under a different set of conditions (e.g., when the administered dose or dosing interval is changed). To illustrate the pharmacokinetic modeling approach for disposition of 7-nitroindazole in the hippocampus during and after repeated intraperitoneal administration of this selective nNOS inhibitor, an empirical model was fit to the extracellular fluid 7-nitroindazole concentrations, corrected for the efficiency of recovery, obtained from the microdialysis probe implanted in the hippocampus. The results of this modeling exercise are displayed graphically in Fig. 1. Numerical estimates of the pharmacokinetic parameters that compose the model, together with the statistical uncertainty associated with those estimates, are recovered from this type of analysis, and can be used for predictive purposes (in extrapolating concentrations that would be produced by other drug administration schedules, for example). The model was capable of describing fluctuations in extracellular fluid 7-nitroindazole concentrations, which occurred due to the fact that the nNOS inhibitor was administered as a series of discrete bolus doses instead of a continuous infusion. By correcting the unbound extracellular fluid concentrations returned from analyzing microdialysate (assumed to be equivalent to unbound serum concentrations) for the presumed degree of protein binding, the model was capable of predicting total 7-nitroindazole concentrations in serum (also displayed in Fig. 1).

In this particular case, the parameters recovered from the pharmacokinetic analysis of the 7-nitroindazole concentration–time data would have no mechanistic utility, and simply provide an empirical means for expressing the shape of the concentration–time profile. However, this step in the analysis was crucial to developing the more important integrated pharmacokinetic–pharmacodynamic model for the influence of 7-nitroindazole on nitric oxide content, determined by indirect analysis, in the hippocampus. Average nitric oxide concentration–time profiles in untreated rats are shown in Fig. 2. Nitric oxide fluctuated reproducibly in individual animals, with a frequency of approximately 7 h.

Understanding the temporal changes in nitric oxide concentrations in control animals was a requisite step in building an integrated pharmacokinetic–pharmacodynamic model for the influence of 7-nitroindazole on nitric oxide concentrations in hippocampal

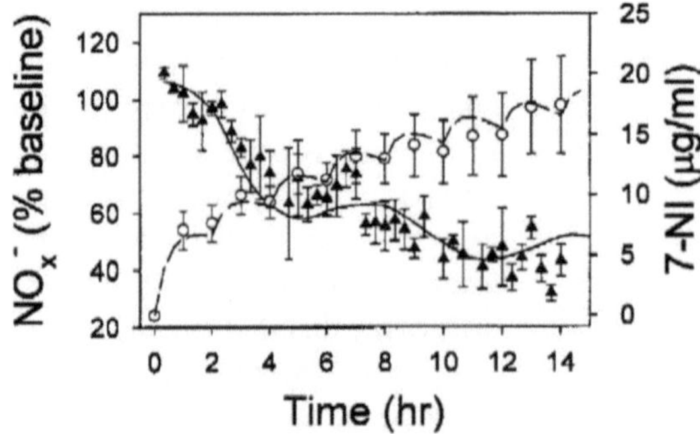

Fig. 2. Concentrations of nitric oxide (NO_x^-; *closed symbols*) and 7-nitroindazole (7-NI; *open symbols*) in hippocampal microdialysate during repeated 7-nitroindazole administration in rats. Symbols represent observed data (mean ± SD; $n = 4$); *lines* indicate the fit of a kinetic-dynamic model (Fig. 3) to the data. These data are consistent with a 7-nitroindazole IC_{50} 4of 17.2 ± 1.1 μg/ml. From Bush and Pollack (53).

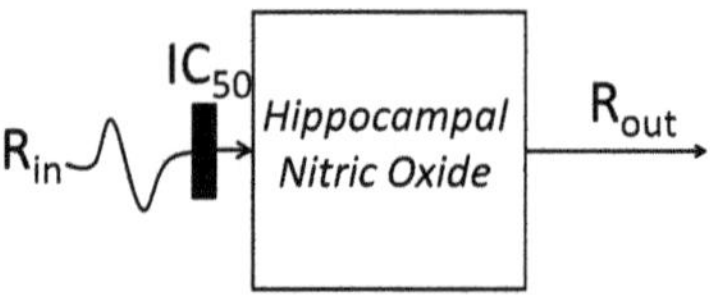

Fig. 3. Schematic of the dynamic model for regulation of nitric oxide content in hippocampal tissue. R_{in} indicates the rate of nitric oxide production, which evidences circadian variation, and R_{out} represents the rate of nitric oxide degradation. Under basal conditions, nitric oxide content reflects the balance of R_{in} and R_{out}. The nNOS inhibitor 7-nitroindazole blocks nitric oxide formation and may be modeled as decreasing R_{in} with a characteristic IC_{50}. Adapted from Bush and Pollack (53).

tissue (Fig. 3). The ability of such a model to describe the data obtained in rats treated with the nNOS inhibitor is displayed in Fig. 2. Nitric oxide concentrations in the hippocampus decreased during the first 2 h of the experiment (prior to administration of 7-nitroindazole) consistent with the cyclic behavior of nitric oxide in control animals. While nitric oxide decreased by approximately 15% in control animals by 4 h, nitric oxide concentrations in 7-nitroindazole-treated rats decreased by more than 40% over the same time frame and by approximately 60% at 12 h. Pharmacodynamic modeling revealed an IC_{50} for 7-nitroindazole in hippocampal extracellular fluid of 17 μg/ml, corresponding to a total (bound plus unbound) serum 7-nitroindazole concentration of 38 μg/ml.

4.2. In Vitro Validation

The two analytical methods (the indirect Greiss method and the specific electrochemical technique) were highly correlated in vitro (Fig. 4), indicating that the degradation of nitric oxide to a mixture of nitrate and nitrite (in the absence of other sources of nitrate/nitrite) allowed a reliable estimation of overall nitric oxide concentration. It must be emphasized, however, that the indirect approach

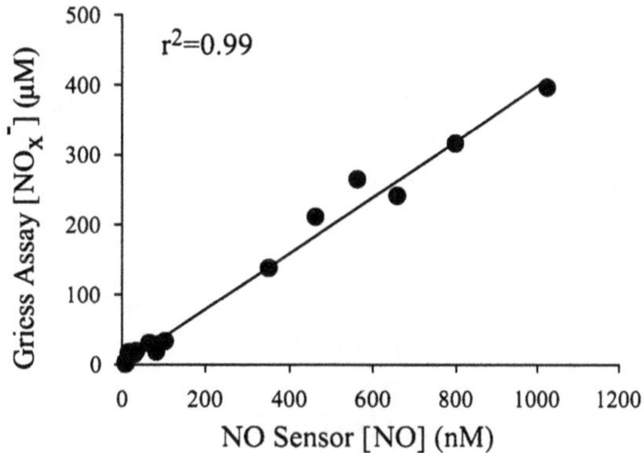

Fig. 4. Comparison of the indirect (Greiss assay) and electrochemical approaches (NO sensor) to determining the concentration of nitric oxide produced in vitro in a solution of Angeli's Salt. From Heinzen and Pollack (54).

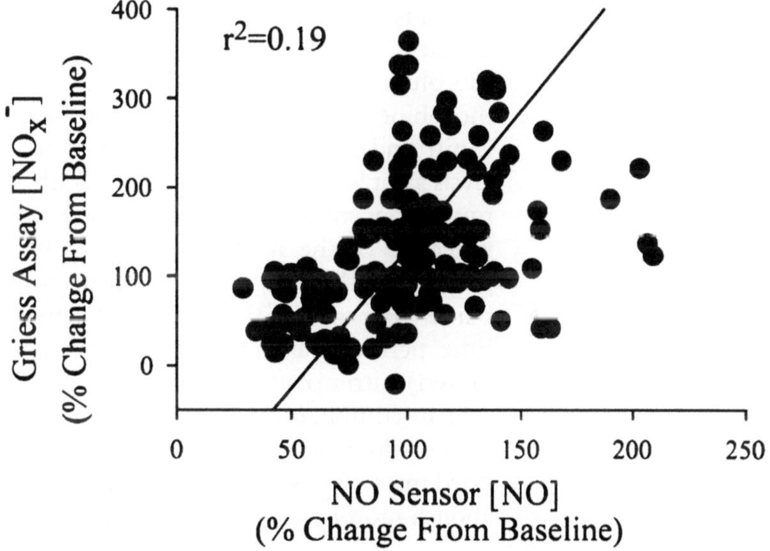

Fig. 5. Comparison of the indirect (Greiss assay) and electrochemical approaches (NO sensor) to determining the concentration of nitric oxide produced in vivo in rats. From Heinzen and Pollack (54).

to determining nitric oxide concentration in vivo is impacted by other (non-nitric oxide-derived) sources of nitrate/nitrite, and the two analytical techniques do not correlate well (Fig. 5). This is not an insurmountable problem with the indirect method, of course. Because nitric oxide is labile, and nitrate/nitrite from other sources are relatively stable, changes in apparent nitric oxide concentration over reasonably short time scales (hours) can be inferred to be the result of altered nitric oxide production. One simply must reference such changes to a baseline value, as described in the preceding section for analysis of the influence of 7-nitroindazole on hippocampal nitric oxide.

4.3. Effect of 7-Nitroindazole on Morphine Tolerance and Brain Nitric Oxide Production

The microdialysis technique is ideal for collecting drug/biochemical data from animals undergoing behavioral tests or observations (such as determination of response to nociceptive stimuli). Serial sampling of the extracellular environment of brain tissue is feasible so that terminal experimentation (one time point per animal), a common aspect of pharmacokinetic/pharmacodynamic experiments with target sites outside of the systemic circulation, is not required.

Morphine administration has been shown to increase nitric oxide production in brain, and that increases in nitric oxide production correspond with loss of morphine-associated antinociceptive activity (37, 51). This study was designed to determine whether 7-nitroindazole can limit increases in hippocampal nitric oxide produced by morphine infusion, and illustrates how microdialysis sampling of a biomarker (nitric oxide) can be integrated with collection of behavioral data (antinociception). The relevant data obtained in this experiment are presented in Fig. 6 and Table 1. Administration of 7-nitroindazole prior to and during morphine infusion decreased hippocampal nitric oxide compared to when morphine was administered alone, consistent with (a) the ability of morphine to elevate brain nitric oxide and (b) the ability of 7-nitroindazole to inhibit nNOS. The data obtained by microdialysis sampling allowed analysis of both peak nitric oxide (the maximum concentration of nitric oxide observed in a given profile, reflective of the magnitude of nNOS activity at a discrete time point) and the area under the concentration–time curve for nitric oxide (reflective of the integrated activity of nNOS over the entire time period evaluated). Both peak nitric oxide and total (integrated) nitric oxide were lower in animals receiving 7-nitroindazole in combination with morphine as compared to rats receiving either saline or morphine alone. Thus, this relatively straightforward microdialysis experiment indicated that 7-nitroindazole could abolish morphine-associated increases in nitric oxide, and suggested

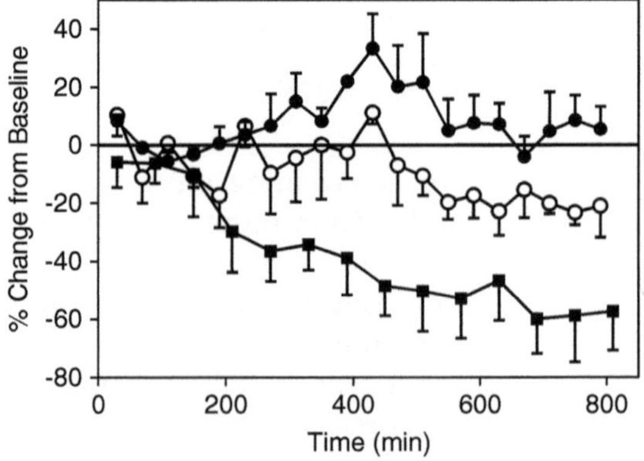

Fig. 6. Time course of hippocampal nitric oxide concentrations during infusion of saline (*open circles*), morphine (*closed circles*), or a combination of 7-nitroindazole and morphine (*closed squares*). Data are presented as mean ± SE, $n = 4$ rats per group.

Table 1
Effect of various treatment regimens on peak nitric oxide concentration and the area under the nitric oxide concentration–time profile (AUC) in hippocampal microdialysate

Treatment	Maximal NO (%)	NO AUC (%h)
Saline	53.1 ± 7.8[a]	−117 ± 86[a]
Morphine	78.8 ± 9.8[b]	106 ± 40[b]
7NI/Morphine	−31.1 ± 10.8[a, b]	−718 ± 153[a, b]
Naloxone/morphine	11.5 ± 11.3[a, b]	−270 ± 74[a]

Data are presented as mean ± SE, $n = 4$ per group
[a]Significantly different from morphine-treated rats, $p < 0.05$
[b]Significantly different from saline-treated rats, $p < 0.05$

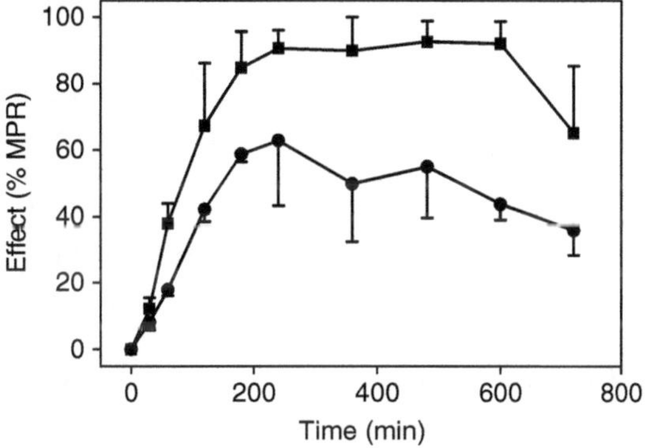

Fig. 7. Time course of antinociceptive effect in rats receiving morphine alone (*circles*) or morphine in combination with 7-nitroindazole (*squares*). Data are presented as mean ± SE ($n = 4$ rats per group). 7-nitroindazole alone did not result in tail-withdrawal latencies that differed from saline-treated rats.

that, if increases in brain nitric oxide were required for the development of tolerance to morphine-associated antinociception, then animals receiving 7-nitroinadazole in combination with morphine should evidence a lower degree of tolerance development than rats receiving morphine alone.

Indeed, coadministration of 7-nitroindazole with morphine produced significantly higher antinociceptive effects during prolonged (12-h) infusions of morphine (Fig. 7, Table 2). Control experiments indicated that neither 7-nitroindazole nor the vehicle used for administration of 7-nitroindazole (peanut oil) produced measurable antinociception, and that peanut oil by itself did not alter the profile of tolerance development in response to morphine infusion. The fact that the antinociceptive effect at the end of the morphine infusion period was not affected by 7-nitroindazole

Table 2
Effect of NOS inhibition on morphine-associated antinociception in rats

Treatment	AUE (%h)[a]	E_{max} (%)[b]	$E_{12\,h}$ (%)[c]
Morphine/peanut oil	501 ± 125[d]	66.9 ± 14.9[d]	35.8 ± 9.14[d]
Morphine/7-nitroindazole	937 ± 58.8[d, e]	100 ± 0[d]	65.1 ± 20.2[d]
Saline/7-nitroindazole	120 ± 23.8	18.2 ± 1.71	10.0 ± 4.23
Saline	76.1 ± 7.8	11.7 ± 1.1	2.59 ± 1.09

Data are presented as mean ± SE, $n = 4$ per group
[a]Area under the effect vs. time curve, 0 through 12 h
[b]Maximum effect observed
[c]Effect observed at 12 h after morphine administration
[d]Significantly different from relevant saline-treated controls, $p < 0.05$
[e]Significantly different from morphine/peanut oil, $p < 0.05$

suggests that the attenuation of tolerance development by inhibition of nNOS was temporary or incomplete. Nevertheless, these data provided compelling evidence that nitric oxide is an important driving force for morphine tolerance, and this hypothesis subsequently was supported by several mechanistic studies (38, 51, 52). Clearly, the key role for nitric oxide in the development of morphine tolerance would have been difficult, if not impossible, to demonstrate without the ability to monitor nitric oxide concentrations in brain via microdialysis sampling.

4.4. Effect of Naloxone on Brain Nitric Oxide Production in Response to Morphine

Although morphine had been shown to increase brain nitric oxide content, the mechanism by which that increased occurred had not been demonstrated. Use of naloxone, a μ-opioid receptor antagonist, allowed evaluation of whether this receptor system was involved in the apparent upregulation of nNOS activity by morphine. Hippocampal nitric oxide was determined during co-infusion of naloxone and morphine. These data are presented in Fig. 8, with companion antinociception data summarized in Table 3. Administration of naloxone with morphine completely inhibited the increases in hippocampal nitric oxide that could be produced by morphine alone and, as anticipated, ablated the pharmacologic effect of the opioid. Peak nitric oxide concentrations, as well as the area under the concentration–time curve for nitric oxide, were lower when morphine was administered in the presence of naloxone than when morphine was administered alone.

While the data did not achieve the level of statistical significance, nitric oxide concentrations in animals receiving morphine and naloxone in combination tended to be lower than nitric oxide concentrations in untreated animals. These data suggest that intact μ-opioid receptor activity is required for morphine to stimulate nitric oxide production and that the μ-opioid receptor system may be involved in basal production of nitric oxide in morphine-naïve

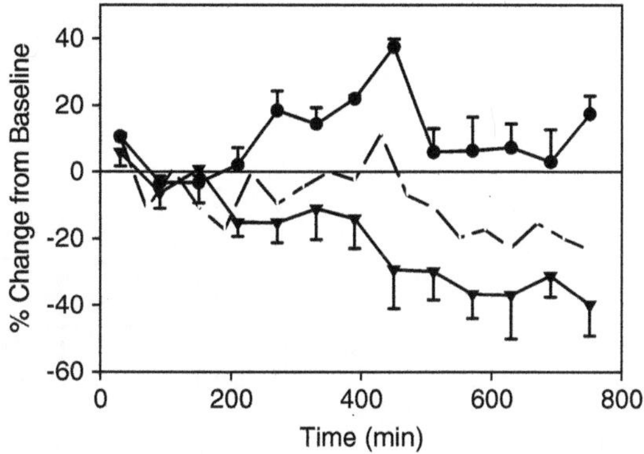

Fig. 8. Time course of changes in hippocampal nitric oxide concentrations (mean ± SE, $n = 4$ per group) during infusion of morphine alone (*circles*) vs. morphine in combination with naloxone (*triangles*). Because nitric oxide content evidences diurnal variations, average data from saline-infused rats are shown for comparison (*broken line*).

Table 3
Effect of μ-receptor blockade on indices of morphine-associated antinociception

Treatment	AUE (%h)[a]	E_{max} (%)[b]	$E_{12\,h}$ (%)[c]
Morphine	591 ± 140[d]	94.1 ± 5.95[d]	52.5 ± 9.51[d]
Morphine/nalorphine	57.7 ± 15.6[e]	9.55 ± 1.54[e]	2.42 ± 1.42[e]
Saline/nalorphine	76.2 ± 5.3	13.9 ± 1.0	6.50 ± 1.74
Saline	120 ± 18	18.2 ± 1.3	10.0 ± 3.2

Data are presented as mean ± SE, $n = 4$ per group
[a] Area under the effect vs. time curve, 0 through 12 h
[b] Maximum effect observed
[c] Effect observed at 12 h after morphine administration
[d] Significantly different from relevant saline-treated controls, $p < 0.05$
[e] Significantly different from morphine/peanut oil, $p < 0.05$

animals. These results provided a framework for understanding morphine tolerance as a biochemical feedback system with morphine agonism of the μ-opioid receptor producing a behavioral effect (antinociception) and a biochemical response (increased nitric oxide), with the biochemical response subsequently down-regulating the behavioral effect (37).

4.5. Effects of L-Arginine-Induced Increases in Brain Nitric Oxide on Morphine-Associated Antinociception

L-Arginine is a substrate for NOS and a precursor of nitric oxide. Increasing L-arginine concentrations in the systemic circulation would therefore be expected to increase brain nitric oxide, assuming that systemically administered L-arginine can cross the blood–brain barrier and gain access to the sites at which nNOS is expressed. L-Arginine does cross the blood–brain barrier, albeit in a capacity-limited manner (52), and so there is a less-than-proportionate

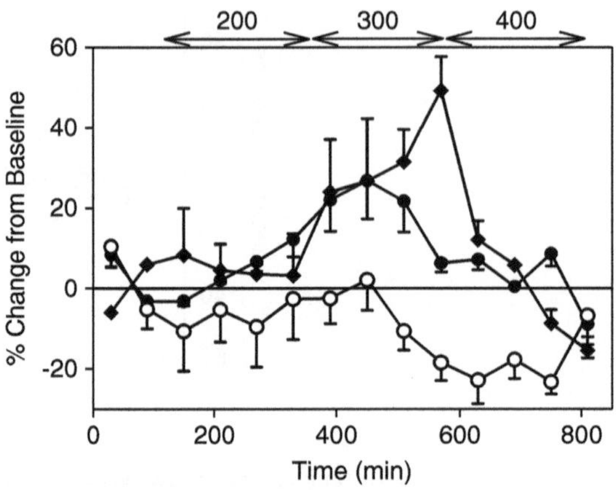

Fig. 9. Elevation of hippocampal nitric oxide concentrations (expressed as change from baseline) during an escalating infusion of L-arginine (*diamonds*). Data obtained in rats infused with morphine (*closed circles*), demonstrating increased nitric oxide concentrations in response to μ-opioid receptor agonism, or saline (*open circles*), demonstrating dirunal fluctuations in hippocampal nitric oxide, are shown for comparison. Data are presented as mean ± SE ($n = 4$ per group).

increase in brain concentration of L-arginine than in systemic concentrations when L-arginine is administered by infusion.

Despite the less-than-proportionate increase in brain tissue L-arginine concentrations as blood L-arginine was increased, infusion of L-arginine resulted in elevated nitric oxide production in the hippocampus (Fig. 9). The L-arginine infusion rate selected (300 mg/kg/h) produced maximal increases in nitric oxide concentrations (49.3 ± 8.0%) that were similar to those observed during morphine treatment (33.4 ± 11.9%). To evaluate the effect of elevated hippocampal nitric oxide on morphine-associated antinociception, rats received a 12-h infusion of L-arginine followed by an 18-h washout to allow nitric oxide concentrations to return to basal levels. Antinociception then was determined in response to a 5-mg/kg subcutaneous dose of morphine. Rats pretreated with L-arginine evidenced significantly lower antinociceptive responses to morphine as compared to control animals (Fig. 10), consistent with the development of functional tolerance to morphine. The area under the effect versus time curve for morphine-associated antinociception, presumably representative of the overall (integrated) antinociceptive response, was statistically ($p < 0.05$) lower in L-arginine-treated rats compared to controls (69.9 ± 31.4 vs. 130 ± 31%MPR min, respectively). L-Arginine did not influence morphine concentrations, and so these data could be interpreted as indicating that increased brain nitric oxide, regardless of the mechanism of increase, is sufficient to impair antinociceptive response to morphine.

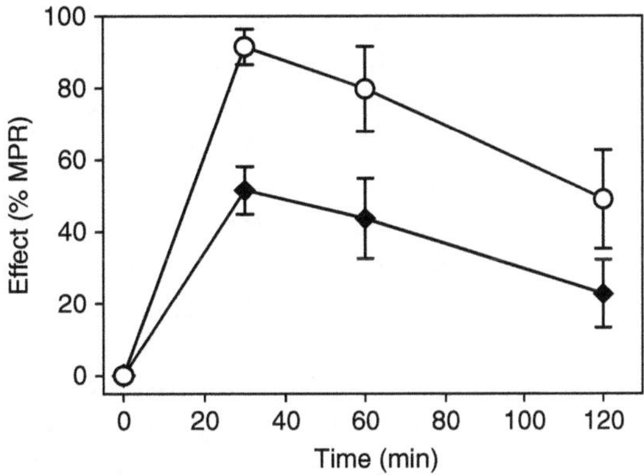

Fig. 10. Antinociception produced by a 5-mg/kg subcutaneous dose of morphine in rats pretreated with 12-h infusion of L-arginine (*closed symbols*) or saline (*open symbols*). Data are presented as mean ± SE (n = 4 per group).

5. Summary and Comments

Nitric oxide is recognized as an important chemical mediator in the brain and other organs and tissues, and plays an integral role in the perception of pain and the response, including the development of tolerance, to opioid analgesics. Understanding the biochemical mechanisms that regulate nitric oxide production is key to exploiting this mediator for therapeutic benefit. Consequently, methods for quantitating nitric oxide concentrations are important experimental tools.

The experiments described in this chapter, focusing on the role of nitric oxide in the development of tolerance to the antinociceptive effects of morphine, were selected to illustrate the utility of microdialysis in CNS pharmacokinetic/pharmacodynamic experiments. Taken together, the results of these experiments indicate that increased production of nitric oxide in the brain is both necessary and sufficient to produce functional tolerance to morphine. The level of contemporary understanding of the role of nitric oxide in opioid tolerance development simply would not have been possible to achieve without the use of microdialysis or the availability of selective inhibitors of the neuronal isoform of nitric oxide synthase.

Because nitric oxide is extraordinarily labile, standard approaches for bioanalysis (e.g., removal of tissue samples with subsequent processing) are not practical. The microdialysis approach for sample collection is ideal in that it obviates the need for removal of tissue or biologic fluid and is amenable to on-line, real-time chemical analyses. Although the data illustrated in this chapter relied on collection and storage of microdialysate samples prior to nitric oxide analysis, it would be feasible to adapt the Greiss assay for on-line conversion of nitric oxide to nitrate and nitrite, with concomitant detection of these breakdown species.

Two caveats must be considered when utilizing microdialysis for the indirect analysis of nitric oxide. The first of these relates to the microdialysis approach itself, and therefore is relevant not only to nitric oxide but to nitric oxide synthase inhibitors as well. Although a variety of techniques (in vitro calibration, retrodialysis) are used to estimate the efficiency of analyte diffusion across the microdialysis membrane with subsequent collection of microdialysate, in reality the actual efficiency of analyte collection from a biologic environment typically is never truly known. Thus, the absolute concentration of the analyte in extracellular fluid, which is expressed as the analyte concentration in microdialysate divided by probe efficiency, cannot truly be known. For pharmacokinetic/pharmacodynamic experiments, the goal of which typically is to determine the *rate of change* of analyte concentrations, an unknown probe efficiency poses no problem, as long as efficiency is stable over the course of the entire experiment. In this regard, it may be more appropriate to report analyte concentrations as "apparent concentrations in extracellular fluid."

The second caution relates to the indirect analysis of nitric oxide, which relies on degradation to a mixture of nitrate and nitrite. These breakdown products of nitric oxide exist in biologic media from a variety of sources, not solely from the degradation of nitric oxide. Therefore, when one compares the analytical results from the indirect method of nitric oxide quantitation with those generated by a specific assay for native nitric oxide, data such as that displayed in Fig. 5 may be observed. The apparent lack of correspondence between the indirect and specific measurements of nitric oxide presents a problem for interpreting nitric oxide concentrations averaged across a group of animals or subjects. However, if one is concerned with the *rate of change* in nitric oxide concentration in an *individual* animal or subject, as is typical for a pharmacokinetic/pharmacodynamic experiment, then the indirect technique provides information that is as valid and useful as that generated by a more specific assay.

References

1. Viswanathan CT, Bansal S, Booth B et al (2007) Quantitative bioanalytical methods validation and implementation: best practices for chromatographic and ligand binding assays. Pharm Res 24:1962–1973
2. Holford NH, Sheiner LB (1981) Pharmacokinetic and pharmacodynamic modeling in vivo. Crit Rev Bioeng 5:273–322
3. Mager DE, Jusko WJ (2008) Development of translational pharmacokinetic-pharmacodynamic models. Clin Pharmacol Ther 83:909–912
4. Zhou Q, Gallo JM (2011) The pharmacokinetic/pharmacodynamic pipeline: translating anticancer drug pharmacology to the clinic. AAPS J 13:111–120
5. Höcht C, Opezzo JA, Bramuglia GF et al (2006) Application of microdialysis in clinical pharmacology. Curr Clin Pharmacol 1:163–183
6. Helmy A, Carpenter KL, Hutchinson PJ (2007) Microdialysis in the human brain and its potential role in the development and clinical

assessment of drugs. Curr Med Chem 14:1525–1537

7. Blakeley J, Portnow J (2010) Microdialysis for assessing intratumoral drug disposition in brain cancers: a tool for rational drug development. Expert Opin Drug Metab Toxicol 6:1477–1491
8. Barbosa RM, Lourenço CF, Santos RM et al (2008) In vivo real-time measurement of nitric oxide in anesthetized rat brain. Methods Enzymol 441:351–367
9. Moncada S, Palmer RM, Higgs EA (1989) Biosynthesis of nitric oxide from L-arginine. A pathway for the regulation of cell function and communication. Biochem Pharmacol 38:1709–1715
10. Bauer V, Sotníková R (2010) Nitric oxide—the endothelium-derived relaxing factor and its role in endothelial functions. Gen Physiol Biophys 29:319–340
11. Stokes GS (2006) Nitrates as adjunct hypertensive treatment. Curr Hypertens Rep 8:60–68
12. Hara S, Mukai T, Kurosaki K, Mizukami H et al (2004) Different response to exogenous L-arginine in nitric oxide production between hippocampus and striatum of conscious rats: a microdialysis study. Neurosci Lett 366:302–307
13. Gören MZ, Arlcioglu-Kartal F, Yurdun T et al (2001) Investigation of extracellular L-citrulline concentration in the striatum during alcohol withdrawal in rats. Neurochem Res 12:1327–1333
14. Nowak P, Brus R, Oswiecimska J et al (2002) 7-Nitroindazole enhances amphetamine-evoked dopamine release in rat striatum. An in vivo microdialysis and voltammetric study. J Physiol Pharmacol 53:251–263
15. Pepicelli O, Raiteri M, Fedele E (2004) The NOS/sGC pathway in the rat central nervous system: a microdialysis overview. Neurochem Int 45:787–797
16. Hong SK, Jung IS, Bang SA et al (2006) Effect of nitric oxide synthase inhibitor and NMDA receptor antagonist on the development of nicotine sensitization of nucleus accumbens dopamine release: an in vivo microdialysis study. Neurosci Lett 409:220–223
17. Chalimoniuk M, Langfort J (2007) The effect of subchronic, intermittent L-DOPA treatment on neuronal nitric oxide synthase and soluble guanylyl cyclase expression and activity in the striatum and midbrain of normal and MPTP-treated mice. Neurochem Int 50: 821–833
18. Raimondi L, Alfarano C, Pacini A et al (2007) Methylamine-dependent release of nitric oxide and dopamine in the CNS modulates food intake in fasting rats. Br J Pharmacol 150:1003–1010
19. Llansola M, Hernandez-Viadel M, Erceg S et al (2009) Increasing the function of the glutamate-nitric oxide-cyclic guanosine monophosphate pathway increases the ability to learn a Y-maze task. J Neurosci Res 87: 2351–2355
20. Li S, Wang W, Wang C et al (2010) Possible involvement of NO/NOS signaling in hippocampal amyloid-β production induced by transient focal cerebral ischemia in aged rats. Neurosci Lett 470:106–110
21. Saulskaya NB, Fofonova NV, Sudorgina PV (2010) Activation of the noergic system of the nucleus accumbens on presentation of contextual danger signals. Neurosci Behav Physiol 40:907–912
22. Kalinchuk AV, McCarley RW, Porkka-Heiskanen T et al (2011) The time course of adenosine, nitric oxide (NO), and inducible NO synthase in the brain with sleep loss and their role in the non-rapid eye movement sleep homeostatic cascade. J Neurochem 116: 260–272
23. Koizumi H, Fujisawa H, Suehiro E et al (2011) Neuroprotective effects of Ebselen following forebrain ischemia: involvement of glutamate and nitric oxide. Neurol Med Chir (Tokyo) 51:337–343
24. Karlsson GA, Chaitoff KA, Hossain S et al (2007) Modulation of cardiovascular responses and neurotransmission during peripheral nociception following nNOS antagonism within the periaqueductal gray. Brain Res 1143: 150–160
25. Zelinski LM, Ohgami Y, Chung E et al (2009) A prolonged nitric oxide-dependent, opioid-mediated antinociceptive effect of hyperbaric oxygen in mice. J Pain 10:167–172
26. Toda N, Kishioka S, Hatano Y et al (2009) Modulation of opioid actions by nitric oxide signaling. Anesthesiology 110:166–181
27. Abdel-Zaher AO, Abdel-Rahman MS, Elwasei FM (2010) Blockade of nitric oxide overproduction and oxidative stress by *Nigella sativa* oil attenuates morphine-induced tolerance and dependence in mice. Neurochem Res 35:1557–1565
28. Horton TL, Pollack GM (1991) Enterohepatic recirculation and renal metabolism of morphine in the rat. J Pharm Sci 80:1147–1152
29. Dagenais C, Zong J, Ducharme J et al (2001) Effect of mdr1a P-glycoprotein gene disruption, gender, and substrate concentration on brain uptake of selected compounds. Pharm Res 18:957–963

30. Dagenais C, Graff CL, Pollack GM (2004) Variable modulation of opioid brain uptake by P-glycoprotein in mice. Biochem Pharmacol 67:269–276
31. Kalvass JC, Olson ER, Cassidy MP et al (2007) Pharmacokinetics and pharmacodynamics of seven opioids in P-glycoprotein-competent mice: assessment of unbound brain EC50, u and correlation of in vitro, preclinical, and clinical data. J Pharmacol Exp Ther 323:346–355
32. Chen C, Pollack GM (1997) Blood-brain disposition and antinociceptive effects of D-penicillamine2,5-enkephalin in the mouse. J Pharmacol Exp Ther 283:1151–1159
33. Chen C, Pollack GM (1998) Altered disposition and antinociception of (D-penicillamine(2,5)) enkephalin in mdr1a-gene-deficient mice. J Pharmacol Exp Ther 287:545–552
34. Bauer B, Yang X, Hartz AM et al (2006) In vivo activation of human pregnane X receptor tightens the blood-brain barrier to methadone through P-glycoprotein up-regulation. Mol Pharmacol 70:1212–1219
35. Ouellet DM, Pollack GM (1995) A pharmacokinetic-pharmacodynamic model of tolerance to morphine analgesia during infusion in rats. J Pharmacokinet Biopharm 23:531–549
36. Ouellet DM, Pollack GM (1997) Pharmacodynamics and tolerance development during multiple intravenous bolus morphine administration in rats. J Pharmacol Exp Ther 281:713–720
37. Heinzen EL, Pollack GM (2004) Pharmacodynamics of morphine-induced neuronal nitric oxide production and antinociceptive tolerance development. Brain Res 1023:175–184
38. Heinzen EL, Pollack GM (2004) The development of morphine antinociceptive tolerance in nitric oxide synthase-deficient mice. Biochem Pharmacol 67:735–741
39. Stamler JS, Feelisch M (1996) Methods in nitric oxide research. Wiley, New York
40. Bush MA, Pollack GM (2000) Pharmacokinetics and protein binding of the selective neuronal nitric oxide synthase inhibitor 7-nitroindazole. Biopharm Drug Dispos 21:221–228
41. Paxinos G, Watson C (1986) The rat brain in stereotaxic coordinates. Academic, New York
42. Colin A-K (1988) Microdialysis user's guide, 4th edn. Carnegie Medicin, Stockholm
43. Maragos CM, Morley D, Wink DA et al (1991) Complexes of NO with nucleophiles as agents for the controlled biological release of nitric oxide. Vasorelaxant effects. J Med Chem 34:3242–3247
44. Fukuto JM, Hobbs AJ, Ignarro LJ (1993) Conversion of nitroxyl (HNO) to nitric oxide (NO) in biological systems: the role of physiological oxidants and relevance to the biological activity of HNO. Biochem Biophys Res Commun 196:707–713
45. Lee S-C, Wang J-J, Ho S-T et al (1997) Nalbuphine coadministered with morphine prevents tolerance and dependence. Anesth Analg 84:810–815
46. Chen SW, Maguire PA, Davies MF et al (1996) Evidence for mu1-opioid receptor involvement in fentanyl-mediated respiratory depression. Eur J Pharmacol 312:241–244
47. Roy S, Liu HC, Loh HH (1998) mu-opioid receptor-knockout mice: the role of mu-opioid receptor in gastrointestinal transit. Brain Res Mol Brain Res 56:281–283
48. Culpepper-Morgan JA, Holt PR, LaRoche D et al (1995) Orally administered opioid antagonists. Neurosci Lett 56:1187–1192
49. Bourne JA (2003) Intracerebral microdialysis: 30 years as a tool for the neuroscientist. Clin Exp Pharmacol Physiol 30:16–24
50. Cano-Cebrián MJ, Zornoza T, Polache A et al (2005) Quantitative in vivo microdialysis in pharmacokinetic studies: some reminders. Curr Drug Metab 6:83–90
51. Heinzen EL, Booth RG, Pollack GM (2005) Neuronal nitric oxide modulates morphine antinociceptive tolerance by enhancing constitutive activity of the mu-opioid receptor. Biochem Pharmacol 69:679–688
52. Heinzen EL, Pollack GM (2003) Pharmacokinetics and pharmacodynamics of L-arginine in rats: a model of stimulated neuronal nitric oxide synthesis. Brain Res 989:67–75
53. Bush MA, Pollack GM (2001) Pharmacokinetics and pharmacodynamics of 7-nitroindazole, a selective nitric oxide synthase inhibitor, in the rat hippocampus. Pharm Res 18:1607–1612
54. Heinzen EL, Pollack GM (2002) Use of an electrochemical nitric oxide sensor to detect neuronal nitric oxide production in conscious, unrestrained rats. J Pharmacol Toxicol Methods 48:139–146

Chapter 15

Determination of Histamine in Microdialysis Samples from the Rodent Brain by Column Liquid Chromatography

Jan Kehr and Takashi Yoshitake

Abstract

Microdialysis sampling in combination with a suitable analytical method enables in vivo monitoring of histamine release and metabolism in selected brain structures of experimental animals including rats and mice. In the alkaline medium, histamine reacts with *o*-phthalaldehyde (OPA) to form fluorescence products. This feature has been utilized for sensitive determination of histamine by HPLC with postcolumn OPA derivatization of histamine in the microdialysates and other biological samples. Even higher sensitivity and specificity could be achieved by precolumn derivatization of histamine with two molecules of a highly fluorescent tag such as 4-(1-pyrene)butyric acid *N*-hydroxysuccinimide ester (PSE), which yields the formation of the fluorescence intramolecular excimer. This chapter provides a detailed description of these two principal derivatization methods used for HPLC determination of histamine in the microdialysis samples and a summary of the most typical applications of monitoring histamine release in experimental neuropharmacology, neuroendocrinology and neurophysiology.

Key words: Microdialysis, In vivo, Histamine, Neurotransmitter, Brain, High-performance liquid chromatography, Fluorescence detection, Derivatization

1. Introduction

Histamine (2-(4-imidazolyl)ethylamine) as a biologically active molecule was discovered more than a century ago by Sir Henry Dale and his collaborators studying the contraction of smooth muscles of the guinea pig ileum and vasodilatation (1). Several monographs and excellent reviews were published on the biological role of histamine in the regulation of a broad spectrum of physiological functions (2–4, 6–10) including secretion of the gastric acid in the stomach, secretion of pituitary hormones, mediating inflammatory reactions and immune responses and acting as a neurotransmitter in the nervous system (for review, see (2–4)). The presence of histamine in the brain was first described by Kwiatkowski

Giuseppe Di Giovanni and Vincenzo Di Matteo (eds.), *Microdialysis Techniques in Neuroscience*, Neuromethods, vol. 75,
DOI 10.1007/978-1-62703-173-8_15,

in 1941 (5). However, since the mammalian brain also contains the histamine-rich mast cells (11), the presence of neuronal histamine and mapping of the brain histaminergic system was finally demonstrated by the use of immunohistochemical methods as late as in the early 1980s (12–14). These studies revealed that histamine is present exclusively in the neurons of the tuberomammillary nucleus from which histamine-containing afferents are diffusely distributed throughout the brain. Histamine together with its four receptor subtypes is implicated in mediation of several physiological and behavioral functions such as sleep-wakefulness, circadian rhythm, thermoregulation, food intake, cortical arousal and cognitive function, affective behavior and nociception (for review, see (8–10)). Consequently, histamine and its receptors have been implicated in the pathophysiology of disorders of the central and peripheral nervous systems, including sleep and eating disorders, mood disorders, neuroinflammation and dementia, pruritus and pain (for review, see (9)).

Determination of histamine in biological samples is traditionally performed by the use of commercially available radioenzymatic assays based on methylation of histamine via methyltransferase and radioactive [^{3}H] S-adenosyl-L-methionine (15). More recent methods are based on enzyme-link immunosorbent assays (ELISA) and utilize nonradioactive labels for spectrophotometric or fluorescence detection of the labeled antibody. Typically the labeled antibody reacts in a competitive way with free histamine in the sample and a fixed amount of histamine molecules covalently bound to the surface of the reaction well of the microtiter plate. However, these assays usually require relatively large sample volumes (>200 μl), the samples should be run in duplicates, and the limit of detection is around 1 nM at best. Therefore, for analysis of trace levels of histamine in limited sample volumes such as microdialysis samples or cell lysates it is more convenient to use high-performance liquid chromatography (HPLC) with fluorescence detection following precolumn or postcolumn derivatization of histamine with a suitable fluorescence yielding reagent. Histamine itself is a weak fluorophore but it contains a reactive primary amine and a secondary amine in the imidazole ring, which both can be used for derivatization reactions. Probably the most common and wide-spread fluorescence derivatization reagent is *o*-phthalaldehyde (OPA). OPA has no intrinsic fluorescence but in the presence of weak nucleophiles such as thiols (e.g., 2-mercaptoethanol) and at mild alkaline conditions, OPA reacts with primary amines and amino acids to yield highly fluorescent substituted isoindoles ((16), for review, see (17)) with maximal excitation (λ_{ex}) and emission (λ_{em}) wavelengths at ≈340 and ≈450 nm, respectively. However, histamine, histidine, arginine, and some polyamines react with OPA in alkaline media to form fluorescence products even in the absence of the thiol nucleophile. This feature has been utilized to achieve higher specificity for the detection of histamine in biological samples and to facilitate chromatographic separation of OPA-histamine derivative, as will be discussed below.

Another strategy to increase the specificity and sensitivity of histamine determination utilizes the conjugation of two fluorescence tags to the amine moieties of the histamine molecule. This will yield the intramolecular excimer fluorescence that would cause a shift to the higher emission wavelengths as compared to the wavelength of the fluorescent label itself as initially described for derivatization of polyamines and amino acids ornithine and lysine (18, 19) and later for histamine (20, 21). OPA and intramolecular excimer fluorescence formation are the two principal derivatization methods used for HPLC determination of histamine in the microdialysis samples as will be discussed below in more detail.

2. Materials

2.1. Determination of Histamine by HPLC with Fluorescence Detection

2.1.1. Derivatization of Histamine with OPA

Chemicals

Histamine dihydrochloride, sodium dihydrogen phosphate, potassium carbonate, sodium 1-octanesulfonate, *o*-phthalaldehyde and methanol were of highest purity grade purchased from Sigma-Aldrich (St. Louis, MO, USA). Deionizer water (more than 18 MΩ cm of specific resistance value) was obtained from a MilliQ water purification system (Millipore, MA, USA).

Instruments

The chromatographic system for postcolumn derivatization of histamine by OPA/carbonate reagent is schematically depicted in Fig. 1. The HPLC system included three isocratic HPLC pumps EP-700 each with an inbuilt degasser unit (Eicom, Kyoto, Japan), a temperature oven ATC-700 (Eicom), a CMA/200 Refrigerated Microsampler (CMA/Microdialysis, Stockholm, Sweden), a fluorescence detector L-7480 (Merck/Hitachi, Ibaraki, Japan), and a computerized data acquisition system CSW (Data Apex, Prague, The Czech Republic). The fluorescence detector was equipped with a 12-μl flow cell and operated at an excitation wavelength of 340 nm and an emission wavelength of 450 nm. Alternative HPLC pumps could be a LC-10AD, (Shimadzu, Kyoto, Japan) or a micro-LC pump (LC-100; ALS, Tokyo, Japan) and a LC-27A Degasser (ALS, Tokyo, Japan). All connecting tubing, fittings and reaction coils were purchased from VICI Jour (Schenkon, Switzerland). A HPLC column EICOMPAK SC-5ODS (3.0 mm, i.d. × 150 mm) and a precolumn PREPAKSET CA-ODS (3.0 mm, i.d. × 4 mm) were purchased from Eicom (Kyoto Japan).

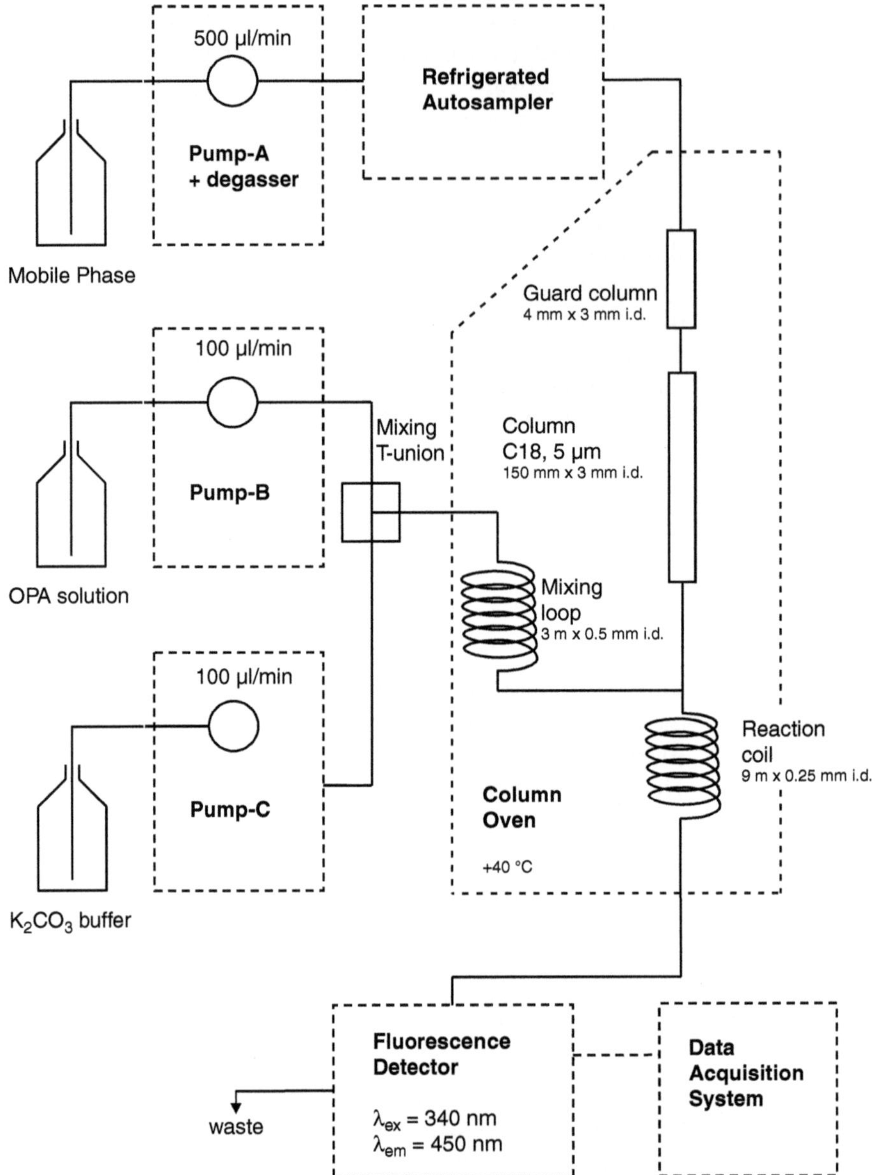

Fig. 1. A scheme of the automated liquid chromatographic system for the postcolumn derivatization of histamine with OPA and fluorescence detection.

The mobile phase (Pump A) was a mixture of 0.1 M NaH_2PO_4 buffer and methanol (9:1 v/v) and contained sodium 1-octanesulfonate at a final concentration of 0.786 mM, the flow-rate was 500 µl/min. The OPA solution (Pump B) was prepared by dissolving in methanol and water to a final concentration of 0.596 mM OPA, the flow-rate was 100 µl/min. Potassium carbonate solution (0.5 M) was pumped by the Pump C at a flow rate of 100 µl/min.

2.1.2. Derivatization of Histamine with PSE or PBC

Chemicals

Histamine dihydrochloride, potassium carbonate and acetonitrile (CHROMASOLV® Plus) were purchased from Sigma-Aldrich (St. Louis, MO, USA). 4-(1-pyrene)butyric acid *N*-hydroxysuccinimide ester (PSE) was purchased from Molecular Probes (Eugene, OR, USA), 4-(1-pyrene) butanoyl chloride (PBC) was purchased from Toronto Research Chemicals, Inc. (North York, Ontario, Canada). The reagents were used without further purification. Deionized water (more than 18 MΩ cm of specific resistance value) was obtained from a MilliQ water purification system (Millipore, MA, USA).

Instruments

The HPLC system included a LC-10AD pump (Shimadzu, Kyoto, Japan), a CMA/200 Refrigerated Microsampler (CMA/ Microdialysis, Stockholm, Sweden), a LC-27A Degasser and a column oven (ALS, Tokyo, Japan), a fluorescence detector L-7480 (Merck/Hitachi, Ibaraki, Japan) and a computerized data acquisition system CSW (Data Apex, Prague, The Czech Republic). The fluorescence detector was equipped with a 12-μl flow cell and operated at an excitation wavelength of 345 nm and an emission wavelength of 500 nm. All connecting tubing, fittings and reaction coils were purchased from VICI Jour (Schenkon, Switzerland). The microbore column (L-column, 150 mm × 1.0 mm i.d., C18 silica, particle size, 5 μm) was purchased from Chemicals Evaluation and Research Institute (Tokyo, Japan) and was kept at temperature of 35°C. The mobile phase was a mixture containing 75% acetonitrile in water and the flow-rate was 50 μl/min.

3. Methods

3.1. Determination of Histamine by HPLC with Fluorescence Detection

3.1.1. Derivatization of Histamine with OPA

Derivatization Reaction

In alkaline media ($pH \approx 12.5$), histamine reacts with OPA forming flourophores through several intermediate steps as revealed in detail by Rönnberg and colleagues (22). These authors proposed that the main and final fluorophore product of the histamine–OPA reaction following its termination by acidification to $pH \approx 2.5$ is the substituted dihydrophenanthroline derivative (F_{acid}) as shown in Fig. 2. The reaction scheme and the kinetics of fluorescence reaction were further elucidated by Yoshimura and collaborators (23). These authors, in agreement with (22) proposed that in the alkaline medium, the primary amine group of histamine first reacts with one of the aldehyde groups of OPA to form a Schiff base that turns to a tetrahydroimidazo[4,5-c]pyridine derivative (Fig. 2). In the presence of excess histamine over OPA, the residual aldehyde group may further react with the amino group of tetrahydropyridine to form a highly fluorescent but unstable isoindole, marked as F_{base} in the reaction scheme. Subsequently, F_{base} degrades to the

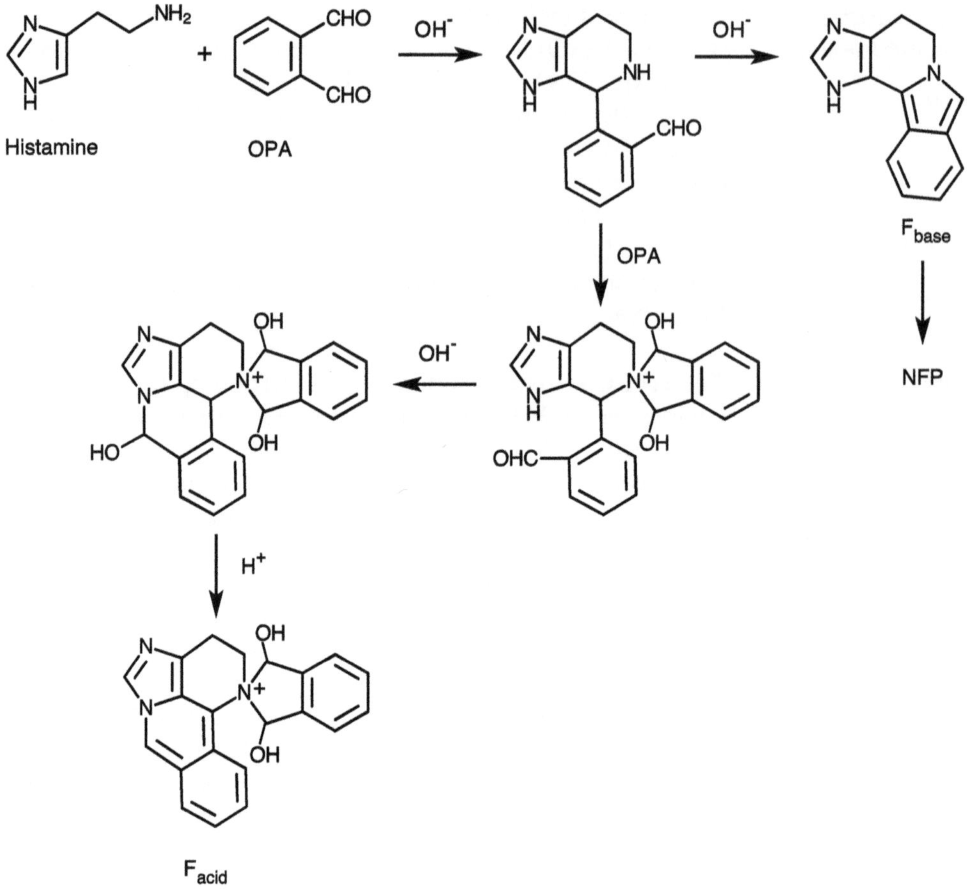

Fig. 2. Reaction scheme for derivatization of histamine with OPA in the alkaline medium (modified from (23)).

nonfluorescence products (NFP) as indicated in Fig. 2. The speed of this degradation when measured as the decrease in the fluorescence intensity, accelerates by increasing the concentration of OPA and prolonging the reaction time (23). On the other hand, in the excess of OPA, the amino group of the tetrahydropyridine ring could react with another OPA to form intermediate products that following acidification yield the final and stable fluorescence dehydrophenanthroline derivative (F_{acid}), as already described by Rönnberg and collaborators (22). Although several steps in the reaction between histamine and OPA are still not fully clarified and have been questioned (17), these findings have important implications when optimizing the OPA derivatization protocol for histamine determination by HPLC.

In general, precolumn derivatization of histamine with OPA followed by separation of the fluorescent derivative on the reversed-phase column offers low sensitivity of about 3 pmol of histamine (24). A possible explanation of this fact could imply a long retention time (>10 min) of the histamine fluorescent derivative causing its partial degradation to the nonfluorescent products before entering the fluorescence detector. A substantial improvement of sensitivity of the precolumn derivatization method of histamine could be achieved by acidification of the final OPA–histamine derivative (25), in line with the theoretical studies. The authors reported the limit of sensitivity at 0.2 pmol histamine in 1 ml standard or plasma samples before a rather complicated extraction procedure comprising several steps (25). The postcolumn derivatization procedure requires several extra pumps, reaction coils and is associated with larger consumption of reagents and organic solvents. On the other hand, the method offers higher sensitivity for determination of histamine in small sample volumes and offers easier optimization of chromatographic separation of nonderivatized histamine from other monoamines, polyamines, amino acids and other possible interferences such as drugs, which are histamine analogs, e.g., (*R*)α-methylhistamine (26). The high-sensitive postcolumn derivatization method originally described for determination of histamine in plasma and brain tissue samples (27) offers a more suitable approach for analysis of histamine in the brain microdialysates (26, 28, 29). Histamine is typically separated in the ion-paired reversed-phase mode, the effluent at the column outlet is mixed with the alkaline OPA reagent, and following the reaction in the mixing coil, optionally acidified with phosphoric acid before entering the fluorescence detector. Several investigators have demonstrated a higher fluorescence yield of the histamine–OPA derivative at a low pH as compared to that of the alkaline solution (27, 30), which is in good agreement with the theoretical studies on OPA–histamine reaction mechanisms discussed above. The detection limits depend on the chromatographic conditions and the sensitivity of the fluorescence detector used, and were reported to range between 250 fmol (30), 50 fmol (27) and 4.5 fmol (26). Although it was reported that postcolumn derivatization of histamine could be performed by the use of only one reagent containing both sodium hydroxide and lower concentration (0.002%) of OPA (31), there is a general agreement that OPA in the alkaline solution is unstable. Therefore, many investigators (26, 28, 29) have suggested to keep the methanolic OPA and alkaline solutions separated and mix them together just before entering the reaction coil with the chromatographic effluent. The present method builds on this principle with a slight modification by replacing the aggressive sodium hydroxide with potassium carbonate solution.

Procedure

1. Preparation of the mobile phase: Weigh accurately 14.04 g of sodium dihydrogen phosphate dihydrate, and dissolve it in deionized water to make exactly 900 ml. Add 100 ml of

methanol to this solution, and finally add 170 mg of sodium 1-octanesulphonate to the mixture (see Note 1).

2. Preparation of the OPA solution: Weigh 16 mg of *o*-phthalaldehyde, and dissolve it in 5 ml of methanol. Add deionized water to this solution to make exactly 200 ml. Store the solution in an amber bottle in the refrigerator (see Note 5).
3. Preparation of the potassium carbonate solution: Weigh 13.82 g of potassium carbonate, and dissolve it in deionized water while stirring and dilute the solution to the final volume of 200 ml with deionized water.
4. Standard solution of histamine: Weigh 1.11 mg of histamine dihydrochloride, dissolve it in 900 μl deionized water and pipette 100 μl of 0.1 M HCl. This will give a standard stock solution of 10 mM histamine in 10 mM HCl. The aliquots of the stock solution are kept frozen (–20°C) in amber-colored test tubes. The working solutions for calibrations of the HPLC system are prepared daily by diluting the stock solution in the microdialysis perfusion medium, e.g., the Ringer solution or the artificial CSF (see Note 8).
5. Preparation and maintenance of the HPLC system: Open the drain valve of the each pump and wash the pump-head assemblies of each pump with at least 5 ml of deionized water using a 10 ml plastic syringe and the connecting tubing (see Notes 2 and 3). Thereafter, fill the pump-heads with at least 5 ml of each respective solvent: the mobile phase for Pump A, the OPA solution for Pump B and the carbonate solution for Pump C (see Note 6). Equilibrate the column with the mobile phase for at least 1 h at the flow rate of 500 μl/min (see Note 4). Pump the reagent solvents through the mixing and reaction coils at least for 1 h at the flow rate of 100 μl/min for the each solution. Set the temperature of the column oven to 40°C, and wait until the display of the temperature shows 40°C constantly. Turn on the power of the fluorescence detector, and check or preset the excitation and emission wavelengths. Wait until the baseline is completely stabilized with no drift before starting the calibration of the HPLC system (see Notes 5 and 9).

Typical Chromatograms and Applications

Typical chromatograms of histamine standard and histamine in the microdialysis samples and the homogenates from the rat hypothalamus and prefrontal cortex derivatized according to the above-described protocol are shown in Fig. 3a–d.

The applicability of the HPLC derivatization method for sensitive determination of histamine in the brain microdialysates was demonstrated in a number of neuropharmacological and neurophysiological studies. A summary of several of the most typical applications is shown in Table 1.

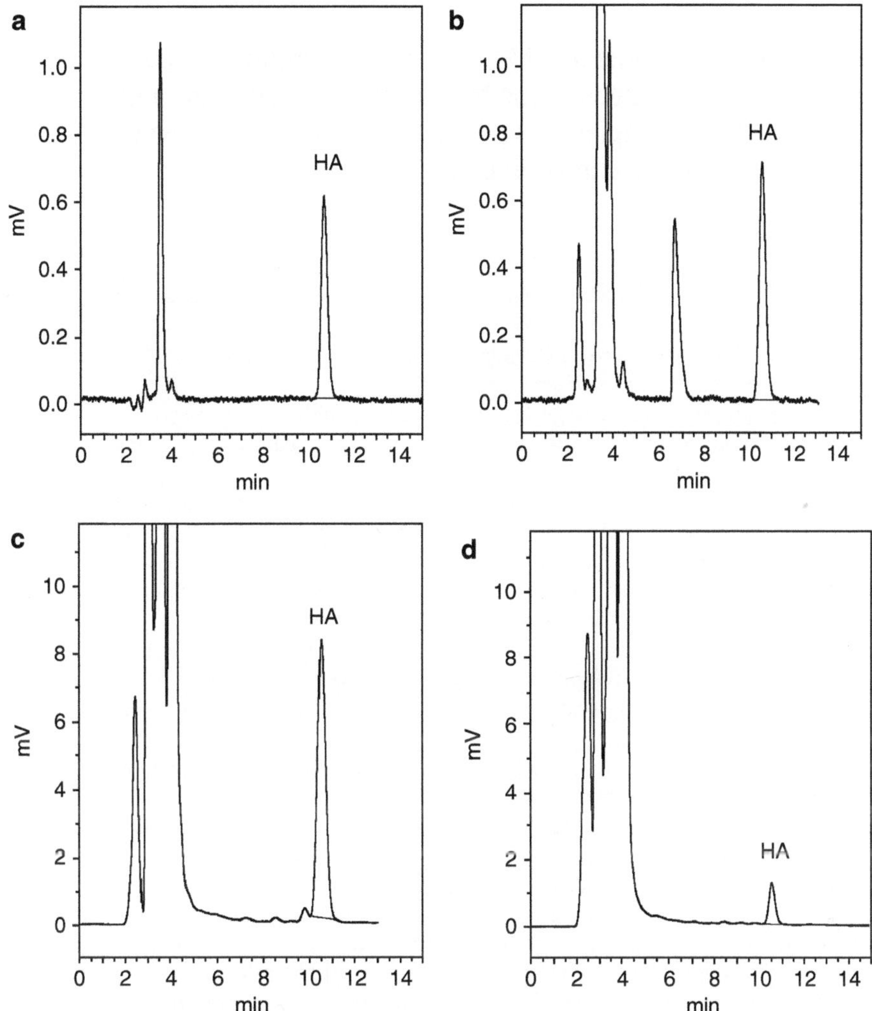

Fig. 3. Representative chromatograms of (**a**) a histamine standard solution at concentration of 1×10^{-9} mol/l; 10 μl corresponding to 10 fmol/injection; (**b**) a microdialysis sample from the hippocampus of the awake rat at basal conditions. The microdialysis samples were collected in 10-min intervals at a flow-rate of 2 μl/min; the injected volume was 20 μl; (**c**) a chromatogram of a sample homogenate from rat hypothalamus (10 μl injection); (**d**) a chromatogram of a sample homogenate from rat prefrontal cortex (10 μl injection). Peaks marked HA correspond to the histamine–OPA derivative.

3.1.2. Derivatization of Histamine and Tele-Methylhistamine with OPA/MCE

An alternative method for histamine derivatization is based on the reaction with OPA in the presence of a nucleophile such as the aliphatic thiol, 2-mercaptoethanol (MCE). This method was originally developed and is routinely used for fluorescence derivatization and detection of primary amino acids (16) in biological samples including brain microdialysates (for review see (64, 65)). Typically, the reagent is a mixture of a methanolic solution of OPA and MCE in the borate buffer (pH ≈ 10), the reaction takes place at room temperature or even at +4°C within 30–60 s, which makes the method especially suitable for automated precolumn derivatization and separation of the fluorescent isoindole derivatives on a

Table 1
A representative list of some most typical neuropharmacological and neurophysiological applications for determination of histamine in brain microdialysis samples using HPLC with the postcolumn OPA derivatization and fluorescence detection

Year	Microdialysis study	Reference
Characterization of neuronal histamine (HA), histamine receptors. Pharmacological characterization		
1991	Effects of, high K^+, electrical stimulation, removal of Ca^{2+} ions on HA levels in hypothalamus of anesthetized rat	(28)
1991	Effects of, high K^+, removal of Ca^{2+} ions, thioperamide, metoprine, alpha-fluoromethylhistidine on HA levels in hypothalamus of anesthetized rat	(29)
1992	Effects of alpha-fluoromethylhistidine, (*R*)-alpha-methylhistamine, metoprine	(26)
1993	Pentobarbital, muscimol, diazepam decrease, reserpine increased striatal HA	(32)
1993	Vestibular stimulation increased HA release in rat hypothalamus	(33)
1994	Morphine increased HA release in rat striatum, blocked by naltrexone	(34)
1995	Alpha 2-adrenoceptor antagonist atipamezole decreased HA release	(35)
1995	5-HT2C/2A antagonist methysergide decreased suprachiasmatic HA release	(36)
1997	Methamphetamine increased HA levels in rat hypothalamus	(37)
1998	H3 agonist immepip decreased and antagonist clobenpropit or thioperamide increased HA release	(38)
2000	Effects of TTX, diltiazem, high K^+, veratridine, ouabain on basal release of histamine in rat cortex	(39)
2002	H3 receptor agonist alpha-methylhistamine attenuated, and H3 antagonist thioperamide potentiated handling stress-induced HA release	(31)
2003	H3 agonist immepip reduced cortical HA release	(40)
2005	H3 receptor agonist methimepip reduced basal HA release in rat brain	(41)
2006	Cannabinoid CB-1 receptor agonist increased HA	(42)
2007	Novel histamine H3 receptor antagonists, increased HA release	(43)
2010	H3 receptor antagonist GSK189254 increased HA release in rat in the tuberomammillary nucleus	(44)
Food intake		
1991	Feeding increased histamine levels in rat hypothalamus	(45)
2008	Effects of H(3)-inverse agonist clobenpropit, leptin on HA in high fat diet-induced obesity mice	(46)
2010	Hard pellets increased, whereas soft pellets did not affect HA release in rat amygdala	(47)
2011	5-HT2 receptor antagonists and antipsychotics risperidone or aripiprazole increased hypothalamic HA release in mice and decreased food intake	(48)
Sleep, wakefulness, circadian rhythm		
1992	HA release was significantly higher during the dark than in the light period	(49)
2001	Orexin A infusion increased wakefulness and HA release	(50)
2003	Modafinil increased HA release in rat hypothalamus	(51)
2003	Prostaglandin E2 promoted wakefulness, increased HA release	(52)
2004	HA in frontal cortex increased in wake periods	(53)
2005	Adenosine A(2A) R agonist CGS21680 promoted sleep, inhibited HA release	(54)
2008	Modafinil increased HA release and locomotor activity	(55)
2008	H3 antagonist ciproxifan increased HA release in H1 receptor KO mice	(56)

(continued)

Table 1 (continued)

Year	Microdialysis study	Reference
Psychiatric disorders: schizophrenia, ADHD		
2007	Methylphenidate and atomoxetine increased HA release in rat PFC	(57)
2008	Atomoxetine increased HA release in SHR model of ADHD	(58)
2010	Ketamine increased HA levels in mPFC, attenuated by mGluR2/3 agonist LY379268	(59)
Neurological disorders: Ischemia, epilepsy		
1992	Histamine levels increased after middle cerebral artery occlusion	(60)
2005	Thioperamide or metoprine reduced infarction size and increased HA release	(61)
2007	Antiepileptic effect of deep brain stimulation in TMN increased HA release in frontal cortex	(62)
Cutaneous microdialysis		
1998	Effects of probe implantation on HA release	(63)
2011	Effects of ovalbumin and histamine liberator compound 48/80 in guinea pig	(73)

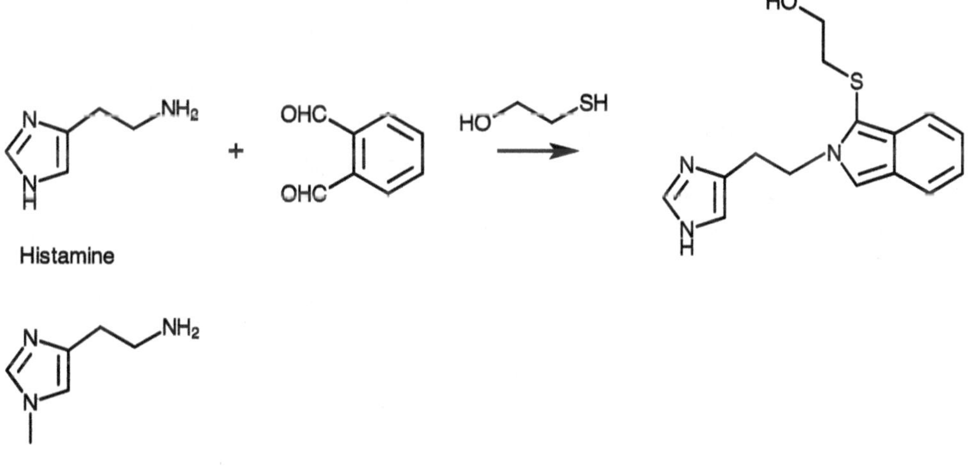

Fig. 4. Derivatization of histamine and *tele*-methylhistamine with OPA/MCE reagent in mild alkaline conditions.

reversed-phase column. The advantage of this derivatization method is that it also enables determination of the major histamine metabolite *tele*-methylhistamine (N^{τ}-methylhistamine; 1-methylhistamine) as shown in the reaction scheme (Fig. 4). The method was developed for determination of histamine and *tele*-methylhistamine in urine (66, 67) and rat brain (66, 68) and for determination of *tele*-methylhistamine for estimation of histamine turnover in rat hypothalamus and amygdala (69). However, the sensitivity of the precolumn derivatization with OPA/MCE reagent is rather law, about 5 pmol for histamine (66) and 0.05 pmol for *tele*-methylhistamine (69). Further modification of the method was described by Saito and collaborators (70) who used on-column derivatization of histamine by adding OPA and *N*-acetyl-L-cysteine acting as a nucleophile in the mobile phase. The detection limit of histamine was about 4.5 pmol. Taken together, the sensitivity of the OPA/thiol-based derivatization protocols seems insufficient enough for the fluorescence determination of histamine in brain microdialysates. Likewise, the detection of histamine–OPA/MCE derivative by HPLC with electrochemical detection offers only a slightly improved detection limit of histamine to 0.45 pmol (71) and 0.1 pmol (72) as discussed in Sect. 3.2.

3.1.3. Derivatization of Histamine with PSE or PBC

Derivatization Reaction

Derivatization of histamine by 4-(1-pyrene)butyric acid *N*-hydroxysuccinimide ester (PSE) (20, 21) or by 4-(1-pyrene) butanoyl chloride (PBC) (73) yields the corresponding dipyrene-labeled histamine derivative which possesses intramolecular excimer fluorescence. Initially, the derivatization chemistry to form the excimer fluorescence products was developed for the determination of polyamines with PSE (18) and with PBC (74). The derivatization protocols were further optimized for determination of histamine in urine (20) and in brain microdialysis samples (21) using the PSE reagent, and recently, the PBC reagent (73). The chemical reaction is very similar for both reagents: one reagent molecule will via its *N*-hydroxysuccinimide (PSE) group or butanoyl chloride (PBC) group conjugate the highly fluorescent pyrene label to the primary amino group of histamine, whereas the second reagent molecule will attack the secondary amine of the imidazole ring of histamine as depicted in the reaction scheme for PSE or PBC derivatization of histamine shown in Fig. 5. The resulting intramolecular excimer causes a shift in the emitted light

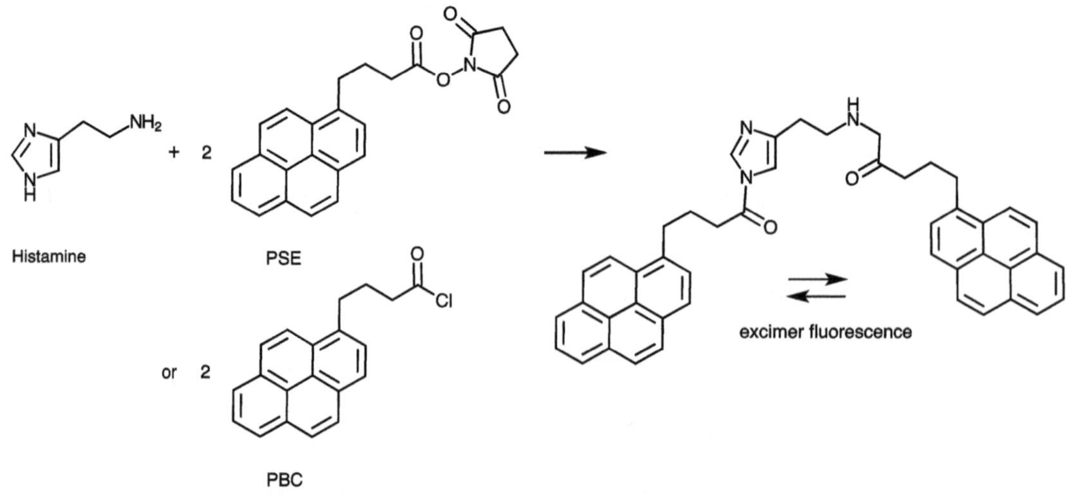

Histamine + 2 PSE or 2 PBC → excimer fluorescence

Fig. 5. Derivatization reaction of histamine with PSE or PBC. Conjugation of two molecules of PSE or two molecules of PBS to histamine yields the dipyridine histamine derivative possessing intramolecular excimer fluorescence (shift to the higher emission wavelength).

towards higher wavelengths ($\lambda_{em} \approx 450$–540 nm) as compared to the pyrene reagent alone or the histamine-pyrene monomer that possess maximum emission at lower wavelengths (≈385 nm), (20). This strategy increases the specificity and sensitivity of histamine determination to a detection limit of 0.3 fmol in 10 μl samples (20 μl derivatized solution injected on-column). Both PBC and PSE are commercially available, however, the PBC compound seems to be offered in a higher purity, thereby significantly reducing the number and the peak heights of fluorescent interferences that appear already in the derivatized blank samples.

Procedure

1. Preparation of the mobile phase: Mix 750 ml of acetonitrile with 250 ml of deionized water.
   Keep the mobile phase in an amber colored bottle (see Note 1).
2. Preparation of the PSE reagent: Weigh 0.92 mg of PSE and dissolve it in 1 ml of acetonitrile, this will provide a 3.0 mM PSE solution.
3. Preparation of 3 mM potassium carbonate solution: Weigh 41.46 mg of potassium carbonate, and dissolve it in 90 ml deionized water while stirring and dilute the solution to the final volume of 100 ml with deionized water. Store the solution in an amber bottle in the refrigerator.
4. Derivatization reagent: Mix 3.0 mM PSE, 3.0 mM potassium carbonate and acetonitrile at a 1:1:18 volume ratio. The reagent solution is stable for at least 1 week when stored at −20°C (see Note 7).
5. Preparation of histamine standard: Weigh 1.11 mg of histamine dihydrochloride, dissolve it in 900 μl deionized water and pipette 100 μl of 0.1 M HCl. This will give a standard stock solution of 10 mM histamine in 10 mM HCl. The aliquots of the stock solution are kept frozen (−20°C) in amber-colored test tubes. The working solutions for calibrations of the HPLC system are prepared daily by diluting the stock solution in the microdialysis perfusion medium, e.g., the Ringer solution or the artificial CSF (see Note 8).
6. Derivatization procedure: To a 10 μl portion of a microdialysis sample or histamine standard solution (corresponding to a 10-min microdialysis sample) pipette 20 μl of the derivatization reagent. The vials are tightly sealed and heated at 100°C for 90 min in an oven or a heated block (e.g., Reacti-Therm Model 18970, Fisher Scientific, Pittsburgh, PA, USA). After cooling in iced water, a 20 μl portion of the reaction mixture is injected onto the chromatograph. For the reagent blank, Ringer solution or artificial cerebrospinal fluid (aCSF) is subjected to the same procedure.

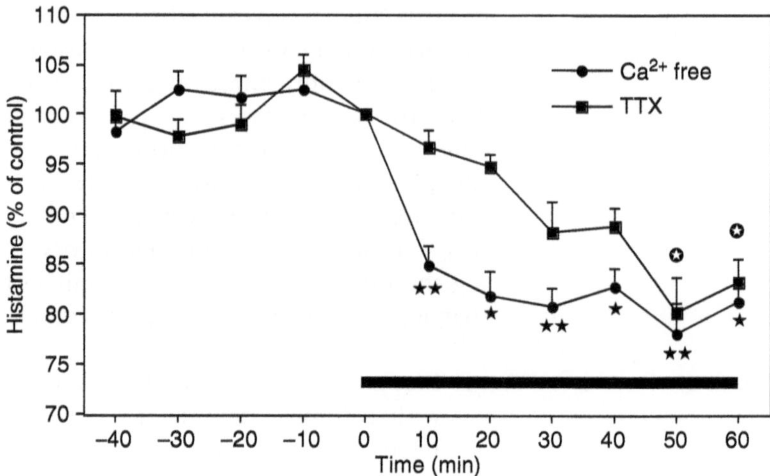

Fig. 6. Effects of the removal of calcium ions from Ringer solution (Ca^{2}-free) and the addition of the sodium channel blocker TTX (1 μM) to the Ringer solution on basal histamine levels in the hypothalamus and striatum, respectively. The *bar* indicates the perfusion of the modified Ringer solution during 60 min. The values are expressed as mean ± S.E.M., $n = 5$ rats, (★★) $P < 0.01$, (✪) $P < 0.05$, paired *t*-test. Reproduced from (21) with permission from Elsevier, The Netherlands.

7. Preparation and maintenance of the HPLC system: Open the drain valve of the HPLC pump and wash the pump-head assemblies of the pump with at least 5 ml each of deionized water, acetonitrile and the mobile phase using a 10 ml plastic syringe and the connecting tubing. Equilibrate the new column with the mobile phase for at least 4–5 h at the flow rate of 50 μl/min. Set the temperature of the column oven to 35°C. Turn on the power of the fluorescence detector, and check or preset the excitation and emission wavelengths. Wait until the baseline is completely stabilized with no drift before starting the calibration of the HPLC system (see Note 9).

Typical Chromatograms, Applications Examples

The PSE derivatization technique optimized for microbore HPLC allows determination of extracellular histamine levels in the microdialysates from different brain regions of awake rats and mice. The method was applied for characterization of neuronally released histamine following perfusions with high potassium, tetrodotoxin (TTX)-containing or calcium-free Ringer solution (Fig. 6) and following forced swim stress (Fig. 7). Typical chromatograms of histamine standard and histamine in the microdialysis samples from different regions of the rat brain are shown in Fig. 8a–e.

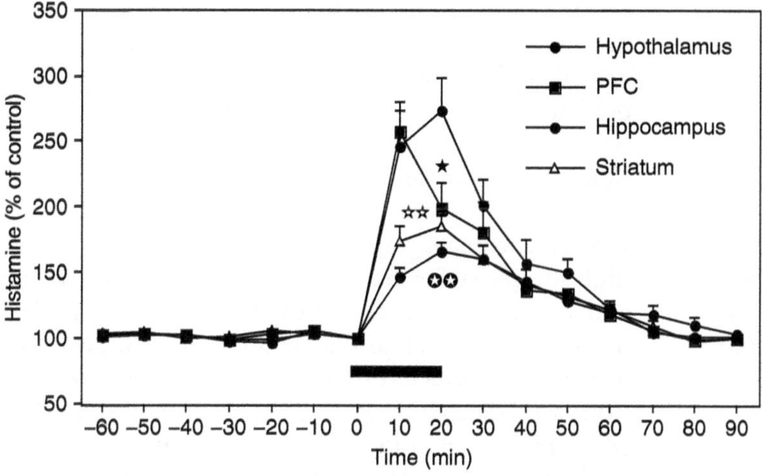

Fig. 7. Effects of forced swimming (20 min) on the release of histamine in the hypothalamus, prefrontal cortex, hippocampus and striatum of the rat. The values are expressed as mean ± S.E.M., $n = 5$ rats, (✪✪), (✪✪) $P < 0.01$, (☆) $P < 0.05$ vs. hypothalamus group, ANOVA, Fisher's PLSD test. Reproduced from (21) with permission from Elsevier, The Netherlands

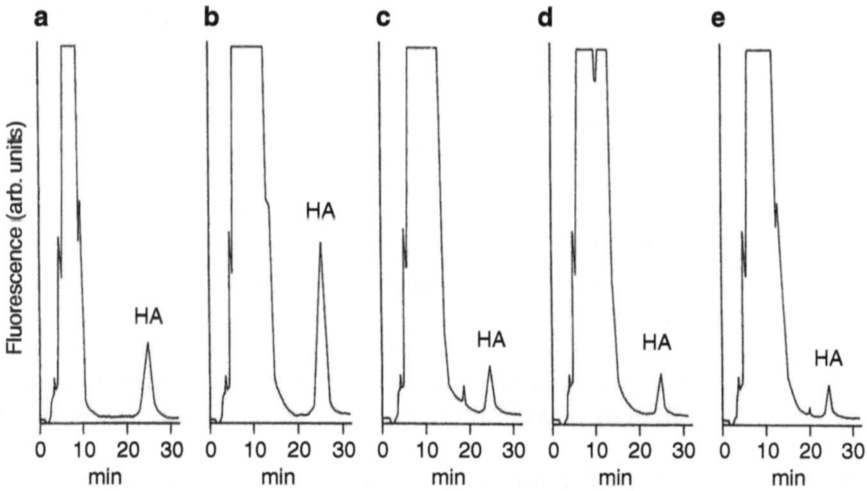

Fig. 8. Chromatograms of a standard solution containing 15 fmol/10 μl histamine (**a**) and microdialysis samples collected at basal conditions from hypothalamus (**b**), prefrontal cortex (**c**), hippocampus (**d**) and striatum (**e**). The basal levels of histamine in the dialysates from hypothalamus, prefrontal cortex, hippocampus, and striatum were 35.45 ± 4.56, 9.05 ± 1.56, 7.83 ± 0.86, and 6.54 ± 0.66 fmol in 10 μl samples, respectively. Peaks marked HA correspond to the histamine-PSE derivative. Reproduced from (21) with permission from Elsevier, The Netherlands.

3.2. Determination of Histamine by Other HPLC Methods

3.2.1. Determination of Histamine by HPLC with Electrochemical Detection

It is well recognized that the OPA/MCE-derivatives of primary amines and amino acids are also electrochemically active and can be detected by HPLC with an electrochemical detector operating in an oxidative mode (75). A similar protocol was described for derivatization of histamine (71, 72), whereas Mine and collaborators mode (76) developed a precolumn derivatization method using the water-soluble Bolton–Hunter reagent that allowed detection of histamine in rat brain nuclei down to 0.1 pmol of histamine and 0.2 pmol of *tele*-methylhistamine. Further improvement of the detection of histamine down to 30 fmol in plasma samples was achieved by the use of HPLC coupled with a pulsed amperometric detector (77). However, any of these methods enables determination of histamine in the microdialysis samples, most likely because of low sensitivity and specificity.

3.2.2. Determination of Histamine with Liquid Chromatography/Mass Spectrometry (LC/MS)

Liquid chromatography coupled to mass spectrometry offers a high specificity for the detection of histamine, *tele*-methylhistamine and related biogenic amines in biological samples (78–81) and food products such as cheese and wine (82, 83). Thus, histamine in wine was detected by LC/MS following precolumn derivatization with dansyl chloride (83), 1,2-naphthoquinone-4-sulfonate (82). Histamine and *tele*-methylhistamine in CSF of narcoleptic subjects were determined following derivatization with 4-bromobenzenesulfonyl chloride (79). The latter assay was linear in the concentration range of 0.05–5 nM for each amine. Similar sensitivity was achieved for determination of histamine and *tele*-methylhistamine in the microdialysis samples from rat prefrontal cortex by the use of hydrophilic interaction liquid chromatography-tandem mass spectrometry following derivatization with propionic anhydride (78). Here, the limits of quantification were 0.75 nM (7.5 fmol/10 μl sample) for histamine and 0.67 nM for *tele*-methylhistamine. In addition, histamine could be detected without derivatization by the use of liquid chromatography-electrospray ionization tandem mass spectrometry (LC-ESI-MS/MS) as described for determination of histamine in the human basophilic cells (80) and determination of histamine, *tele*-methylhistamine and *tele*-methylimidazolacetic acid in rat cerebrospinal fluid (81). The sensitivity of the latter method was 0.5 nM for histamine and its metabolites, which would allow using this assay also for the analysis of typical microdialysis samples from the rat brain.

3.3. Conclusions

Determination of histamine in the microdialysis samples from the rodent brain requires to apply a technique which offer high sensitivity, at least to 1 nM for the detection of histamine in 10–20 μl sample volumes. Liquid chromatography techniques developed for the analysis of histamine in brain samples offer many advantages over the conventional immunoenzymatic assays or gas chromatographic methods. A simplest, most convenient and sufficiently sensitive HPLC method for the determination of histamine in brain microdialysis samples is based on postcolumn derivatization of histamine with OPA and following fluorescence detection.

At optimized conditions as described in Sect. 3.1.1 and illustrated in Fig. 3, histamine can be detected down to 2 fmol/10 μl, which is sufficient for determination of histamine concentrations in typical microdialysis samples from the rodent brain, where the basal levels range between 5 and 40 fmol/10 μl, depending on brain structure and the microdialysis protocol used. The LC/MS/MS methods offer similar sensitivity (≈5 fmol/10 μl) for the detection of histamine and in addition, a possibility to simultaneously measure both histamine and its major metabolites, *tele*-methylhistamine and *tele*-methylimidazolacetic acid. Finally, the precolumn derivatization protocol utilizing the formation of intramolecular excimer fluorescence with PSE (Sect. 3.1.3) or PBC reagents is the method of choice for applications where even higher sensitivity, down to the sub-fmol range (30 pM or 0.3 fmol/10 μl) is required. Typical examples of such applications are the microdialysis samples from nuclei of the mouse brain or skin microdialysates in models of inflammation and neuropathic pain.

4. Notes

1. It is rather easy to obtain false data, just because of using glassware, chemicals and other accessories contaminated with biological material. Always use the highest purity water for preparation of HPLC buffers.
2. Use new tubing, frits, ferrules and fittings when setting up the HPLC system for the postcolumn OPA derivatization method.
3. Wash the pump-head assemblies of the each pump with methanol followed by deionized water before installing the mobile phase and reagent solutions.
4. Longer equilibration time is needed when using a new column. Pump the mobile phase at least for 4–5 h at the flow rate of 500 μl/min.
5. The OPA reagent ages rather fast, therefore it is important to calibrate the system regularly and to prepare the fresh reagent every day.
6. A special care must be taken when handling the alkaline solution and waste. Control regularly the performance of the pump used for delivery of the carbonate buffer solution, check there is no leakage of the solution from the pump head and the connecting tubing. Use gloves and safety glasses.
7. The major difference between the protocols applying the postcolumn OPA derivatization and the precolumn derivatization with PSE for high-sensitive analysis of histamine is associated with the presence of contaminants. Both PSE and PBC reagents are commercially available but their purity may vary between the suppliers and from batch to batch. It should be noted that even at 99.9% purity, these reagents contain trace levels of fluorescence impurities, which give a number of peaks in the chromatogram and increase the risk of interfering with the elu-

tion peak of histamine derivative, particularly at the high amplification rate of the detector.

8. Use sterile Ringer or aCSF solutions for the microdialysis experiments.
9. Always run a series of blank samples (water, Ringer, aCSF or even the mobile phase itself) to be sure that there is no interfering peak eluting at the position of the histamine derivative.

Appendix

List of the Main Suppliers and Their Respective Webpage Addresses

Supplier	Product	Webpage addresses
ALS Co.	HPLC pump	www.als-japan.com
CMA/Microdialysis	Microinfusion pump	www.microdialysis.se
DataApex	Data acquisition system	www.dataapex.com
Eicom Corp.	Microdialysis system, HPLC	www.eicom.co.jp
Hitachi	HPLC, detectors	www.hitachi-hitec.com
Microbiotech/se	Microdialysis accessories	www.microbiotech.se
Molecular Probes	PSE reagent	www.invitrogen.com
Shimadzu	HPLC pumps	www.shimadzu.com
Sigma-Aldrich	OPA reagent, chemicals	www.sigmaaldrich.com
Toronto research Chemicals	PBC reagent	www.trc-canada.com
VICI Jour	HPLC accessories	www.vici.com

References

1. Dale HH, Laidlaw PP (1910) The physiological action of imidazolethyl-amine. J Physiol 41:318–344
2. Khardori N, Shahid M, Ali Khan R, Tripathi T (2010) Biomedical aspects of histamine: current perspectives. Springer, Heidelberg
3. Thurmond RL (2010) Histamine in inflammation (advances in experimental medicine and biology). Springer, Heidelberg
4. Falus A, Grosman N, Darvas Z (2004) Histamine: biology and medical aspects. S Karger Pub, Basel
5. Kwiatkowski H (1943) Histamine in nervous system. J Physiol 102:32–41
6. Watanabe T, Wada H (1991) Histaminergic neurons: morphology and function. CRC, Boca Raton
7. Wada H, Inagaki N, Yamatodani A, Watanabe T (1991) Is the histaminergic neuron system a regulatory center for whole-brain activity? Trends Neurosci 14:415–418
8. Haas H, Panula P (2003) The role of histamine and the tuberomamillary nucleus in the nervous system. Nat Rev Neurosci 4: 121–130
9. Haas HL, Sergeeva OA, Selbach O (2008) Histamine in the nervous system. Physiol Rev 88:1183–1241

10. Alvarez EO (2009) The role of histamine on cognition. Behav Brain Res 199:183–189
11. Oishi R, Nishibori M, Saeki K (1983) The simultaneous assay of histamine and tele-methylhistamine in the rat brain after microwave irradiation and immersion in liquid nitrogen: contamination by extraencephalic mast-cell histamine. Brain Res 261:180–183
12. Watanabe T, Taguchi Y, Shiosaka S, Tanaka J, Kubota H, Terano Y, Tohyama M, Wada H (1984) Distribution of the histaminergic neuron system in the central nervous system of rats; a fluorescent immunohistochemical analysis with histidine decarboxylase as a marker. Brain Res 295:13–25
13. Panula P, Yang HY, Costa E (1984) Histamine-containing neurons in the rat hypothalamus. Proc Natl Acad Sci USA 81:2572–2576
14. Steinbusch MWM, Mulder AH (1984) Immunohistochemical localization of histamine in neurons and mast cells in the rat brain. In: Björklund A, Hökfelt T, Kuhar MJ (eds) Classical transmitters and transmitter receptors in the CNS, part II. Handbook of chemical neuroanatomy, vol 3. Elsevier, Amsterdam
15. Strecker RE, Nalwalk J, Dauphin LJ, Thakkar MM, Chen Y, Ramesh V, Hough LB, McCarley RW (2002) Extracellular histamine levels in the feline preoptic/anterior hypothalamic area during natural sleep-wakefulness and prolonged wakefulness: an in vivo microdialysis study. Neuroscience 113:663–670
16. Roth M (1971) Fluorescence reaction for amino acids. Anal Chem 43:880–882
17. Zuman P (2004) Reactions of orthophthalaldehyde with nucleophiles. Chem Rev 104: 3217–3238
18. Nohta H, Satozono H, Koiso K, Yoshida H, Ishida J, Yamaguchi M (2000) Highly selective fluorometric determination of polyamines based on intramolecular excimer-forming derivatization with a pyrene-labeling reagent. Anal Chem 72:4199–4204
19. Yoshida H, Nakano Y, Koiso K, Nohta H, Ishida J, Yamaguchi M (2001) Liquid chromatographic determination of ornithine and lysine based on intramolecular excimer-forming fluorescence derivatization. Anal Sci 17:107–112
20. Yoshitake T, Ichinose F, Yoshida H, Todoroki K, Kehr J, Inoue O, Nohta H, Yamaguchi M (2003) A sensitive and selective determination method of histamine by HPLC with intramolecular excimer-forming derivatization and fluorescence detection. Biomed Chromatogr 17:509–516
21. Yoshitake T, Yamaguchi M, Nohta H, Ichinose F, Yoshida H, Yoshitake S, Fuxe K, Kehr J (2003) Determination of histamine in microdialysis samples from rat brain by microbore column liquid chromatography following intramolecular excimer-forming derivatization with pyrene-labeling reagent. J Neurosci Methods 127:11–17
22. Rönnberg AL, Hansson C, Drakenberg T, Håkanson R (1984) Reaction of histamine with o-phthalaldehyde: isolation and analysis of the fluorophore. Anal Biochem 139:329–337
23. Yoshimura T, Kaneuchi T, Miura T, Kimura M (1987) Kinetic analysis of the fluorescence reaction of histamine with orthophthalaldehyde. Anal Biochem 164:132–137
24. Egger D, Reisbach G, Hültner L (1994) Simultaneous determination of histamine and serotonin in mast cells by high-performance liquid chromatography. J Chromatogr B Biomed Appl 662:103–107
25. Tsuruta Y, Kohashi K, Ohkura Y (1978) Determination of histamine in plasma by high-speed liquid chromatography. J Chromatogr 146:490–493
26. Itoh Y, Oishi R, Adachi N, Saeki K (1992) A highly sensitive assay for histamine using ion-pair HPLC coupled with postcolumn fluorescent derivatization: its application to biological specimens. J Neurochem 58:884–889
27. Yamatodani A, Fukuda H, Wada H, Iwaeda T, Watanabe T (1985) High-performance liquid chromatographic determination of plasma and brain histamine without previous purification of biological samples: cation-exchange chromatography coupled with post-column derivatization fluorometry. J Chromatogr 344:115–123
28. Mochizuki T, Yamatodani A, Okakura K, Takemura M, Inagaki N, Wada H (1991) In vivo release of neuronal histamine in the hypothalamus of rats measured by microdialysis. Naunyn Schmiedebergs Arch Pharmacol 343:190–195
29. Itoh Y, Oishi R, Nishibori M, Saeki K (1991) Characterization of histamine release from the rat hypothalamus as measured by in vivo microdialysis. J Neurochem 56:769–774
30. Allenmark S, Bergström S, Enerbäck L (1985) A selective postcolumn o-phthalaldehyde-derivatization system for the determination of histamine in biological material by high-performance liquid chromatography. Anal Biochem 144:98–103
31. Westerink BH, Cremers TI, De Vries JB, Liefers H, Tran N, De Boer P (2002) Evidence for activation of histamine H3 autoreceptors during handling stress in the prefrontal cortex of the rat. Synapse 43:238–243
32. Chikai T, Oishi R, Saeki K (1993) Microdialysis study of the effects of sedative drugs on extracellular histamine in the striatum of freely moving rats. J Pharmacol Exp Ther 266:1277–1281

33. Horii A, Takeda N, Matsunaga T, Yamatodani A, Mochizuki T, Okakura-Mochizuki K, Wada H (1993) Effect of unilateral vestibular stimulation on histamine release from the hypothalamus of rats in vivo. J Neurophysiol 70:1822–1826
34. Chikai T, Oishi R, Saeki K (1994) Increase in the extracellular histamine concentration in the rat striatum by mu-opioid receptor activation. J Neurochem 62:724–729
35. Laitinen KS, Tuomisto L, MacDonald E (1995) Effects of a selective alpha 2-adrenoceptor antagonist, atipamezole, on hypothalamic histamine and noradrenaline release in vivo. Eur J Pharmacol 285:255–260
36. Laitinen KS, Tuomisto L, Laitinen JT (1995) Endogenous serotonin modulates histamine release in the rat hypothalamus as measured by in vivo microdialysis. Eur J Pharmacol 285:159–164
37. Ito C, Onodera K, Yamatodani A, Watanabe T, Sato M (1997) The effect of methamphetamine on histamine release in the rat hypothalamus. Psychiatry Clin Neurosci 51:79–81
38. Jansen FP, Mochizuki T, Yamamoto Y, Timmerman H, Yamatodani A (1998) In vivo modulation of rat hypothalamic histamine release by the histamine H3 receptor ligands, immepip and clobenpropit. Effects of intrahypothalamic and peripheral application. Eur J Pharmacol 362:149–155
39. Washington B, Shaw JB, Li J, Fisher B, Gwathmey J (2000) In vivo histamine release from brain cortex: the effects of modulating cellular and extracellular sodium and calcium channels. Eur J Pharmacol 407:117–122
40. Lamberty Y, Margineanu DG, Dassesse D, Klitgaard H (2003) H3 agonist immepip markedly reduces cortical histamine release, but only weakly promotes sleep in the rat. Pharmacol Res 48:193–198
41. Kitbunnadaj R, Hashimoto T, Poli E, Zuiderveld OP, Menozzi A, Hidaka R, de Esch IJ, Bakker RA, Menge WM, Yamatodani A, Coruzzi G, Timmerman H, Leurs R (2005) N-substituted piperidinyl alkyl imidazoles: discovery of methimepip as a potent and selective histamine H3 receptor agonist. J Med Chem 48:2100–2107
42. Cenni G, Blandina P, Mackie K, Nosi D, Formigli L, Giannoni P, Ballini C, Della Corte L, Mannaioni PF, Passani MB (2006) Differential effect of cannabinoid agonists and endocannabinoids on histamine release from distinct regions of the rat brain. Eur J Neurosci 24:1633–1644
43. Harusawa S, Kawamura M, Araki L, Taniguchi R, Yoneyama H, Sakamoto Y, Kaneko N, Nakao Y, Hatano K, Fujita T, Yamamoto R, Kurihara T, Yamatodani A (2007) Synthesis of novel 4(5)-(5-aminotetrahydropyran-2-yl)imidazole derivatives and their in vivo release of neuronal histamine measured by brain microdialysis. Chem Pharm Bull(Tokyo) 55:1245–1253
44. Giannoni P, Medhurst AD, Passani MB, Giovannini MG, Ballini C, Corte LD, Blandina P (2010) Regional differential effects of the novel histamine H3 receptor antagonist 6-((3-cyclobutyl-2,3,4,5-tetrahydro-1H-3-benzazepin-7-yl)oxy)-N-methyl-3-pyridinecarboxamide hydrochloride (GSK189254) on histamine release in the central nervous system of freely moving rats. J Pharmacol Exp Ther 332:164–172
45. Itoh Y, Oishi R, Saeki K (1991) Feeding-induced increase in the extracellular concentration of histamine in rat hypothalamus as measured by in vivo microdialysis. Neurosci Lett 125:235–237
46. Ishizuka T, Hatano K, Murotani T, Yamatodani A (2008) Comparison of the effect of an H(3)-inverse agonist on energy intake and hypothalamic histamine release in normal mice and leptin resistant mice with high fat diet-induced obesity. Behav Brain Res 188:250–254
47. Ishizuka T, Sako N, Murotani T, Morimoto A, Yamatodani A, Ohura K (2010) The effect of hardness of food on amygdalar histamine release in rats. Brain Res 1313:97–102
48. Murotani T, Ishizuka T, Isogawa Y, Karashima M, Yamatodani A (2011) Possible involvement of serotonin 5-HT2 receptor in the regulation of feeding behavior through the histaminergic system. Neuropharmacology 61:228–233
49. Mochizuki T, Yamatodani A, Okakura K, Horii A, Inagaki N, Wada H (1992) Circadian rhythm of histamine release from the hypothalamus of freely moving rats. Physiol Behav 51:391–394
50. Huang ZL, Qu WM, Li WD, Mochizuki T, Eguchi N, Watanabe T, Urade Y, Hayaishi O (2001) Arousal effect of orexin A depends on activation of the histaminergic system. Proc Natl Acad Sci USA 98:9965–9970
51. Ishizuka T, Sakamoto Y, Sakurai T, Yamatodani A (2003) Modafinil increases histamine release in the anterior hypothalamus of rats. Neurosci Lett 339:143–146
52. Huang ZL, Sato Y, Mochizuki T, Okada T, Qu WM, Yamatodani A, Urade Y, Hayaishi O (2003) Prostaglandin E2 activates the histaminergic system via the EP4 receptor to induce wakefulness in rats. J Neurosci 23:5975–5983
53. Chu M, Huang ZL, Qu WM, Eguchi N, Yao MH, Urade Y (2004) Extracellular histamine level in the frontal cortex is positively correlated

with the amount of wakefulness in rats. Neurosci Res 49:417–420

54. Hong ZY, Huang ZL, Qu WM, Eguchi N, Urade Y, Hayaishi O (2005) An adenosine A receptor agonist induces sleep by increasing GABA release in the tuberomammillary nucleus to inhibit histaminergic systems in rats. J Neurochem 92:1542–1549
55. Ishizuka T, Murakami M, Yamatodani A (2008) Involvement of central histaminergic systems in modafinil-induced but not methylphenidate-induced increases in locomotor activity in rats. Eur J Pharmacol 578:209–215
56. Huang ZL, Mochizuki T, Qu WM, Hong ZY, Watanabe T, Urade Y, Hayaishi O (2006) Altered sleep-wake characteristics and lack of arousal response to H3 receptor antagonist in histamine H1 receptor knockout mice. Proc Natl Acad Sci USA 103:4687–4692
57. Horner WE, Johnson DE, Schmidt AW, Rollema H (2007) Methylphenidate and atomoxetine increase histamine release in rat prefrontal cortex. Eur J Pharmacol 558:96–97
58. Liu LL, Yang J, Lei GF, Wang GJ, Wang YW, Sun RP (2008) Atomoxetine increases histamine release and improves learning deficits in an animal model of attention-deficit hyperactivity disorder: the spontaneously hypertensive rat. Basic Clin Pharmacol Toxicol 102:527–532
59. Fell MJ, Katner JS, Johnson BG, Khilevich A, Schkeryantz JM, Perry KW, Svensson KA (2010) Activation of metabotropic glutamate (mGlu)2 receptors suppresses histamine release in limbic brain regions following acute ketamine challenge. Neuropharmacology 58:632–639
60. Adachi N, Itoh Y, Oishi R, Saeki K (1992) Direct evidence for increased continuous histamine release in the striatum of conscious freely moving rats produced by middle cerebral artery occlusion. J Cereb Blood Flow Metab 12:477–483
61. Motoki A, Adachi N, Semba K, Liu K, Arai T (2005) Reduction in brain infarction by augmentation of central histaminergic activity in rats. Brain Res 1066:172–178
62. Nishida N, Huang ZL, Mikuni N, Miura Y, Urade Y, Hashimoto N (2007) Deep brain stimulation of the posterior hypothalamus activates the histaminergic system to exert antiepileptic effect in rat pentylenetetrazol model. Exp Neurol 205:132–144
63. Groth L, Jørgensen A, Serup J (1998) Cutaneous microdialysis in the rat: insertion trauma and effect of anaesthesia studied by laser Doppler perfusion imaging and histamine release. Skin Pharmacol Appl Skin Physiol 11:125–132
64. Kehr J (1999) Monitoring chemistry of brain microenvironment: biosensors, microdialysis and related techniques (Chapter 41). In: Windhorst U, Johansson H (eds) Modern techniques in neuroscience research. Springer, Heidelberg
65. Kehr J, Yoshitake T (2006) Monitoring brain chemical signals by microdialysis. In: Grimes CA, Dickey EC, Pishko MV (eds) Encyclopedia of sensors, vol 6. American Scientific Publishers, Valencia
66. Tsuruta Y, Kohashi K, Ohkura Y (1981) Simultaneous determination of histamine and N tau-methylhistamine in human urine and rat brain by high-performance liquid chromatography with fluorescence detection. J Chromatogr 224:105–110
67. Granerus G, Wass U (1984) Urinary excretion of histamine, methylhistamine (1-MeHi) and methylimidazoleacetic acid (MeImAA) in mastocytosis: comparison of new HPLC methods with other present methods. Agents Actions 14:341–345
68. Saeki K, Oishi R, Nishibori M (1985) Simultaneous determination of histamine and *tele*-methylhistamine in the mammalian central nervous system and histamine turnover measurements. Adv Biosci 51:173–184
69. Itoh Y, Oishi R, Nishibori M, Saeki K (1989) *tele*-Methylhistamine levels and histamine turnover in nuclei of the rat hypothalamus and amygdala. J Neurochem 53:844–848
70. Saito K, Horie M, Nose N, Nakagomi K, Nakazawa H (1992) High-performance liquid chromatography of histamine and 1-methylhistamine with on-column fluorescence derivatization. J Chromatogr 595:163–168
71. Harsing LG Jr, Nagashima H, Vizi ES, Duncalf D (1986) Electrochemical determination of histamine derivatized with o-phthalaldehyde and 2-mercaptoethanol. J Chromatogr 383:19–26
72. Jensen TB, Marley PD (1995) Development of an assay for histamine using automated high-performance liquid chromatography with electrochemical detection. J Chromatogr B Biomed Appl 670:199–207
73. Yoshitake T, Ijiri S, Yoshitake S, Todoroki K, Yoshida H, Kehr J, Nohta H, Yamaguchi M (2012) Determination of histamine in microdialysis samples from guinea pig skin by high-performance liquid chromatography with fluorescence detection. Skin Pharmacol Physiol 25(2):65–72
74. Yoshida H, Harada H, Nakano Y, Nohta H, Ishida J, Yamaguchi M (2004) Liquid chromatographic determination of polyamines in

human urine based on intramolecular excimer-forming fluorescence derivatization using 4-(1-pyrene)butanoyl chloride. Biomed Chromatogr 18:687–693

75. Joseph MH, Davies P (1983) Electrochemical activity of o-phthalaldehyde-mercaptoethanol derivatives of amino acids. Application to high-performance liquid chromatographic determination of amino acids in plasma and other biological materials. J Chromatogr 277:125–136
76. Mine K, Jacobson KA, Kirk KL, Kitajima Y, Linnoila M (1986) Simultaneous determination of histamine and N tau-methylhistamine with high-performance liquid chromatography using electrochemical detection. Anal Biochem 152:127–135
77. Washington B, Smith MO, Robinson TJ, Olubadewo JO, Ochillo R (1991) A rapid qualitative and quantitative method of assaying histamine in small plasma volume. J Liq Chromatogr 14:2189–2200
78. Bourgogne E, Mathy FX, Boucaut D, Boekens H, Laprevote O (2011) Simultaneous quantitation of histamine and its major metabolite 1-methylhistamine in brain dialysates by using precolumn derivatization prior to HILIC-MS/MS analysis. Anal Bioanal Chem 402(1):449–59
79. Croyal M, Dauvilliers Y, Labeeuw O, Capet M, Schwartz JC, Robert P (2011) Histamine and tele-methylhistamine quantification in cerebrospinal fluid from narcoleptic subjects by liquid chromatography tandem mass spectrometry with precolumn derivatization. Anal Biochem 409:28–36
80. Koyama J, Taga S, Shimizu K, Shimizu M, Morita I, Takeuchi A (2011) Simultaneous determination of histamine and prostaglandin D(2) using an LC-ESI-MS/MS method with positive/negative ion-switching ionization modes: application to the study of anti-allergic flavonoids on the degranulation of KU812 cells. Anal Bioanal Chem 401(4):1385–92
81. Zhang Y, Tingley FD 3rd, Tseng E, Tella M, Yang X, Groeber E, Liu J, Li W, Schmidt CJ, Steenwyk R (2011) Development and validation of a sample stabilization strategy and a UPLC-MS/MS method for the simultaneous quantitation of acetylcholine (ACh), histamine (HA), and its metabolites in rat cerebrospinal fluid (CSF). J Chromatogr B Analyt Technol Biomed Life Sci 879:2023–2033
82. García-Villar N, Hernández-Cassou S, Saurina J (2009) Determination of biogenic amines in wines by pre-column derivatization and high-performance liquid chromatography coupled to mass spectrometry. J Chromatogr A 1216:6387–6393
83. Konakovsky V, Focke M, Hoffmann-Sommergruber K, Schmid R, Scheiner O, Moser P, Jarisch R, Hemmer W (2011) Levels of histamine and other biogenic amines in high-quality red wines. Food Addit Contam Part A Chem Anal Control Expo Risk Assess 28:408–416

Chapter 16

Intracerebral Microdialysis in the Study of Limbic Seizure Mechanisms and Antiepileptic Drug Action Using Freely Moving Rats

Jeanelle Portelli and Ilse Smolders

Abstract

In vivo microdialysis has become a widely used tool to help elucidate the neurochemical alterations accompanying epilepsy, as well as the mechanisms of action of both old and new generation antiepileptic drugs. With the need of novel antiepileptic drugs for the high percentage of pharmacoresistant patients, microdialysis serves both as a drug delivery tool and/or sampling method which is often combined with other complementary neurotechniques in order to discover potential drug leads. An advantage microdialysis presents in epilepsy-related investigations is that seizures are focally evoked in conscious animals that have minimal restrictions on their movements and general behavior. This allows for the analysis of seizure behavior manifestations by visual scoring of seizure severity in combination with electrical activity monitoring via electrocorticography. In this chapter we will demonstrate how one can use microdialysis to identify and further investigate known and potential anticonvulsant drugs, in addition to the better understanding of seizure mechanisms. Major focus will be given towards the focal pilocarpine model of limbic seizures.

Key words: Microdialysis, Seizures, Epilepsy, Pilocarpine, Hippocampus, Electrocorticography

1. Introduction

Epilepsy is one of the most common serious neurological disorders. Around 50 million persons worldwide have active epilepsy with recurrent seizures and, in spite of the medical advances over the years, 30% of these patients remain drug resistant (1). Major importance is directed towards the identification of compounds that act in new ways and on novel molecular targets (2). Animal models are still essential in the discovery of new antiepileptic drugs that do not fall under the "me-too" category and that offer better tolerability, less drug interactions, and improved pharmacokinetics (3).

The pilocarpine rat model for temporal lobe epilepsy is considered to be highly isomorphic with the human disorder since it produces

Giuseppe Di Giovanni and Vincenzo Di Matteo (eds.), *Microdialysis Techniques in Neuroscience*, Neuromethods, vol. 75,
DOI 10.1007/978-1-62703-173-8_16,

morphological changes, altered membrane properties and altered synaptic responses of hippocampal neurons which are similar to what has been observed in epileptic human brains (4, 5). In this post-status epilepticus model, pilocarpine is administered systemically and usually presents in three phases: the status epilepticus phase, the latent/silent phase, and the spontaneous seizures phase (6). It is possible to perform microdialysis in this chronic model in order to investigate neurochemical alterations taking place in these different phases. In this chapter however we will focus on a variant of the aforementioned pilocarpine rat model, namely the focally evoked pilocarpine-induced limbic seizure model in which pilocarpine is administered locally into the hippocampus via a microdialysis probe (7, 8).

Focally evoked pilocarpine-induced seizures are known to mimic the generation of complex partial seizures closer than limbic epilepsy that is induced via systemic administration of the chemoconvulsant pilocarpine (7). The focal pilocarpine model in freely moving animals has been used in our laboratory for pharmacological screening experiments since 1997 (7, 9–29). In all these years, intrahippocampal pilocarpine perfusion has proven to be a reproducible way to induce limbic seizures with comparable severity.

We have characterized our model using electrographic monitoring in the past (7, 15, 17–19, 26). More recently we combined video monitoring with electrocorticographical (ECoG) recordings (9). In this model, none of the pilocarpine control animals develop status epilepticus and mortality rate is very low. Moreover, animals normally have time to recover between consecutive seizures. As such, the focal pilocarpine model can be considered to be a model for repeated limbic seizures with and without secondary generalization, rather than a model for status epilepticus. Further in this chapter we explain the scoring system that we use to determine the severity of the behavioral seizures, together with a more detailed explanation on at which stage the rat is considered to be experiencing secondary generalization.

Pilocarpine-induced seizures are initiated by M_1 muscarinic receptors and maintained by NMDA receptor activation (7, 30). By administering pilocarpine to the animal by intracerebral microdialysis there is the added advantage that seizure-related changes in extracellular neurotransmitter or neuropeptide concentrations can be observed throughout the experiment. Focus has been directed towards the hippocampus (hence intrahippocampal microdialysis) due to its involvement in the pathophysiology of temporal lobe epilepsy (7, 21). Thus intrahippocampal microdialysis is used both as a tool for drug delivery and as a sampling technique.

Since the microdialysis technique for the monitoring of in vivo extracellular neurotransmitter concentrations in brain regions has been discussed in other chapters of this book, in this chapter we will solely focus on describing how to perform the focal pilocarpine model for limbic seizures. In this model pilocarpine is perfused, via a microdialysis probe, locally into the hippocampus. Compounds can be tested for their effect on seizures by means of intrahippocampal administration, as well as via intracerebroventricular (icv) and systemic administration. In this chapter we describe the appropriate surgical procedures for both intrahippocampal and icv probe localization. Apart from looking at neurochemical and behavioral seizure alterations, one can also observe brain electrical changes. This can be performed by ECoG recordings, thus we illustrate how to implant a radiotelemetric transmitter in the rat. We also describe in detail how to set up the microdialysis sampling experiment itself in awake rats together with practical information on what to look out for. A more comprehensive overview on microdialysis as a beneficial tool in epilepsy research can be found at (31).

2. Materials

2.1. Perfusion Medium Composition

Modified Ringer's solution is used as the perfusion fluid for microdialysis, containing 147 mM NaCl, 4 mM KCl, and 2.3 mM $CaCl_2$. The following salts are weighed and dissolved in 250 mL ultrapure water (Arium pro UV).

- 2.15 g NaCl
- 0.075 g KCl
- 0.126 g $CaCl_2 \cdot 6H_2O$

The solution is then filtered through a 0.2 μm membrane filter under vacuum and is kept at a temperature of 4°C for 1 week.

2.2. Materials Needed for a Microdialysis Experiment

1. Laboratory rat.
2. Weighing scale.
3. Anesthesia: mixture of 1.5 mL ketamine, 1.5 mL diazepam, and 1.0 mL physiological saline (0.9% NaCl solution). Syringe to administer anesthesia intraperitoneally (ip).
4. Stereotaxic atlas of the rat brain (32).
5. Stereotaxic frame with heating plate and stereotaxic manipulator.
6. Surgery lamp to illuminate surgery field.
7. Intracerebral steel guide cannulas (CMA/12).
8. Hemostats.
9. Cotton swabs.
10. Dental drill.
11. Jeweler forceps, jeweler screwdriver, anchor screws.
12. 21 gauge × 2 in. needle (green needle) and 18 gauge × 1.5 in. needle (pink needle).

13. Carbon steel surgical blades (#24).
14. Acrylic dental cement.
15. Ketoprofen (4 mg/kg).
16. Microdialysis probe (CMA/12, 3 mm membrane length).
17. Microinfusion pumps (CMA/100).
18. Raturn microdialysis stand-alone system that includes return controller, stainless steel platform with support stand, motor, balance arm and tether assembly, bowl cage.
19. Rat collar.
20. FEP tubing (0.12 mm inner diameter, internal volume of 1.2 μL/100 mm length).
21. Tubing adapters to connect FEP tubing to probes and syringes.
22. Pilocarpine hydrochloride (minimum 99% titration).
23. Gas-tight syringes 1.0 and 2.5 mL.

2.3. Materials Needed for Video-ECoG Monitoring

1. Shaver to shave abdomen
2. Fine scissors
3. Adson tissue forceps
4. Metal surgical cannula
5. Sterile disposable scalpel blade (#22)
6. Thread for sutures (4-0 Perma-Hand silk)
7. Sterile drape (Foliodrape®)
8. Adhesive op-tape (Barrier®)
9. Iso-Betadine®
10. Amoxicillin (1 mg/kg)
11. Sterilized radiotelemetric transmitter (F20-EET, Data Sciences International®)
12. Radiotelemetric receiver (PhysioTel™ Receiver Model RPC-1, Data Sciences International®)
13. Computer with acquisition software installed (Notocord-hem Evolution®)

3. Methods

3.1. Surgical Procedure for Microdialysis Probe and ECoG Transmitter Implantation

Following are the detailed step-by-step protocols on rats to implant intracerebral guide cannulas in the hippocampus only, or in the hippocampus and the lateral cerebral ventricle (icv), in the presence/absence of ECoG radiotelemetric transmitter implantation (Table 1).

Table 1
A clearer representation on which steps should be followed in this methods Sect. 3.1 depending on the surgery to be performed

	Intrahippocampal guide cannula alone	Intrahippocampal and icv guide cannula	Intrahippocampal guide cannula + ECoG	Intrahippocampal and icv guide cannulas + ECoG
Anesthesia	X	X	X	X
ECoG transmitter implantation			X	X
Intrahippocampal guide cannula implantation	X	X	X	X
Icv guide implantation		X		X
Postoperative treatment	Analgesia	Analgesia	Analgesia + antibiotics	Analgesia + antibiotics

3.1.1. Anesthesia

1. Anesthetize rats using a mixture of ketamine and diazepam (start dose of 90.5:4.5 mg/kg). This mixture is injected ip in the lower-right quadrant of the rat's abdomen. If following the start dose the rats are not fully anesthetized after 4 min, an additional injection of 20:1 mg/kg should be administered. The latter procedure can be repeated every 4 min until the rat is sufficiently sedated. The maximum total amount of anesthesia administered must not exceed 250:14.5 mg/kg ketamine:diazepam. Ketamine is used as the anesthetic while diazepam functions as a muscle relaxant to counteract the increased muscle tone induced by ketamine. Surgical anesthesia is judged by the lack of a withdrawal reflex when the hind toes are pinched, regular respiratory rate at 30% below normal rate, loss of corneal reflexes (by verification of retraction of the globe and blinking in response to gently touching the cornea), the loss of jaw reflexes (by verification of head shaking, attempts to masticate and movement of the tongue), and no reaction to skin pinch over area to be incised. Weigh rat using a weighing scale.

3.1.2. ECoG Transmitter Implantation

2. Shave the lower central part of the rat's abdomen. Rinse the shaved area thoroughly and disinfect with Iso-Betadine®. Place the rat on a sterile drape. Using a scalpel (#24), perform an incision from a few mm posterior to the eyes back to a few mm behind the interaural line. Use a cotton swab to clean any blood. The rat does not need to be on the stereotaxic frame at this point. Position rat to be on its back, and perform an incision in the lower central part of the abdomen. Tent the skin with

Adson forceps to retract it from the viscera and using a #22 sterile scalpel nick through the skin, passing through the abdominal wall and the adherent parietal peritoneum. An incision slightly longer than the width of the radiotelemetric transmitter can be performed using this scalpel or else by using the tip of the scissors, keeping the skin tented at all times. Special attention must be taken in order not to damage the underlying organs. Implant the radiotelemetric transmitter into the abdominal space, and place the leads in the cavity of the metal surgical cannula. Push this cannula, subcutaneously, towards the incision present at the top of the head of the rat. Retract the cannula slowly from the head so that both measuring and reference electrodes of the transmitter are positioned on the rat's skull. Close the incision in the abdomen by performing, for instance, vertical or horizontal mattress sutures—it is important that the abdomen is sufficiently closed. Disinfect the skin again, and place a clean cotton swab on the stitches which is kept in place by covering it with adhesive op-tape.

3. Place the rat on the stereotaxic frame, using a heating plate as support. It is important that the head is tightly secured and in position. A detailed explanation on how to place the rat on the stereotaxic apparatus can be found elsewhere in this book (c.f. Chap. 9).
4. Retract the skin from each side of the incision using hemostats. Gently scrape the periosteum using the belly of the blade. Dry the bone using a cotton swab and mark the location of bregma using a fine tip marker. The measuring electrode of the transmitter should be positioned above the right hippocampus (+4.6 mm lateral and 5.6 mm anterior according to bregma), whereas the reference electrode is positioned above the cerebellum (1 mm anterior according to lambda). These coordinates are determined using the stereotaxic atlas as well as taking into consideration the weight of the rat. The coordinates mentioned in this chapter are best appropriate for rats weighing 270–310 g. The electrodes are screwed just until the tip of the screw touches the dura mater.

3.1.3. Intrahippocampal Guide Cannula Implantation

5. For precise intracranial steel guide cannula implantation in the left CA1–CA3 hippocampal area, the coordinates mentioned in point 4 should be adequately corrected (−4.6 mm lateral and 5.6 mm anterior according to bregma). Before drilling, two additional screws should be anchored on opposite sides of the skull. The green needle is used to create the depth of the hole (without damaging the underlying dura mater), while the pink needle is then used to make the hole slightly wider. Screws are anchored using jeweler forceps and a jeweler screwdriver.

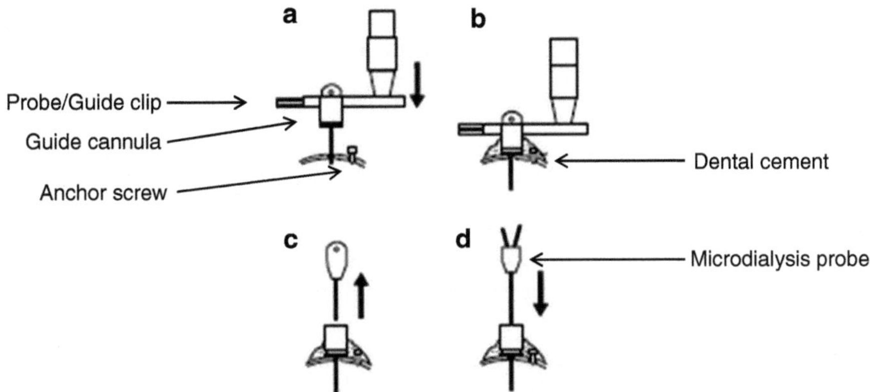

Fig. 1. Implantation of the guide cannula and microdialysis probe. (**a**) Insert the guide cannula in the hole; (**b**) apply dental cement to fix guide cannula in place; (**c**) remove hippocampal guide cannula obturator when cement is dry; (**d**) insert microdialysis probe carefully without hitting the probe membrane (Illustration with permission from N. Grimberg).

6. Mark the location of the coordinates using a fine tip marker and using the dental drill, drill the bone taking care not to damage the dura mater. The guide cannula is then positioned on top of the dura mater and lowered into the brain to 3 mm (if a 3 mm membrane length probe is used) above the actual probe membrane position in the left CA1–CA3 hippocampal area (4.6 mm ventral from dura). Once in place, apply dental cement to surround and cover the guide cannula and anchor screws. Remove hemostats while applying cement, making sure that the cement extends to the skin. Figure 1 illustrates the implantation of the guide cannula into the brain.

3.1.4. Intracerebroventricular Guide Implantation

7. To additionally implant a guide in the lateral ventricle, mark the location of the icv coordinates (−1.4 mm lateral and 0.9 mm anterior according to bregma) at the same time the coordinates for intrahippocampal implantation are marked. The bone above this brain location is also drilled together with the location above the hippocampus. Since the volume of the left ventricle is much smaller than that of the hippocampus, for precise icv guide cannula implantation the cannula is first implanted in the hippocampus and covered sufficiently with dental cement so that it stays in place, without covering the hole for icv cannula implantation and bregma. Do not remove hemostats at this point. After the dental cement is dry and the intrahippocampal guide cannula is securely in place, a second guide cannula is placed on the guide holder and the coordinates for icv cannula implantation are reread taking into consideration that minute but still significant movements could have occurred. The guide cannula is then positioned just above the dura mater and lowered into the brain to 1 mm above the location of the actual tip of the probe (3.5 mm ventral from dura). The rest of the bare

skull is covered with dental cement, hemostats removed, and allowed to dry. When cement is dry, free rat from the stereotaxic apparatus and transfer to the microdialysis cage.

3.1.5. Postoperative Treatment

8. Rat is administered analgesia ip (4 mg/kg ketoprofen) at the end of the surgical procedure for postoperative pain control.
9. If a radiotelemetric transmitter is also implanted, the rat is administered antibiotics subcutaneously (amoxicillin 1 mg/kg) to prevent bacterial infections.

3.2. Microdialysis Cage Setup

1. If recording ECoG, the microdialysis setup should take place where there is also a video-ECoG monitoring unit equipped. The radiotelemetric receiver, which is coupled to the acquisition software, can be placed under the stainless steel platform with support stand. The video recorder is positioned in front of the microdialysis cage for optimal recording of the rat throughout the experiment.
2. Fill a 2.5 mL microdialysis glass syringe with ultrapure water and mount it in the microinfusion pump. Mount the microdialysis probe in a probe/guide clip, and place the probe in a vial filled with ultrapure water. Tubing adapters are used to connect the FEP tubing with the syringe and with the probe. First flush the probe with ultrapure water at a flow of 8–10 μL/min for 5 min. Replace the syringe with one containing modified Ringer's solution, and after flushing again at a flow of 8–10 μL/min for 5 min, start perfusing the probe at a rate of 2 μL/min.
3. Fill the cage with tissue paper, food pellets and a bottle of water, and place the rat in the cage. A rat collar should be adjusted loosely on the neck of the rat. The hippocampal guide cannula obturator is removed and replaced by a microdialysis probe (CMA/12; 3 mm membrane length) (Fig. 1c, d), which is continuously perfused with Ringer's solution at 2 μL/min. Attach the rat collar with the tether. The rat is allowed to recover from the anesthesia and surgery, and experiments are normally started 24 h after surgery.

3.3. The Focal Pilocarpine Model for Limbic Seizures

Individual sampling for the analysis of neurotransmitters and neuropeptides can be performed using this setup. Since these protocols have been covered extensively in other chapters of this book, we did not provide details on dialysate collection and analysis in this chapter. In our laboratory, dialysate collections are normally collected every 20 min at a flow rate of 2 μL/min, thus obtaining 40 μL dialysate in every collection. We normally then split the dialysate in two: 15 μL in order to analyze for glutamate or GABA and 25 μL where we can analyze for noradrenaline, dopamine, and serotonin. To the latter, 6.25 μL antioxidant mixture is added to prevent monoamine oxidation. For neuropeptide

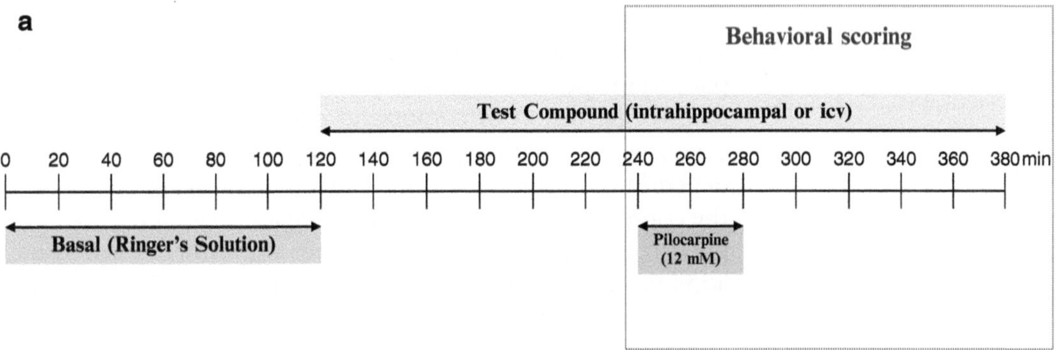

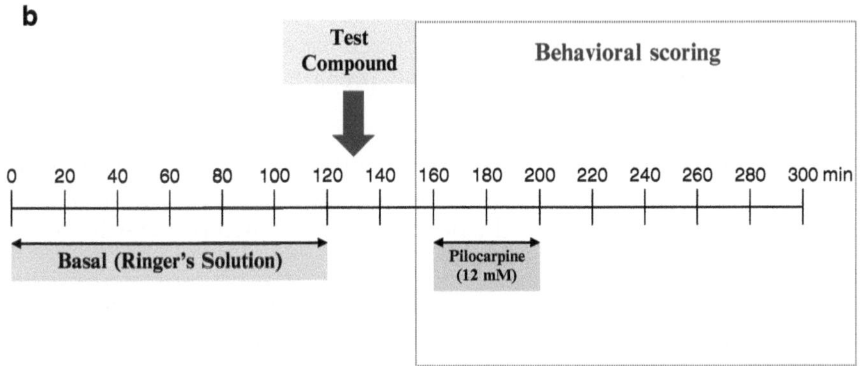

Fig. 2. Experimental microdialysis protocol for (**a**) intrahippocampal or icv administration of test compound and (**b**) systemic administration of test compound. In the latter example, the test compound is systemically administered 30 min prior pilocarpine administration.

assessment, the amount of dialysate required for analysis depends on the loop volume of the LC-MS/MS (33).

1. The microdialysis experiment is initiated the morning after surgery. The radiotelemetric transmitter in the rat is activated, and modified Ringer's solution is perfused for the first 2 h of the experiment. Intrahippocampal basal samples can be collected during this period. If compounds are tested, these can be administered either intrahippocampally, systemically (if they cross the blood brain barrier) or else via icv (Fig. 2).
2. For intrahippocampal compound administration, the test compound is dissolved in modified Ringer's solution and perfused through the hippocampal microdialysis probe for 2 h prior to administration of the chemoconvulsant pilocarpine. Pilocarpine is coadministered with the test compound, and perfusion of the test compound continues till the end of the experiment.
3. Test compounds can also be administered systemically. We usually administer the test compound ip around 30–60 min prior pilocarpine administration. The time is decided taking into account pharmacokinetic properties of the test compound (e.g., C_{max}) if known.

4. If compound is administered continuously throughout the experiment via icv, compound is administered for 2 h prior pilocarpine perfusion. Pilocarpine is perfused in the hippocampus. Icv perfusion of compound continues till the end of the experiment. If the compound is administered as a bolus dose, depending on the compound and the aim of the experiment, the compound is administered via the icv probe a few min prior pilocarpine administration. In our experiments we usually administer bolus dose volumes of 2–3 μL over 2 min, and we recommend not to exceed a volume of 5 μL. If compound is administered to the lateral ventricle, first the tubing and the icv probe need to be prepared.

Since icv probes are only required for compound administration and not for sampling, we normally prepare icv probes as following:

5. Remove 2 mm from the 3 mm semipermeable membrane length of a used CMA/12 microdialysis probe. Remove the rest of the membrane, and with a pair of pliers remove the outer steel shaft of the probe. This modified probe will serve as a tool for drug administration.
6. Replace the icv guide cannula obturator with the icv probe, and then start administration of the compound. If compound is administered via bolus dose, do not remove the icv probe after total compound volume administration to prevent any backflow of the compound from the ventricle.

3.3.1. Pilocarpine Administration and Seizure Severity Assessment

To invoke limbic seizures, 12 mM pilocarpine in modified Ringer's solution is perfused for 40 min (two collection samples) into the hippocampus. Pilocarpine-induced seizures are characterized by a sequential development of typical behavioral patterns and electrographic activity. The behavioral phenomena are observed and scored in all rats from the initiation of pilocarpine administration till the end of the experiment, amounting to seven collection periods in total. Seizure severity is rated on a modified seizure severity scale based on Racine's scale (23, 34). This scale was adapted by our laboratory in order to include all behavioral changes observed in focal limbic seizure models. The scale is as follows.

(0) Normal, non-epileptic activity
(1) Mouth and facial movements, hyperactivity, grooming, sniffing, scratching, wet-dog shakes
(2) Staring, head nodding, tremor
(3) Forelimb clonus, forelimb extension
(4) Rearing, salivating, tonic-clonic activity
(5) Falling

For each animal, the total seizure severity score (TSSS) is calculated as the sum of the seizure severity scores during the seven

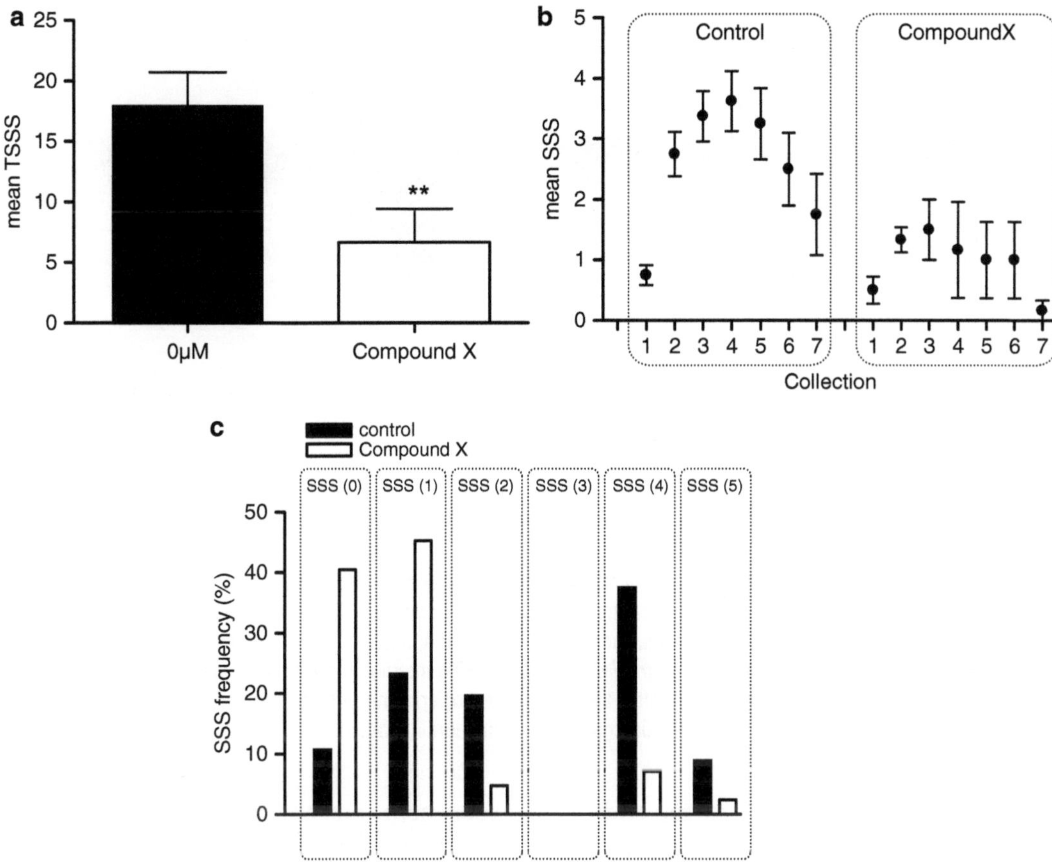

Fig. 3. (**a**) An example of mean TSSS graph. The *black bar* denotes the control group, whereas the *white bar* denotes an anticonvulsant compound using the focal pilocarpine model for limbic seizures. (**b**) The mean SSS attributed at each collection period following intrahippocampal administration of pilocarpine until the end of the experiment for both the control and test compound groups. (**c**) The SSS frequency in percentage for each score for both the control and test compound groups. The SSS (3) is not often observed in this model.

collection periods following the start of convulsant administration (Fig. 3a). Rats administered pilocarpine intrahippocampally normally show the same characteristic seizure severity profile with a gradual increase, a peak around collections 3 and 4, followed by a gradual decrease in seizure severity (Fig. 3b). Rats experiencing seizure behaviors as enlisted in scores (1) and (2) are considered to experience partial limbic seizures without secondary generalization according to Racine's original scale (34) (Fig. 3c). Progression to secondary generalization is attributed to behaviors detailed from scores (3) till (5). The ECoG recordings throughout the microdialysis experiment serve as a way to monitor ECoG activity prior and following pilocarpine administration, as well as to ensure whether seizure activity is generalized or not. An example of a full ECoG recording in a control rat can be observed in Fig. 4.

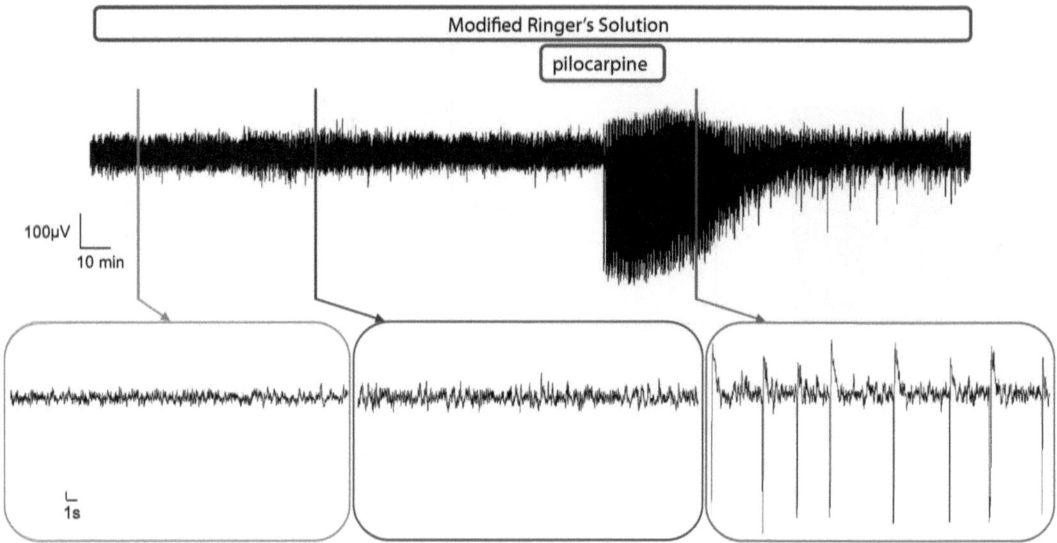

Fig. 4. Representative ECoG recording of a control rat that undergoes pilocarpine administration. The ECoG recording represents 2 h 30 min of modified Ringer's perfusion, 40 min administration of pilocarpine, and 2 h 40 min perfusion of modified Ringer's solution alone till end of experiment. The insets of magnified windows represent EEG samples for 30 s in each highlighted time window. The *box on the left* indicates a 30 s interval in the first 15 min of the experiment; the *middle box* indicates a 30 s interval, 80 min after the start of the experiment; and the *box on the right* indicates a 30 s interval, 50 min following initiation of intrahippocampal pilocarpine administration.

At the end of the experiment, rats are killed by an overdose of pentobarbital (Nembutal®), and probe localization and tissue damage is histologically verified and evaluated postmortem.

4. Notes

1. The Wistar rat strain is normally the best suited strain to perform pilocarpine-induced limbic seizures (5). For years we have used in our laboratory the focal pilocarpine model as an acute limbic seizures model. At a certain point, due to breeding problems at the vendor and apparent changes in pilocarpine-induced seizure susceptibility, we were forced to change breeding locations and vendors over a period of 2 years. We experienced significant intrastrain variations on ketamine dosing to establish anesthesia, on pilocarpine-induced seizure susceptibility, and on basal extracellular hippocampal noradrenaline, dopamine, serotonin, GABA, and glutamate levels of all pilocarpine-treated rats included in our studies (23). This clearly illustrates that intrastrain differences do exist from one vendor/breeding location to another, or even between rats from the same breeding location. Thus when performing a study, rats from the same stock should be

preferentially used and control rats should be performed at random and on a regular basis, i.e., at least every fortnight.

2. We have checked variations of the suggested modified Ringer's solution and found that this composition is best suited for the focal pilocarpine model of limbic seizures. In our laboratory we have investigated alterations of this composition by decreasing the Ca^{2+} content and/or adding Mg^{2+}. We established that Mg^{2+} ions invoke an anticonvulsant effect, which can be explained by the block of ion channels of NMDA receptors by Mg^{2+} ions. Reducing Ca^{2+} ion concentration affected neurotransmitter release, especially that of dopamine (unpublished data).
3. Acrylic dental cement is incapable to stick to bone. This is why at least two anchor screws are placed on the skull. It is imperative that the screws are positioned securely to the skull, and at the same time half of the length is still visible on the skull surface. This allows for the dental cement to also attach to the part under the head of the screw. Ideally screws should be positioned on opposite sides of the skull for better anchorage of the cement. Without screws, the cement "head" is unable to remain attached on the rat's head till the day of the experiment.
4. Especially when the rat is implanted with the radiotelemetric transmitter, in order to avoid the rat from becoming dehydrated, around 5 mL of physiological saline is administered subcutaneously to the rat.
5. To avoid the rat's eyes from being dry and irritated during surgery, one drop of 0.3% hypromellose (Tears Naturale®) is administered to the eyes of the rat after the rat is secured to the stereotaxic frame.
6. Do not scratch the periosteum with the tip of the blade since it could become difficult to establish the exact location of bregma.
7. On certain occasions, blood flow persists following the removal of the periosteum. This makes it difficult to detect bregma and the coordinates of interest, and also prolongs surgery time thus risking that the animal will wake up before surgery is complete. In order to help stopping the unwanted blood flow, one can wipe the skull with 10% $FeCl_3$ since this accelerates the coagulation rate of blood resulting in an attenuation of bleeding.
8. The skull bone overlying the hippocampus is thinner that the bone overlying the lateral ventricle. If drilling of the icv probe location is performed first, take care not to apply the same force when drilling above the hippocampal location in order not to risk removing part of the cortex in the process.
9. When drilling the hole for the icv probe, take note not to drill part of the suture above it since this could lead to heavy blood flow due to underlying vessels.

10. Cage setup can be performed before or after the surgical procedure of the rat. Since it is easier to insert the microdialysis probe in the rat's brain while the rat is still anesthetized, it is best that the microdialysis setup preparation takes place prior to the surgical procedure.
11. Especially when performing both icv and intrahippocampal probe experiments, the raturn setup comes very handy. This is due to the lack of swivels, thus allowing for continual access of the FEP tubing from the syringe to the probe. This is advantageous since it prevents dead volume in tubings; also this setup avoids twisting of inlet and outlet tubing, which could lead to a discontinuation of flow. Apart from that, the lack of swivels prevents any contamination or oxidation that could take place in swivels. For very slow flow, as in continuous icv administration, dead volumes could pose a problem thus this setup radically reduces this nuisance.
12. The perfusion fluid, in this case modified Ringer's solution, should be at room temperature when filling the syringe since this will lead to less air bubbles. Also, gently tap the syringe to remove any remaining air bubbles since this can lead to air bubbles in the microdialysis probe thus interfering with drug administration and sampling process.
13. Run the pump in order to be sure that liquid leaves the outlet of the microdialysis membrane prior to inserting the probe into the rat. At a high flow (8–10 μL/min), the membrane may appear to "sweat." This is no problem since it is due to ultrafiltration of fluid through the membrane. However if this "sweating" also occurs when the flow is at 2 μL/min, the probe should be discarded due to faults in the membrane. "Sweating" at this flow rate could lead to more compound and chemoconvulsant reaching the hippocampus than desired.
14. Prior to starting the microdialysis experiment, especially if collecting dialysate samples to be analyzed, it is recommended to cut the end part of the outlet FEP tubing. The reason is that salt crystal formation, as a result of 24 h of modified Ringer's solution flow, could occur at the end of the tubing. This could result in flow problems and contamination of the sample dialysate.
15. An hour before icv drug administration, connect the icv probe with FEP tubing and perfuse ultrapure water or modified Ringer's solution at the low flow to be used for drug administration in order to check whether the solution appears at the lower end of the metal shaft since low flows can be sometimes problematic. When the icv probe is successfully tested, start

perfusing with the compound in question for at least a certain amount of min. The minimal amount of time can be determined by calculating the exact dead volume of the inlet FEP tubing together with the microdialysis probe inlet and shaft by looking at their length and internal volume, and correlate it with the administered flow rate. This is performed to ensure that the test compound and not the previous solution is administered to the rat's brain.

16. Following pilocarpine administration, it is ideal to remove the water bottle that is hanging from the cage. As seen from the seizure severity scale, an epileptic rat could experience slavering. Sometimes slavering is not easily detected in the rat, and often the rat experiences a wet-dog shake after the slavering episode. This would lead to saliva drops to be transferred to the wall of the cage. By removing the water bottle, one can be assured that the transparent drops are those coming from the rat and not from the water bottle.
17. Check the icv probe at the end of the experiment to ensure that it did not get blocked during the experiment. Also check the microdialysis probe to ascertain that the membrane is not damaged. If treated delicately, microdialysis probes can be used for an average of three experiments. Rinse all tubings and probes with ultrapure water at a high flow at the end of the experiment. Keep microdialysis probe in ultrapure water, cover with parafilm, and store at 4°C. To prevent any bacterial contamination in the FEP tubings in-between experiments, flush the tubings with 70% ethanol. Note that in the following experiment, tubings should be flushed with ultrapure water before connecting tubings with the microdialysis probe since 70% ethanol can damage the microdialysis membrane.

Acknowledgments

We would like to thank G. De Smet for his assistance in formulating this chapter, as well as our colleagues that have contributed in optimizing and validating the techniques described.

Appendix

List of main suppliers and web page address

Supplier	Product	Web page address
Sigma-Aldrich	Salts for modified Ringer's solution, pilocarpine	www.sigmaaldrich.com
David Kopf instruments	Stereotaxic frame	www.kopfinstruments.com
Narishige International Ltd	Stereotaxic manipulator	www.narishige.co.jp
Harvard Apparatus	Surgical instruments	www.harvardapparatus.com
Aurora Borealis Control	Guide cannulas	www.microdialysis.se
Aurora Borealis Control	Microdialysis probes	www.microdialysis.se
Aurora Borealis Control	FEP tubing	www.microdialysis.se
Aurora Borealis Control	Anchor Screws	www.microdialysis.se
Aurora Borealis Control	Tubing Adapters	www.microdialysis.se
Aurora Borealis Control	Microinfusion pump	www.microdialysis.se
Sartorius STEDIM biotech	Ultrapure water	www.sartorius-stedim.com
Data Sciences International	ECoG apparatus	www.datasci.com
BASi	Microdialysis Raturn	www.basinc.com

References

1. Pitkanen A, Lukasiuk K (2011) Mechanisms of epileptogenesis and potential treatment targets. Lancet Neurol 10:173–186
2. Rogawski MA (2006) Molecular targets versus models for new antiepileptic drug discovery. Epilepsy Res 68:22–28
3. Rogawski MA (2006) Diverse mechanisms of antiepileptic drugs in the development pipeline. Epilepsy Res 69:273–294
4. Isokawa M, Mello LE (1991) NMDA receptor-mediated excitability in dendritically deformed dentate granule cells in pilocarpine-treated rats. Neurosci Lett 129:69–73
5. Curia G, Longo D, Biagini G et al (2008) The pilocarpine model of temporal lobe epilepsy. J Neurosci Methods 172:143–157
6. Morimoto K, Fahnestock M, Racine RJ (2004) Kindling and status epilepticus models of epilepsy: rewiring the brain. Prog Neurobiol 73:1–60
7. Smolders I, Khan GM, Manil J et al (1997) NMDA receptor-mediated pilocarpine-induced seizures: characterization in freely moving rats by microdialysis. Br J Pharmacol 121:1171–1179
8. Millan MH, Chapman AG, Meldrum BS (1993) Extracellular amino acid levels in hippocampus during pilocarpine-induced seizures. Epilepsy Res 14:139–148
9. Aourz N, De Bundel D, Stragier B et al (2011) Rat hippocampal somatostatin sst3 and sst4 receptors mediate anticonvulsive effects in vivo: indications of functional interactions with sst2 receptors. Neuropharmacology 61:1327–1333
10. Clinckers R, Gheuens S, Smolders I et al (2005) In vivo modulatory action of extracellular glutamate on the anticonvulsant effects of hippocampal dopamine and serotonin. Epilepsia 46:828–836
11. Clinckers R, Smolders I, Meurs A et al (2004) Anticonvulsant action of GBR-12909 and citalopram against acute experimentally induced limbic seizures. Neuropharmacology 47: 1053–1061
12. Clinckers R, Smolders I, Meurs A et al (2005) Quantitative in vivo microdialysis study on the influence of multidrug transporters on the blood-brain barrier passage of oxcarbazepine: concomitant use of hippocampal monoamines

as pharmacodynamic markers for the anticonvulsant activity. J Pharmacol Exp Ther 314:725–731

13. Clinckers R, Zgavc T, Vermoesen K et al (2010) Pharmacological and neurochemical characterization of the involvement of hippocampal adrenoreceptor subtypes in the modulation of acute limbic seizures. J Neurochem 115:1595–1607
14. De Bundel D, Aourz N, Kiagiadaki F et al (2010) Hippocampal sst(1) receptors are autoreceptors and do not affect seizures in rats. Neuroreport 21:254–258
15. Khan GM, Smolders I, Ebinger G et al (2000) Anticonvulsant effect and neurotransmitter modulation of focal and systemic 2-chloroadenosine against the development of pilocarpine-induced seizures. Neuropharmacology 39:2418–2432
16. Khan GM, Smolders I, Ebinger G et al (2000) Flumazenil prevents diazepam-elicited anticonvulsant action and concomitant attenuation of glutamate overflow. Eur J Pharmacol 407:139–144
17. Khan GM, Smolders I, Ebinger G et al (2001) 2-Chloro-N(6)-cyclopentyladenosine-elicited attenuation of evoked glutamate release is not sufficient to give complete protection against pilocarpine-induced seizures in rats. Neuropharmacology 40:657–667
18. Khan GM, Smolders I, Lindekens H et al (1999) Effects of diazepam on extracellular brain neurotransmitters in pilocarpine-induced seizures in rats. Eur J Pharmacol 373:153–161
19. Lindekens H, Smolders I, Khan GM et al (2000) In vivo study of the effect of valpromide and valnoctamide in the pilocarpine rat model of focal epilepsy. Pharm Res 17:1408–1413
20. Meurs A, Clinckers R, Ebinger G et al (2007) Sigma 1 receptor-mediated increase in hippocampal extracellular dopamine contributes to the mechanism of the anticonvulsant action of neuropeptide Y. Eur J Neurosci 26: 3079–3092
21. Meurs A, Clinckers R, Ebinger G et al (2008) Seizure activity and changes in hippocampal extracellular glutamate, GABA, dopamine and serotonin. Epilepsy Res 78:50–59
22. Meurs A, Portelli J, Clinckers R et al (2012) Neuropeptide Y increases in vivo hippocampal extracellular glutamate levels through Y1 receptor activation. Neurosci Lett 510:143–147
23. Portelli J, Aourz N, De Bundel D et al (2009) Intrastrain differences in seizure susceptibility, pharmacological response and basal neurochemistry of Wistar rats. Epilepsy Res 87:234–246
24. Smolders I, Bortolotto ZA, Clarke VR et al (2002) Antagonists of GLU(K5)-containing kainate receptors prevent pilocarpine-induced limbic seizures. Nat Neurosci 5:796–804
25. Smolders I, Clinckers R, Meurs A et al (2008) Direct enhancement of hippocampal dopamine or serotonin levels as a pharmacodynamic measure of combined antidepressant-anticonvulsant action. Neuropharmacology 54:1017–1028
26. Smolders I, Khan GM, Lindekens H et al (1997) Effectiveness of vigabatrin against focally evoked pilocarpine-induced seizures and concomitant changes in extracellular hippocampal and cerebellar glutamate, gamma-aminobutyric acid and dopamine levels, a microdialysis-electrocorticography study in freely moving rats. J Pharmacol Exp Ther 283:1239–1248
27. Smolders I, Lindekens H, Clinckers R et al (2004) In vivo modulation of extracellular hippocampal glutamate and GABA levels and limbic seizures by group I and II metabotropic glutamate receptor ligands. J Neurochem 88:1068–1077
28. Smolders I, Van Belle K, Ebinger G et al (1997) Hippocampal and cerebellar extracellular amino acids during pilocarpine-induced seizures in freely moving rats. Eur J Pharmacol 319:21–29
29. Stragier B, Clinckers R, Meurs A et al (2006) Involvement of the somatostatin-2 receptor in the anti-convulsant effect of angiotensin IV against pilocarpine-induced limbic seizures in rats. J Neurochem 98:1100–1113
30. Maslanski JA, Powelt R, Deirmengiant C et al (1994) Assessment of the muscarinic receptor subtypes involved in pilocarpine-induced seizures in mice. Neurosci Lett 168:225–228
31. Smolders I, Clinckers R, Michotte Y (2011) Microdialysis as a tool to unravel neurobiological mechanisms of seizures and antiepileptic drug action. In: Tsai TH (ed) Applications of microdialysis in pharmaceutical science. Wiley, New York
32. Paxinos G, Watson C (1986) The rat brain in stereotaxic coordinates, 2nd edn. Academic, San Diego
33. Lanckmans K, Stragier B, Sarre S et al (2007) Nano-LC-MS/MS for the monitoring of angiotensin IV in rat brain microdialysates: limitations and possibilities. J Sep Sci 30:2217–2224
34. Racine RJ (1972) Modification of seizure activity by electrical stimulation. II. Motor seizure. Electroencephalogr Clin Neurophysiol 32:281–294

Index

Giuseppe Di Giovanni and Vincenzo Di Matteo (eds.), *Microdialysis Techniques in Neuroscience*, Neuromethods, vol. 75, DOI 10.1007/978-1-62703-173-8, © Springer Science+Business Media, LLC 2013

I

L

M

N

O

P

Q

R

S

U

ProductSafety@springernature.com
Springer Nature Customer Service Center GmbH
Europaplatz 3, 69115 Heidelberg, Germany

MIX
Papier aus verantwortungsvollen Quellen
Paper from responsible sources
FSC® C105338

If you have any concerns about our products,
you can contact us on
ProductSafety@springernature.com

In case Publisher is established outside the EU,
the EU authorized representative is:
Springer Nature Customer Service Center GmbH
Europaplatz 3, 69115 Heidelberg, Germany

Printed by Libri Plureos GmbH
in Hamburg, Germany